Artificial Intelligence and Machine Learning in Sports Science

Daniel Memmert
Editor

Artificial Intelligence and Machine Learning in Sports Science

Springer

Editor
Daniel Memmert
Institut für Trainingswissenschaft und Sportinformatik
Deutsche Sporthochschule Köln
Köln, Germany

ISBN 978-3-662-70154-6 ISBN 978-3-662-70155-3 (eBook)
https://doi.org/10.1007/978-3-662-70155-3

Translation from the German language edition: "Künstliche Intelligenz und maschinelles Lernen in der Sportwissenschaft" by Daniel Memmert, © Springer Verlag GmbH, DE 2025. Published by Springer Verlag GmbH, DE. All Rights Reserved.

This Springer imprint is published by the registered company Springer-Verlag GmbH, DE, part of Springer Nature.
The registered company address is: Heidelberger Platz 3, 14197 Berlin, Germany

With love Kim, Lina, and Ute.

Preface

This specialist book is one of the first German-language publications to provide a comprehensive overview of how artificial intelligence (AI) and machine learning (ML) are used in the context of sport science research and practice.

In addition to the basics of AI and ML, various applications are described, including self-learning algorithms for analyzing athletes' movement patterns and intelligent wearables that provide real-time data. By integrating big data, game results, fitness parameters, and individual performance can be analyzed in depth, leading to new developments in research. From performance analysis to injury prevention and personalized training methods, there are many opportunities for future research and practice.

More than 30 experts contribute to covering the wide range of topics related to AI and ML and concisely summarize the latest state of research. Various topics are clustered in overarching book parts, including general basics, metrics in team sport, metrics in individual sport, and applications in sport science.

An outlook also addresses future issues relating to the use of AI and ML in sport and their responsible application. Overall, specialists and researchers in sport informatics, sport technology, training science, and sport medicine receive a comprehensive reference work with practical examples of an innovative field of research.

Designing a reference book requires a good mix of intrinsically motivated and sometimes hard work, which a large number of people have done excellently. I would like to take this opportunity to express my sincere thanks to these colleagues. First and foremost, we would like to thank the authors[1] of the book chapters for their willingness to contribute their expertise to our textbook and for their constant desire to improve. Many thanks for the consistently good and friendly cooperation.

I would also like to thank Ms. Erika Graf for her ever-vigilant support of the book and for her many comments and advice. I would also like to thank our student assistants for adopting a critical student perspective, for their constructive feedback to the authors, and for their numerous contributions to smooth out rough edges.

Finally, my sincere thanks go to the constant, very friendly, and always extremely competent support for our book project from the staff at Springer-Verlag. First and

[1] To improve readability and avoid unnecessary repetition, this book omits gender-specific differentiations. All person designations and pronouns apply equally to all genders. This includes women, men, and nonbinary individuals and aims to ensure inclusive and respectful language.

foremost, Ken Kissinger (program planning), who invested a lot of time, commitment, and energy in the book. This cannot be taken for granted, nor can the speed of his process. Without his expertise, this would not have been possible and I would like to express my sincere thanks to him. I would also like to mention Meike Barth (project management) from Springer-Verlag, who successfully oversaw the book project right up to the handover of production. I would also like to thank everyone else involved in the production process (copy editing, typesetting), for their professional cooperation.

I hope you enjoy reading it and gain a lot of new insights.

Köln, Germany Daniel Memmert

Contents

Part V Outlook

Part I

Basics

AI and ML: Definitions and Basics 1

Daniel Memmert and Leily Bakhtiar

Abstract

This chapter focuses on the definitions and basics of artificial intelligence (AI) and machine learning (ML) in science in general and sports science in particular. After the classification and structuring of AI and ML in general science, the ML research landscape is then reflected on with a special focus on sports practice and sports science. It is shown that ML-based approaches already exist in numerous sports such as soccer, volleyball, field hockey, tennis, badminton and basketball, but have not yet found their way into sports practice to any significant extent. Although event and position data are now generated as standard in many sports, they are used almost exclusively for basic research purposes. On the one hand, however, performance-relevant metrics for training and competition could already be developed on the basis of ML. On the other hand, ML can contribute to the development and testing of theories in various areas of sports science and sports informatics in the future.

1.1 What Does AI and ML Mean?

The concept of human intelligence is probably one of the most controversial topics in psychology (Hagemann et al., 2022). Contemporary definitions of human intelligence agree that intelligence encompasses mental abilities such as logic,

D. Memmert (✉)
Institut für Trainingswissenschaft und Sportinformatik, Deutsche Sporthochschule Köln, Köln, Germany
e-mail: memmert@dshs-koeln.de

L. Bakhtiar
Institute of Exercise Training and Sport Informatics, German Sport University Cologne, Köln, Germany

D. Memmert (ed.), *Artificial Intelligence and Machine Learning in Sports Science*,
https://doi.org/10.1007/978-3-662-70155-3_1

reasoning, problem solving and planning. For years, scientists have endeavored to transform these concepts in a non-human, or rather artificial, way, so that these attempts have led to the foundation of the science of artificial intelligence (AI).

Definition and Goals of Artificial Intelligence (AI) Artificial intelligence (AI) is a field of computer science that deals with the development of computer systems that can perform tasks that normally require human intelligence.

The goal of artificial intelligence is to create computers or machine systems that can think, learn and make decisions in a similar way to humans. AI systems use various approaches and techniques, including machine learning, neural networks, genetic algorithms and natural language processing. These systems can analyze large amounts of data, recognize patterns and correlations, and then use these insights to make decisions or perform specific tasks. AI has numerous areas of application, including speech recognition, image recognition, machine translation, automation of tasks and much more (see Fig. 1.1).

John McCarthy was the first to use the term AI in 1956 when he organized the first academic conference on the subject, while the idea of AI originated thousands of years ago (Toosi et al., 2021). Some researchers believe that it all started with ancient Greece, where inventors created objects called "automata" that worked autonomously and moved independently of human intervention. In Greek mythology, there are many references to amazing automatons, the oldest of which is Talos, a 10-m-long giant bronze automaton that circled the island's shores three times a day to protect Europeans from pirates and invaders.

Years later, technical drawings made by Leonardo da Vinci, the famous Italian inventor, in the late fifteenth century made it possible to create a mechanical humanoid automaton. It was a robotic knight that could stand, lie down, raise its visor and walk independently, which was very advanced and human-like for the time. Figure 1.2 shows an exact replica of Leonardo's robot or mechanical knight based on his drawings (Werthner et al., 2022).

Although the original idea of machines with human-like behavior dates back to ancient times, we want to focus on the development of artificial intelligence since the twentieth century, when engineers and scientists began to blaze the trail of modern artificial intelligence.

As mentioned earlier, the process of the emergence and development of artificial intelligence goes back thousands of years, but the birth of artificial intelligence, the concept we are dealing with today, was in 1950 when Alan Turing, a young British polymath, published an article entitled "Computer Machinery and Intelligence" in which he proposed how to build intelligent machines and test their intelligence (Turing, 2012). Five years later, John McCarthy held a workshop on "AI" at Dartmouth, which led to the word "AI" entering common parlance. The development of AI can be divided into different time periods, which are summarized in Fig. 1.3.

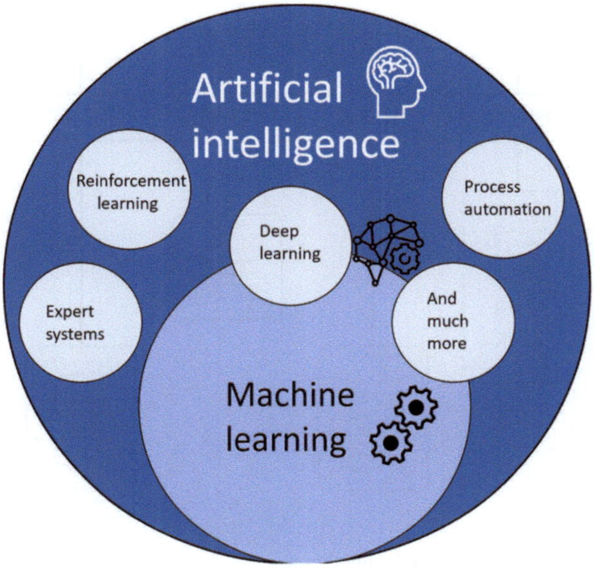

Fig. 1.1 The relationship between AI and ML. (Adapted from Goodfellow et al., 2016)

Fig. 1.2 Model of a robot based on the drawings by Leonardo da Vinci (Werthner et al., 2022)

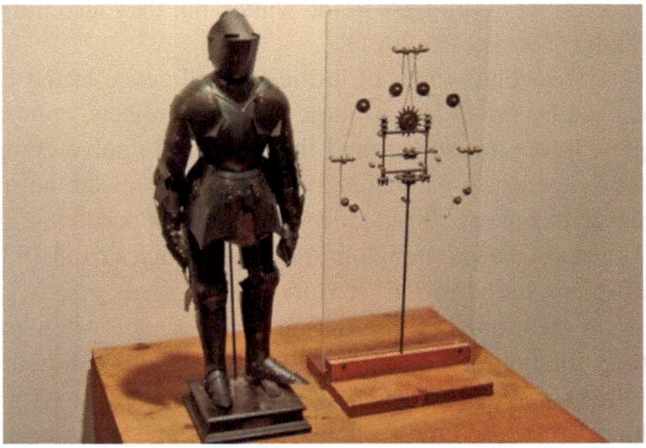

Most people believe that AI has always made rapid progress over the decades, but like other sciences, AI has experienced setbacks and recessions during its development, known as AI winters. It can be said that AI has generally experienced two winters in its timeline. The first winter began around 1973, when Professor James Lighthill of Cambridge University reported in his essay "Artificial Intelligence: A General Survey" that " in no part of the field have discoveries made so far produced the major impact that was then promised" (Agar, 2020), prompting the British government to withdraw support for AI research from most UK universities.

To survive the AI winter, AI researchers had to rename their projects to other terms that implied AI, such as "machine learning", "computer science", "knowledge-based systems", and "pattern recognition". These vague terms that were proposed

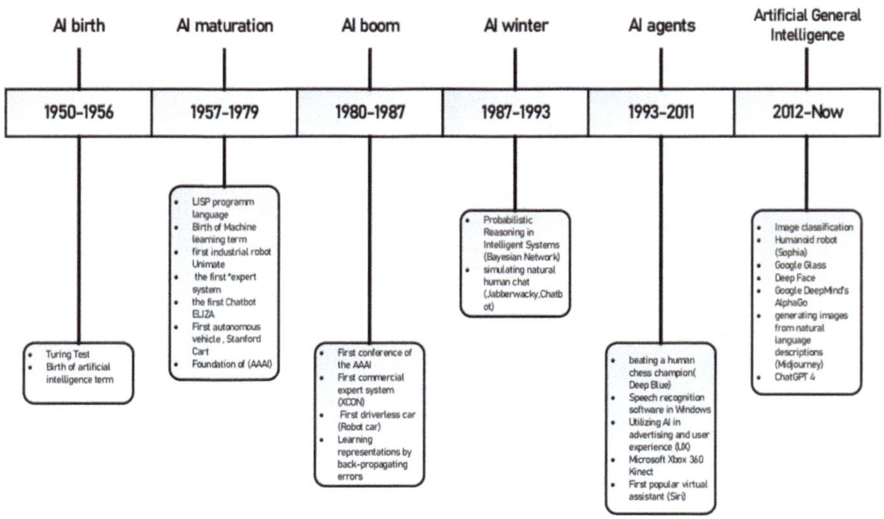

Fig. 1.3 Historical timeline of artificial intelligence

allowed AI to progress in the winter, but they also hid the progress of AI from the public (Toosi et al., 2021).

The second winter of artificial intelligence lasted from 1987 to 1993 and was caused by the failure of large-scale AI projects, competition with other technologies, lower investment and the inability to solve complex problems. During this time, interest in AI declined as expectations were not met. The AAAI conference, which had over 6000 attendees in 1986, dropped to just 2000 by 1991. The New York Times also saw a decline in AI-related articles from 1987 to 1995, reaching its lowest point that year (Fast & Horvitz, 2017).

As mentioned earlier, during the AI winter, when governments had severely cut their funding for this field, researchers published their work under different titles. Furthermore, over time and as knowledge about AI has progressed, many topics have been raised in this field of science and much research has been conducted on it, so that today it is difficult to divide AI into completely separate subfields and define clear boundaries between them (see Fig. 1.1). Technically, all of the aforementioned subfields of artificial intelligence overlap in terms of goals and tools, but they also have some differences and distinctions that are significant. In academia, machine learning (ML) is the best known, as it is used in various disciplines.

▶ **Definition and Goals of Machine Learning (ML)** Machine learning (ML) is a branch of artificial intelligence that deals with the development of algorithms and techniques that enable computers to learn from experience and master tasks without being explicitly programmed.

In ML, algorithms are developed that can analyze data, recognize patterns and learn from them. They are trained on large datasets to create models that can make predictions or recognize patterns. These models can then be applied to new data to make predictions or decisions. There are different types of machine learning, including supervised learning, unsupervised learning and semi-supervised learning. In supervised learning, models are trained using labeled training data (data where the desired or correct outcome is known), while unsupervised learning identifies patterns and structures in unlabeled data. Reinforcement learning deals with the interaction of an agent with an environment, where the agent learns through rewards and punishments based on its actions. Machine learning is used in various application areas, such as image and speech recognition, medical diagnosis, finance, autonomous driving and many other areas where large amounts of data need to be analyzed and patterns recognized.

Arthur Samuel, a pioneer of IBM research in the field of artificial intelligence, first coined the term "ML" in 1959 for playing computer checkers at championship level. He developed an evaluation function that used the positions of the pieces on the board so that the computer program could learn to distinguish correct moves from bad ones. The basis and core of the concept of ML lies in the word "learning", which means that machines can learn from data and experience, just like humans. For example, remember how you learned numbers when you were a child. From birth, a child gradually learns arithmetic and counting by observing and experiencing numbers in games and activities, hearing their parents count fingers and toes, and seeing the shape of numbers using objects such as clocks and phones. The basic idea of machine learning is also based on this principle.

If we want a machine to recognize numbers from each other, we have to convert all the steps above that human intelligence does to a computer program. First, we need a set of observations and experiences called a dataset, which consists of a large number of images in the form of numbers. The aforementioned labeled dataset plays the role of childhood experiences and parents' teachings. The next step is to develop an algorithm that works similarly to the human brain and finds the common features of the images in the dataset with the same label. These algorithms have different implementations and can range from simple linear functions to more complex optimization functions, which will be explained in the following section. All the steps mentioned above are called training. In the final phase, the output of the trained machine must be tested. This test is performed on a new dataset (test dataset) that is not included in the data set used to train the machine. The test dataset is also labeled, but during the test step, they are fed to the machine without labels. Finally, the output of the machine learning model is compared with the labels of the test data to determine the performance of the implemented model.

Various sources cite different classifications for ML methods. In general, however, they can be divided into three main types: supervised, unsupervised and semi-supervised methods (see Chap. 2 in this book).

In supervised learning, the data from which the machine learning model learns is labeled, i.e. the input and output are known and provided by the user. The example above is an example of a supervised learning technique. In contrast, unsupervised methods use unlabeled data, i.e. the input is known, but the output is unknown and is discovered by the algorithm. Clustering algorithms are considered an unsupervised approach, e.g. social network analysis, anomaly detection, DNA pattern clustering, to name just a few examples.

Semi-supervised learning is a hybrid technique that combines both labeled and unlabeled data to train a model. It is useful when a lot of unlabeled data is available, but labeling it is costly or difficult. Semi-supervised learning can improve the performance of the model compared to using only labeled data by exploiting the hidden patterns and structures in the unlabeled data. For example, in text classification, we can use a small amount of labeled texts and a large amount of unlabeled texts to train a model that can assign categories to new texts.

As already mentioned, there is a wide range of algorithms that extract features from the input data (datasets). The core of these algorithms are optimization functions that aim to find the optimal values for the parameters of the machine learning model.

Depending on the learning task, ML offers different algorithms that can be divided into two main approaches: classical machine learning and deep learning (Mukhamediev et al., 2021). Classical ML models are based on traditional statistical and mathematical methods such as linear regression, logistic regression, support vector machine, decision trees, k-nearest neighbors and naive Bayes classifiers.

The algorithms mentioned have common features. They generally perform better with small datasets and require fewer computing resources for execution. In addition, their results are easier to interpret and understand for non-ML specialists. However, these methods need to extract features from the dataset before training. They usually perform poorly on large datasets, and another disadvantage of this category is the lack of ability to learn complex features.

To explain the history of deep learning, we need to mention the year 1959, when Frank Rosenblatt conducted the first mathematical studies on the "perceptron", which became the basis of artificial neural networks (ANN). In simple terms, a "perceptron" is a building block that receives an input, performs linear and non-linear calculations with it and forwards it to the output (Fradkov, 2020). "Perceptrons", also known as nodes or neurons, are then organized into layers to form a network. The architecture of a basic ANN therefore consists of three layers: the input layer, the hidden layer and the output layer. Simply put, the concept of deep learning is to learn deep hierarchies of features with a densely interconnected neural network with many layers, in other words, a deep neural network (DNN) is an ANN that has more hidden layers between the input and output layers, which gives it more depth.

Deep learning methods, which have recently experienced significant growth, are suitable for highly complex problems, have high accuracy compared to classic methods, support large amounts of data and are able to learn complex features. However, a major problem with these models is that they act as a black box, i.e. the end user does not know how a decision was made. In addition, training these models requires powerful computing resources such as a graphical programming unit (GPU).

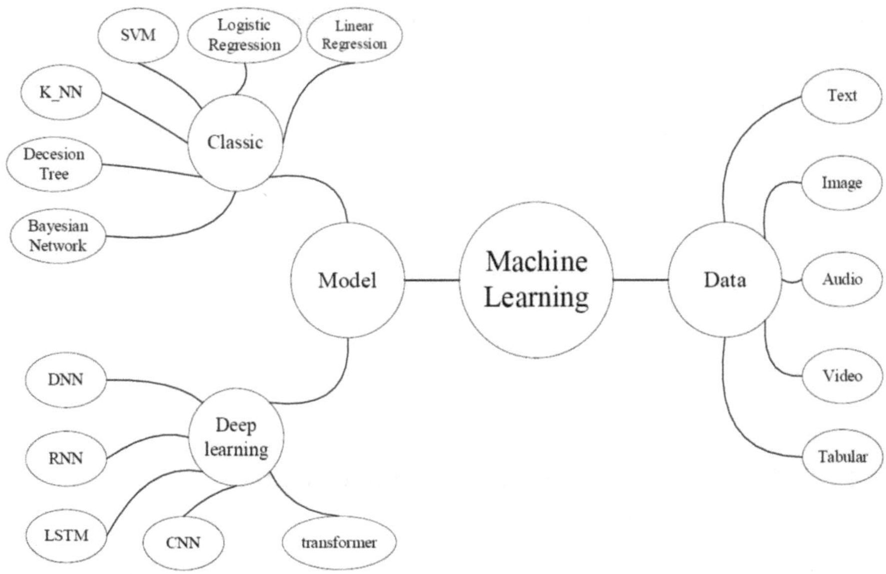

Fig. 1.4 Machine learning models and their data types

Depending on the dataset, a collection of data used to train the model can be divided into text, image, audio, video and tabular data (see for a summary in sports, Memmert, 2023). It should be noted that all input data should be converted into numeric fields (matrix) before being fed into the model, which is the most important step of preprocessing. Figure 1.4 illustrates the summary of ML models and different types of datasets.

In summary, it can be said that each category of ML models is suitable for certain problems and data types. Nevertheless, according to the No-Free-Lunch (NFL) theorem (Wolpert & Macready, 1997), there is no universal optimization strategy that is suitable for all problems. In simple words, no optimization algorithm (ML models) is better than another when we compare its average performance for all types of objective functions. For example, deep learning performs very well for image classification, while linear regression may be the better choice for house price prediction. As shown in Fig. 1.5, the amount of data affects the performance of classical ML models and deep learning. The diagram shows that classical methods perform better when the amount of data is small. Conversely, deep learning can recognize complex patterns in large datasets better than the classical methods (Mukhamediev et al., 2021).

The amount of data generated daily is increasing, and the resources to store and process this data are also improving, leading to greater demand and interest in models based on deep learning. Therefore, it can be said that deep learning has significantly revolutionized ML.

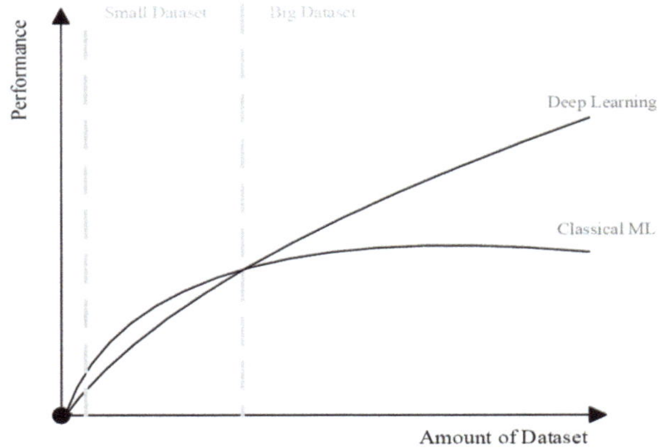

Fig. 1.5 Comparison of the performance of deep learning and classical ML based on the amount of available data. (Based on Mukhamediev et al., 2021)

1.2 ML: From the Academic World to Industry and Commercialization

Here are some of the innovations in deep learning that have taken ML from academia to industry and commercial use.

1.2.1 Convolutional Neural Network (CNN)

Convolutional Neural Network, or CNN for short, is a type of deep learning model that is recognized as the basis and foundation for computer vision systems. The structure of the CNN was developed for processing data that has a grid pattern, such as images, to learn spatial hierarchies of features. A CNN typically consists of three layers, all of which perform specific mathematical operations. The first two layers, the convolutional layer and the pooling layer, perform feature extraction, while the fully connected layer transforms the extracted features into the final output (Yamashita et al., 2018).

CNNs offer numerous applications in various fields and have achieved expert-level performance. In medical science, CNNs can visually detect the presence or absence of cancer cells in images by analyzing thousands of pathology reports. Recently, numerous publications have been published on the application of CNNs for COVID-19 detection in X-ray images (Gupta et al., 2023). In addition, automated driving relies on CNNs to accurately recognize signs or other objects and make decisions based on the results.

1.2.2 Generative Adversarial Networks (GANs)

Goodfellow et al. (2014) introduced Generative Adversarial Networks (GANs), a powerful class of neural networks used for unsupervised learning. GAN networks essentially consist of two main components of neural networks that compete with

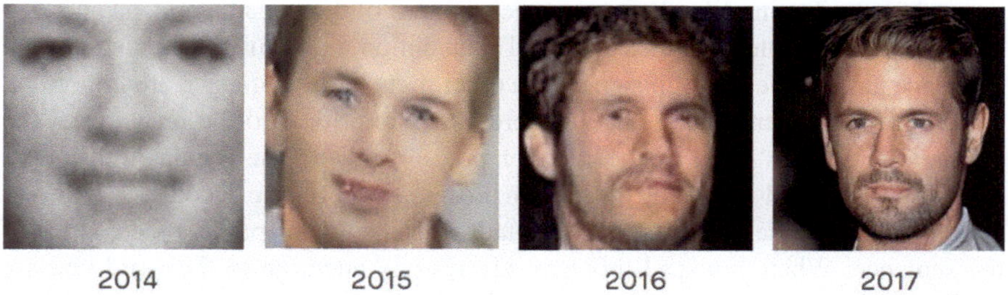

Fig. 1.6 The progress of GANs to generate human fake faces (Brundage et al., 2018)

each other. A neural network that generates data instances is called a generator, and the one that evaluates their realism is called a discriminator. The discriminator model attempts to recognize whether a data instance looks real or fake. The generator model aims to fool the discriminator and trains with more data to achieve convincing results (Goodfellow et al., 2014).

One of the interesting applications of GANs is the generation of images that can serve as patterns for image datasets when there is not enough data for processing, e.g. photos of human faces, realistic photos, and cartoon characters. Figure 1.6 shows the rapid progress of GANs from 2014 to 2017 in the generation of fake human faces (Brundage et al., 2018).

1.2.3 Recurrent Neural Network (RNN)

Another architecture of deep learning that forms the basis for many advances is recurrent neural networks (RNN). In ordinary neural networks, all inputs and outputs are independent of each other; previous input data can no longer influence other, subsequent inputs. However, for a range of applications, the inputs are interdependent and it is important to consider the effects of earlier or even later data for each input. RNNs have implemented the concept of "memory", which allows them to store as much important information as possible from the information or states of previous inputs to produce the next output of the sequence.

In simle words, RNNs implement the memory by use of a feedback loop, in such that it considers the current input and also what it has learned from the former inputs in the previous step. There are different versions of RNNs, such as LSTM, BRNN or GRU, which are used in many natural language processing (NLP) models.

1.2.4 Transformer

In 2017, a new architecture called Transformer was presented with the title "Attention Is All You Need" (Vaswani et al., 2017). In it, the authors proposed a state-of-the-art method for deep learning tasks such as machine translation, text generation and sentiment analysis that can effortlessly handle wide-ranging dependencies. To simplify, the transformers consist of three main parts: an encoder, a decoder and an attention mechanism. The attention mechanism looks at an input sequence and decides at each step which other parts of the sequence are important. The encoder segment retains the keywords that are critical to the meaning of the sentence and sends them to the decoder part along with the regular translation. These new keywords make the translation process easier for the decoder, as it knows which parts of the sentence are relevant and which keywords provide the context of the sentence. When you read this text, always pay attention to the word you are reading while your mind is still memorizing the main idea of the text to provide context. Also, write down the important keywords instead of just writing the translation of the sentence word for word. By looking at the key words, you can always remember the full meaning of the sentence.

The best-known applications that use Transformers are Large Language Models (LLMs) such as BERT and ChatGPT, which were developed by Google and OpenAI respectively.

1.2.5 Generative Artificial Intelligence (GAI)

Generative Artificial Intelligence (GAI) is an unsupervised or partially supervised machine learning concept that can generate novel content such as text, images, music or videos. In fact, it could be suggested that we are now at a tipping point where GAI will change our lives spectacularly. BERT and ChatGPT are other examples of applications of GAI that can perform translations or create new context through the use of transformers. A category of GAIs are also platforms that transform the text into images, models where the input is a text request and the output is an image. Stable Diffusion, Midjourney and DALL-E are text-to-image diffusion models released in 2021. Neural Radiance Field (NeRF) is another platform for GAI that is a type of deep neural network and can generate highly realistic views of complex 3D scenes based on a partial set of 2D images. The model uses a set of images from a scene and interpolates between them to synthesize a complete scene.

GitHub Copilot, an AI programming partner that can create code quickly and efficiently, was launched in 2021. It is powered by OpenAI's GPT-3 AI model, which has been trained on natural language text and source code from various publicly available sources such as GitHub's public repositories. GitHub Copilot can provide full lines of code or entire functions based on how you program. It can also write code from user comments and derive your code from the function name you declare.

1.3 What Significance Does ML Have in Sports Science?

What positions do the players take when in possession of the ball and what other phases of the game usually result from this? When and where is the opponent put under pressure? What system of play, including running routes, has the team practiced? Which moves lead to goals? To answer these questions, data is analyzed using ML approaches (Memmert, 2021). In many professional sports games such as soccer or handball, more and more data has to be processed and interpreted in less and less time in scouting and match analysis (Memmert, 2023). The manual assessment of video recordings would take many hours to record such performance parameters. On the basis of so-called position data, however, this is possible in seconds (Memmert & Raabe, 2024). The precise recording of the positions of each player and possibly other materials thus enables significantly more complex analyses with new types of performance indicators or metrics today (cf. Memmert, 2024). Positional data is now recorded in various sports (cf. Memmert, 2023; Memmert et al., 2017; Low et al. 2020).

Multimodal Analysis for Sports Analytics

The Institute for Exercise Training and Sport Informatics at the German Sport University Cologne (DSHS) and the L3S Research Centre at Leibniz University Hannover have now successfully completed the project "MM4SPA—Multimodal Analysis for Sports Analytics: Intelligent Synchronization and Semantic Enrichment of Position and Video Data for the Analysis of Sports Game Data", funded by the German Federal Ministry of Education and Research (BMBF) (https://mm4spa.github.io/project/). The project aimed to develop machine learning methods for the multimodal semantic analysis of sports games from different domains to enable high-quality analyses by experts (see Fig. 1.7). The project was positively assessed as part of the "Application of Artificial Intelligence Methods in Practice" funding line and financed over a project period of 2 years (2020–2023). With

Fig. 1.7 Concept of the project "MM4SPA—Multimodal Analysis for Sports Analytics: Intelligent Synchronization and Semantic Enrichment of Positional and Video Data for the Analysis of Sports Game Data" (https://mm4spa.github.io/project/)

this funding line, the BMBF supports projects that aim to transfer innovative research results from the field of artificial intelligence into practice.

In particular, the exact synchronization of event, video and position data could be further developed, which is essential for the development and applicability of multimodal and ML approaches (Biermann et al., 2021). In addition to more granular cross-sport definitions and solutions for event detection, an open science framework for processing sports data was also implemented (https://github.com/floodlight-sports/floodlight; s. Raabe et al., 2022).

Another project partner is the Munich-based technology company KINEXON, an expert in real-time collection and processing of sports performance data. In the context of the current project, successful integration of machine learning methods into KINEXON's existing analysis platform has been achieved. The developed procedures were prototypically incorporated into the software. The research center L3S, located at Leibniz University Hannover, contributed the necessary expertise in handling video data and machine learning for the project. The research group explores and develops methods for automatic video analysis and artificial vision using machine learning. Precise field registration, serving as the foundation for the semantic linkage of video and positional data, has allowed for the advancement of methods for recognizing game phases, such as pressing, for further game analysis (Theiner et al., 2022).

Based on video, event, and positional data, scientists have developed and validated various machine learning tools that allow for the complete automated, extremely fast, and reliable extraction and evaluation of sports game data (Memmert, 2023). Therefore, methods of machine learning have become indispensable in sports and are increasingly gaining importance. For example, technological advancements in sensor technology and (semi-)automated data collection enable comprehensive analyses in various sports such as soccer, basketball, or handball (Memmert & Raabe, 2024). Modern tracking systems, for instance, are employed in both competition and training for load management purposes.

Gender Differences in Professional Soccer 2 [Start]

In order to ignore different physiological and anatomical characteristics between women and men and to avoid gender-specific biases when assessing soccer games at the highest level, the tactical performance of both genders was evaluated based on event data (Garnica-Caparrós & Memmert, 2021). The objective KPIs used included: artificial neural networks. The analysis of passing pressure efficiency measures, various pressing indices and various spatial control parameters reveals that women and men have comparable values in all tactical variables. In summary, it was shown that, in contrast to previous video-based analyses, there were no significant differences in soccer-specific tactical performance between women and men in high-performance when using "defaced" position data, where no conclusions can be drawn about gender, as well as objective KPIs are recognizable.

In recent years there have been the first overview articles that have focused primarily on AI and therefore naturally (as included) also on ML. These range from narrative overviews of the influence of AI in sport (Hammes et al., 2022) to bibliometric overviews (Dindorf et al., 2022) and systematic overviews in soccer (Rico-González et al., 2023) to review Paper on forecasting models in sport (Bunker & Susnjak, 2022). Table 1.1 gives a, by no means comprehensive, overview of various studies that have been published in sports science in recent years and have used ML approaches.

In the evaluation and summary of the presented studies, it becomes evident that the majority of the studies addressed inquiries and data related to soccer ($n = 47$), followed by basketball ($n = 6$), Injury prediction ($n = 6$), Load management ($n = 6$), Result Prediction ($n = 5$), American Soccer ($n = 4$) and Hockey ($n = 4$). Among the predominantly used ML approaches are Artificial Neural Networks ($n = 31$), Regression Analysis ($n = 23$), Decision Trees ($n = 22$), Support Vector Machines ($n = 21$), Random Forests ($n = 20$), Naive Bayes ($n = 18$) and xGBoosting ($n = 17$). Publications utilizing ML approaches have significantly increased in recent years (2022 ($n = 12$), 2019 ($n = 12$), 2021 ($n = 11$), 2020 ($n = 7$), 2018 ($n = 7$) and 2016 ($n = 7$)).

Metrics in the German Bundesliga

In an ML study, a total of 50 men's Bundesliga games from the 2014/15 season (2 teams, 2 halves, a total of 200 observations) were analyzed based on position data with various KPIs (Memmert et al., 2017; Schlenger et al., 2023) is automatically evaluated and validated. The focus was on the self-developed analysis tool SOCCER (Perl et al., 2013), which combines conventional data analysis, dynamic state-event modeling and artificial neural networks. The systematic analysis yields relevant results for further research on soccer-specific metrics: First, the magnitude of the overall correlation coefficients was higher for relative metrics than for absolute metrics. Secondly, the correlation of the key figures with the strength of a team is stronger than the correlation with the playing success of a team. Third, the correlation analysis with team strength indicated more positive associations, while the correlation analysis with success was most likely confounded by the interim score of a game and revealed more negative associations.

Counterattack in Professional Soccer

Biermann et al. (2023) used an ML approach to analyze counterattack play in soccer. A counterattack is a tactical team/group strategy to generate scoring opportunities from a stable defensive position. Conversely, teams on offense must be careful not to take too much risk of losing the ball while passing. In order to assess the benefits and risks, it is important to understand the

Table 1.1 A selection of studies in sports science based on ML approaches is shown

Type of sport	Year	Authors	ML
American Soccer	2012	Delen, Cogdell, Kasap	ANN, DT, SVM
American Soccer	2011	David, Pasteur, Ahmad, Janning, M. C	ANN
American Soccer	2003	Kahn	ANN
American Soccer	1996	Purucker	ANN
Baseball	2016	Valero	ANN, SVM, DT
Basketball	2021	Tsai, Lin, Jeng, Liao	SVM
Basketball	2021	Mahmood, Daud, Abbasi	SVM, NB, ANN
Basketball	2019	Thabtah, Zhang, Abdelhamid	NB, ANN, DT
Basketball	2013	Shi, Moorthy, Zimmermann	DT, Rule learners, ANN, NB, RF
Basketball	2012	Cao	RA, ANN, SVM, NB
Basketball	2010	Ivankovic, Rackovic, Markoski, Radosav, Ivkovic	ANN
Basketball	2010	Miljkovic, Gajic, Kovacevic, Konjovic	RA, NB
Basketball	2009	Loeffelholz, Bednar, Bauer	ANN
Beach Volleyball	2020	Wenninger, Link, Lames	CNN, RNN, xGBoost
Biathlon	2018	Maier, Meister, Trösch, Wehrlin	RA, xGBoost, ANN
Boulder Climbing	2021	Duh, Chang	RNN
Boulder Climbing	2019	Kempen	DE-CTW
Cricket	2016	Pathak, Wadhwa	NB, SVM, RF
Fall Detection	2019	Lu, Wu, Feng, Song	CNN
Golf	2020	Kim, Park	LSTM, CNN
Golf	2017	König, Johansson, Riveiro, Brattberg	Cost-sensitive DT, RF
Handball	2022	Kato, Kameda, Yamada, Fujimoto, Aida	DT, RF, xGBoost
Handball	2021	van den Tillaar, Bhandurge, Stewart	RF, xGBoost, SVM
Handball	2023	Bassek, Raabe, Banning, Memmert, Rein	GNN
Handball	2023	Bassek, M., Raabe, D., Memmert, D., & Rein, R.	ANOVA
Hockey	2019	Gu, Foster, Shang, Wei,	SVM
Hockey	2018	Liu, Schulte	DRL
Hockey	2018	Mehrasa, Zhong, Tung, Bornn, Mori	DL
Hockey	2014	Weissbock, Inkpen	ANN, NB, SVM, DT, Rule Learners, RA
Hockey	2013	Weissbock, Viktor, Inkpen	ANN, NB, SVM, DT

(continued)

Table 1.1 (continued)

Type of sport	Year	Authors	ML
Injury prediction	2022	Diniz, Abreu, Lacerda, Martins, Pereira, Ferreira	k-means, XGBoost
Injury prediction	2020	Oliver, Ayala, De Ste. Croix, Lloyd, Myer, Read	DT, RA
Injury prediction	2020	DiCesare, Green, Yuan, Diekfuss, Barber Foss KD, Dudley	xGBoost
Injury prediction	2020	Rommers, Rössler, Verhagen, Vandecasteele, Verstockt, Vaeyens	xGBoost
Injury prediction	2019	Ayala, López-Valenciano, Gámez Martín, De Ste Croix, Vera-Garcia, García-Vaquero	DT
Injury prediction	2019	Rodas, Osaba, Arteta, Fernández, Lucía	RA
Injury prediction	2018	Rossi, Pappalardo, Cintia, Iaia, Fernàndez, Medina	DT, RFECV
Injury prediction	2018	López-Valenciano, Ayala, Puerta, De Ste. Croix, Vera-Garcia, Hernández-Sánchez	AdaBoost, DT, RF
Injury prediction	2016	Tixier, Hallowell, Rajagopalan, Bowman	RF, SGTB
Learning	2012	Guha, Ward	K-SVD
Load management	2021	Campbell, Stewart, Sirotic, Drovandi, Foy, Minett	RA, DT, RF
Load management	2021	Perri, Carlo, Rossi, Trecroci, Alberti, Laia	RA
Load management	2019	Op De Beéck, Jaspers, Brink, Frencken, Staes, Davis	GBRT, NB
Load management	2019	Zago, Sforza, Dolci, Tarabini, Galli	RA, SVM, DT, ANN
Load management	2017	Bartlett, O'Connor, Pitchford, Torres-Ronda, Robertson	ANN
Movement recognition	2019	Cust, Sweeting, Ball, Robertson	CNN, DL, SVM, NB, DT, kNN, Gaussian
Player tracking	2013	Lu, Ting, Little, Murphy	SSL, EM
Player tracking	2011	Xing, Ai, Liu, Lao	Bayesian nets
Result prediction	2019	Keshtkar Langaroudi, Yamaghani	ANN, Bayesian Nets, RA, SVM, FuzzySystem
Result prediction	2019	Bunker, Thabtah	ANN
Result prediction	2013	Haghighat, Rastegari, Nourafza	ANN, SVM, Bayesian Nets, DT, FuzzySystem, RA,
Result prediction	2009	Zdravevski, Kulakov	ANN, NB, SVM, DT
Result prediction	2008	McCabe, Trevathan	ANN
Result prediction	2005	Reed, O'Donoghue	RA, ANN

(continued)

Table 1.1 (continued)

Type of sport	Year	Authors	ML
Return to play prediction	2022	Valle, Mechó, Alentorn-Geli, Järvinen, Lempainen, Pruna	LR, RF, xGBoost
Rugby	2016	O'Donoghue, Ball, Eustace, McFarlan, Nisotaki	RA
Rugby	2004	O'Donoghue, Williams	ANN
Soccer	2023	Cao, Zhao, Xiaodong, Tang, Fan, Zereg	ANN
Soccer	2023	Mohandas, Ahsan, Haider	RA, DT, KNN, SVM, NB, RF
Soccer	2023	Raabe, Nabben, Memmert	TGNets
Soccer	2022	Akif Md Ali, Muhalith, Sahran, Zubli, Kamarudin, Rahman	k-Means, HC
Soccer	2022	Al-Asadi, Tasdemir	RA, DT, RF
Soccer	2022	Anzer, Bauer	XGBoost
Soccer	2022	Kelly, Williams, Cook, Sáiz, Wilson	CVLR
Soccer	2022	Kern, Lober, Hermsdörfer	LSTM
Soccer	2022	Lang, Wild, Isenko, Link	RA, DT, RF, AdaBoost
Soccer	2022	Phatak, Mehta, Wieland, Jamil, Connor, Bassek, Memmert	RA, xGBoost
Soccer	2022	Rodrigues, Pinto	NB, KNN, RF, SVM, DT, xGBoost, RA, ANN
Soccer	2022	Tümer, Akyıldız, Güler, Saka, Ievoli, Palazzo, Clemente	ANN, RA, RBFN
Soccer	2022	Vijay Fidelis, Karthikeyan	ANN
Soccer	2021	Ćwiklinski, Giełczyk, Choraś	RF, NB, AdaBoost
Soccer	2021	García-Aliaga, Marquina, Coterón, Rodríguez-González, Luengo-Sánchez	RIPPER
Soccer	2021	Sheng, Li, Zhang, Mao, Chen	BLS
Soccer	2021	Fernández, Bornn, Cervone	CNN, DL, RA
Soccer	2021	Garnica-Caparrós, Memmert	RA, DT, ANN
Soccer	2021	Inan, Cavas	ANN
Soccer	2021	Jamil, Phatak, Mehta, Beato, Memmert, Connor	RA, xGBoost, RF
Soccer	2020	Goes, Brink, Elferink-Gemser, Kempe, Lemmink	HC, k-means
Soccer	2020	Bongiovanni, Trecroci, Cavaggioni, Rossi, Perri, Pasta	DT
Soccer	2020	Almulla, Alam	kNN, RF, DT, RA, SVM, XGBoost
Soccer	2019	Bilek, Ulas	k-means, DT
Soccer	2019	Dick, Brefeld	DL
Soccer	2019	Geurkink, Vandewiele, Lievens, de Turck, Ongenae, Matthys	xGBoost

(continued)

Table 1.1 (continued)

Type of sport	Year	Authors	ML
Soccer	2019	Leser, Hoch, Tan, Moser, Kellermayr, Baca	UML
Soccer	2019	Berrar, Lopes, Dubitzky	KNN, xGBoost
Soccer	2019	Constantinou	Bayesian Nets
Soccer	2019	Hubacek, Sourek, Zelezny	xGBoost
Soccer	2018	Barron, Ball, Robins, Sunderland	ANN
Soccer	2018	Pappalardo, Cintia	RA, RF
Soccer	2018	Danisik, Lacko, Farkas	RNNs, LSTM
Soccer	2018	Razali, Mustapha, Utama, Din	Bayesian Nets
Soccer	2017	Link, Hoernig	NB
Soccer	2016	Bialkowski, Lucey, Carr, Matthews, Sridharan, Fookes	k-means
Soccer	2016	Brooks, Kerr, Guttag	KNN
Soccer	2016	Knauf, Memmert, Brefeld	CNN
Soccer	2016	Szczepański, McHale	GAMM
Soccer	2016	Prasetio	RA
Soccer	2016	Rein, Memmert	Maximization algorithm, spatio-temporal kernel algorithm, multiscale comparison technique, Bag-of-Words algorithm, ANN
Soccer	2015	Montoliu, Martín-Félez, Torres-Sospedra, Martínez-Usó	KNN, SVM, RF
Soccer	2015	Tax, Joustra	NB, LogitBoost, ANN, RF, DT
Soccer	2012	Odachowski, Grekow	Bayesian Net, SMO, DT
Soccer	2011	Hucaljuk, Rakipovic	NB, Bayessian nets, LogitBoost, kNN,RF, ANN
Soccer	2010	Buursma	RA, NB
Soccer	2010	Huang, Chang	ANN
Soccer	2006	Joseph, Fenton, Neil	Bayesian Nets, DT, NB, kNN
Table Tennis	2020	Tabrizi, Pashazadeh, Javani	SVM, LSTM, CNN
Table Tennis	2015	Draschkowitz, Draschkowitz, Hlavacs	RT, DT, Kstar, NB, SVM
Table Tennis	2013	Mülling, Kober, Kroemer, Peters	MoMP
Tennis	2020	Giles, Kovalchik, Reid	RF, xGBoost
Tennis	2017	Whiteside, Reid	DT, k-means
Text mining and forecasting	2010	Li, Wu,	k-means, SVM
Track and Field	2021	Cui, Jiang	KDE, k-means, DBSCAN
Track and field	2018	Peterson	DT, RA, ANN, SVM
Training Science	2018	Jaspers, De Beéck, Brink, Frencken, Staes, Davis	ANN

(continued)

Table 1.1 (continued)

Type of sport	Year	Authors	ML
Wearables	2018	Nweke, Teh, Al-garadi, Alo	DL
Wearables	2016	Ravi, Wong, Lo, Yang	DL
Wearables	2014	Barshan, Yuksek	NB, ANN, GMM, SVM

AdaBoost Adaptive Boosting, *ANN* Artificial Neural Network, *ANOVA* ANOVA, *BLS* Broad Learning Systems, *Bayesian Nets* Bayesian Network, *CNN* Convolutional Neural Network, *DBSCAN* Density-Based Spatial Clustering of Applications with Noise, *CVLR* cross-validated Lasso regression, DE-CTW Depth-bounded Context Tree Weighting, *DL* Deep Learning, *DT* Decision Trees, *DRL* Deep Reinforcement Learning, *EM* Expectation-Maximization, *FuzzySystem* FuzzySystem, *GAMM* generalized additive mixed model, *GBRT* gradient-boosted regression trees, *GMM* Gaussian mixed models, *GNN* Graphical Neural Network, *HC* Hierarchical Clustering, *KDE* Gaussian Kernel Density Estimation, *KNN* k nearest neighbor, *k-means* k-means, *Kstar* Kstar, *LSTM* Long Short-Term Memory, *LogitBoost* LogitBoost, *MOMP* Mixture of motor primitives, *NB* Naive Bayes, *RA* Regression Analyses, *RBFN* Radial Basis Function, *RF* Random Forest, *RFECV* Recursive Feature Elimination with Cross-Validation, *RIPPER* Repeated Incremental Pruning to Produce Error Reduction, *RNN* Recurrent Neural Networks, *Rule learners* Rule learners, *SGTB* Stochastic Gradient Tree Boosting, *SSL* Semi-supervised-learning, *SVM* Support Vector Machine, *TGNets* Tactical Graph Networks, *UML* unsupervised machine learning, *XGBoost* extreme Gradient Boosting

mechanisms that lead to when and how ball losses lead to counterattacks. The study developed a framework to predict counterattacks before they happen. For this purpose, ball losses in soccer games (N = 289 games, N = 72,367 ball losses, approx. 250 per game) were considered and temporal-spatial labels (markings) were created. In addition, features were constructed based on soccer-specific assumptions and their influence on counterattacks was evaluated. Finally, these functions were used to predict counterattacks using logistic regression models. The results show that players behind the ball are the most important predictors and that if a team loses the ball in the center and there are no more than two players behind the ball, they allow a counterattack almost 30% of the time. This underlines the importance of avoiding losing the ball when building up the game. In a follow-up study (Biermann et al., 2024) it was also proven that overloading on the wings and a 40 m distance from the last line of defense (sweet spot for the defense) can each reduce the probability of counterattacks by more than 50%.

1.4 Conclusion

GAN, RNN, CNN, Transformer or other applications of GAI can effectively and accelerate various processes in daily life, business and science. Nevertheless, we should be aware that GAI is a double-edged sword that also poses many potential dangers. An example of this is deepfake technology, which can create fake or harmful content, which can lead to the spread of misinformation and influence public opinion. The loss of jobs, lack of control over AI or the emergence of biases in AI algorithms are challenges that should be given serious attention.

GAI can solve problems and unlock opportunities and there is no doubt that this breakthrough technology is unstoppable. Nevertheless, as we become increasingly reliant on AI, there should be a moral compass, rules or clear framework to protect people from its potential harms.

With the help of ML approaches in sport, dynamic relationships can be examined on the basis of video, event or position data by taking greater account of situational and therefore context-specific references as well as inter-individual differences (cf. Rein & Memmert, 2016). In the future, training and competition data can also be linked more closely. For this purpose, well-founded theories or models are absolutely necessary (Memmert et al., 2019; Rein et al., 2017).

References

Agar, J. (2020). What is science for? The Lighthill report on artificial intelligence reinterpreted. *The British Journal for the History of Science, 53*(3), 289–310.

Al-Asadi, M. A., & Tasdemır, S. (2022). Predict the value of football players using FIFA video game data and machine learning techniques. *IEEE Access, 10*, 22631–22645.

Ali, M. F. A. M., Muhalith, M. A. F., Sahran, M. S. F., Zubli, Z. Q., Kamarudin, S. N. K., & Rahman, S. A. (2022, September). Clustering in game analysis on FIFA22 official players data. In *2022 3rd International Conference on Artificial Intelligence and Data Sciences (AiDAS)* (pp. 49–54). IEEE.

Almulla, J., & Alam, T. (2020). Machine learning models reveal key performance metrics of football players to win matches in Qatar stars league. *IEEE Access, 8*, 213695–213705.

Anzer, G., & Bauer, P. (2022). Expected passes: Determining the difficulty of a pass in football (soccer) using spatio-temporal data. *Data Mining and Knowledge Discovery, 36*(1), 295–317.

Ayala, F., López-Valenciano, A., Martín, J. A. G., Croix, M. D. S., Vera-Garcia, F. J., del Pilar García-Vaquero, M., et al. (2019). A preventive model for hamstring injuries in professional soccer: Learning algorithms. *International Journal of Sports Medicine, 40*(05), 344–353.

Baboota, R., & Kaur, H. (2019). Predictive analysis and modelling football results using machine learning approach for English Premier League. *International Journal of Forecasting, 35*(2), 741–755.

Barron, D., Ball, G., Robins, M., & Sunderland, C. (2018). Artificial neural networks and player recruitment in professional soccer. *PLoS One, 13*(10), e0205818.

Barshan, B., & Yüksek, M. C. (2014). Recognizing daily and sports activities in two open source machine learning environments using body-worn sensor units. *The Computer Journal, 57*(11), 1649–1667.

Bartlett, J. D., O'Connor, F., Pitchford, N., Torres-Ronda, L., & Robertson, S. J. (2017). Relationships between internal and external training load in team-sport athletes: Evidence for an individualized approach. *International Journal of Sports Physiology and Performance, 12*(2), 230–234.

Bassek, M., Raabe, D., Memmert, D., & Rein, R. (2023). Analysis of motion characteristics and metabolic power in elite male handball players. *Journal of Sports Science & Medicine, 22*(2), 310.

Berrar, D., Lopes, P., & Dubitzky, W. (2019). Incorporating domain knowledge in machine learning for soccer outcome prediction. *Machine Learning, 108*, 97–126.

Bialkowski, A., Lucey, P., Carr, P., Matthews, I., Sridharan, S., & Fookes, C. (2016). Discovering team structures in soccer from spatiotemporal data. *IEEE Transactions on Knowledge and Data Engineering, 28*(10), 2596–2605.

Biermann, H., Theiner, J., Bassek, M., Raabe, D., Memmert, D., & Ewerth, R. (2021, October). A unified taxonomy and multimodal dataset for events in invasion games. In *Proceedings of the 4th international workshop on multimedia content analysis in sports* (pp. 1–10).

Biermann, H., Wieland, F. G., Timmer, J., Memmert, D., & Phatak, A. (2022, September). Towards expected counter-using comprehensible features to predict counterattacks. In *International workshop on machine learning and data mining for sports analytics* (pp. 3–13). Springer.

Biermann, H., Yang, W., Wieland, F. G., Timmer, J., & Memmert, D. (2024). Quantification of turnover danger with xCounter. In U. Brefeld, J. Davis, J. Van Haaren, & A. Zimmermann (Eds.), *Machine learning and data mining for sports analytics. MLSA 2023. Communications in computer and information science* (Vol. 1784). Springer.

Bilek, G., & Ulas, E. (2019). Predicting match outcome according to the quality of opponent in the English premier league using situational variables and team performance indicators. *International Journal of Performance Analysis in Sport, 19*(6), 930–941.

Bongiovanni, T., Trecroci, A., Cavaggioni, L., Rossi, A., Perri, E., Pasta, G., et al. (2021). Importance of anthropometric features to predict physical performance in elite youth soccer: A machine learning approach. *Research in Sports Medicine, 29*(3), 213–224.

Brooks, J., Kerr, M., & Guttag, J. (2016). Using machine learning to draw inferences from pass location data in soccer. *Statistical Analysis and Data Mining: The ASA Data Science Journal, 9*(5), 338–349.

Brundage, M., Avin, S., Clark, J., Toner, H., Eckersley, P., Garfinkel, B., et al. (2018). The malicious use of artificial intelligence: Forecasting, prevention, and mitigation. *arXiv preprint arXiv:1802.07228.*

Bunker, R., & Susnjak, T. (2022). The application of machine learning techniques for predicting match results in team sport: A review. *Journal of Artificial Intelligence Research, 73*, 1285–1322.

Bunker, R. P., & Thabtah, F. (2019). A machine learning framework for sport result prediction. *Applied Computing and Informatics, 15*(1), 27–33.

Buursma, D. (2010). Predicting sports events from past results. In *14th Twente Student Conference on IT* (Vol. 21).

Campbell, P. G., Stewart, I. B., Sirotic, A. C., Drovandi, C., Foy, B. H., & Minett, G. M. (2021). Analysing the predictive capacity and dose-response of wellness in load monitoring. *Journal of Sports Sciences, 39*(12), 1339–1347.

Cao, C. (2012). *Sports data mining technology used in basketball outcome prediction.* Masters Dissertation. Technological University Dublin.

Cao, X., Zhao, X., Tang, H., Fan, N., & Zereg, F. (2023). Football players' strength training method using image processing based on machine learning. *PLoS One, 18*(6), e0287433.

Constantinou, A. C. (2019). Dolores: A model that predicts football match outcomes from all over the world. *Machine Learning, 108*(1), 49–75.

Cui, D., & Jiang, Z. (2021, March). A champion model of men's 100m sprint based on functional data analysis. In *2021 IEEE 6th International Conference on Big Data Analytics (ICBDA)* (pp. 103–107). IEEE.

Cust, E. E., Sweeting, A. J., Ball, K., & Robertson, S. (2019). Machine and deep learning for sport-specific movement recognition: A systematic review of model development and performance. *Journal of Sports Sciences, 37*(5), 568–600.

Ćwiklinski, B., Giełczyk, A., & Choraś, M. (2021). Who will score? A machine learning approach to supporting football team building and transfers. *Entropy, 23*(1), 90.

Danisik, N., Lacko, P., & Farkas, M. (2018, August). Football match prediction using players attributes. In *2018 World Symposium on Digital Intelligence for Systems and Machines (DISA)* (pp. 201–206). IEEE.

David, J. A., Pasteur, R. D., Ahmad, M. S., & Janning, M. C. (2011). NFL prediction using committees of artificial neural networks. *Journal of Quantitative Analysis in Sports, 7*(2), 9. https://doi.org/10.2202/1559-0410.1327

De Beéck, T. O., Jaspers, A., Brink, M. S., Frencken, W. G., Staes, F., Davis, J. J., & Helsen, W. F. (2019). Predicting future perceived wellness in professional soccer: The role of preceding load and wellness. *International Journal of Sports Physiology and Performance, 14*(8), 1074–1080.

Delen, D., Cogdell, D., & Kasap, N. (2012). A comparative analysis of data mining methods in predicting NCAA bowl outcomes. *International Journal of Forecasting, 28*(2), 543–552.

DiCesare, C. A., Green, B., Yuan, W., Diekfuss, J. A., Barber Foss, K. D., Dudley, J., et al. (2020). Machine learning classification of verified head impact exposure strengthens associations with brain changes. *Annals of Biomedical Engineering, 48*, 2772–2782.

Dick, U., & Brefeld, U. (2019). Learning to rate player positioning in soccer. *Big Data, 7*(1), 71–82.

Dindorf, C., Bartaguiz, E., Gassmann, F., & Fröhlich, M. (2022). Conceptual structure and current trends in artificial intelligence, machine learning, and deep learning research in sports: A bibliometric review. *International Journal of Environmental Research and Public Health, 20*(1), 173.

Diniz, P., Abreu, M., Lacerda, D., Martins, A., Pereira, H., Ferreira, F. C., et al. (2022). Pre-injury performance is most important for predicting the level of match participation after Achilles tendon ruptures in elite soccer players: A study using a machine learning classifier. *Knee Surgery, Sports Traumatology, Arthroscopy, 30*(12), 4225–4237.

Draschkowitz, L., Draschkowitz, C., & Hlavacs, H. (2015). Using video analysis and machine learning for predicting shot success in table tennis. *EAI Endorsed Transactions on Creative Technologies, 2*(5), 150096. https://doi.org/10.4108/eai.20-10-2015.150096

Duh, Y. S., & Chang, R. (2021). Recurrent neural network for moonboard climbing route classification and generation. *arXiv preprint arXiv:2102.01788.*

Fast, E., & Horvitz, E. (2017, February). Long-term trends in the public perception of artificial intelligence. *Proceedings of the AAAI Conference on Artificial Intelligence, 31*(1). https://doi.org/10.1609/aaai.v31i1.10635

Fernández, J., Bornn, L., & Cervone, D. (2021). A framework for the fine-grained evaluation of the instantaneous expected value of soccer possessions. *Machine Learning, 110*(6), 1389–1427.

Fradkov, A. L. (2020). Early history of machine learning. *IFAC-PapersOnLine, 53*(2), 1385–1390.

García-Aliaga, A., Marquina, M., Coteron, J., Rodriguez-Gonzalez, A., & Luengo-Sanchez, S. (2021). In-game behaviour analysis of football players using machine learning techniques based on player statistics. *International Journal of Sports Science & Coaching, 16*(1), 148–157.

Garnica-Caparrós, M., & Memmert, D. (2021). Understanding gender differences in professional European football through machine learning interpretability and match actions data. *Scientific Reports, 11*(1), 10805.

Geurkink, Y., Vandewiele, G., Lievens, M., De Turck, F., Ongenae, F., Matthys, S. P., et al. (2019). Modeling the prediction of the session rating of perceived exertion in soccer: Unraveling the puzzle of predictive indicators. *International Journal of Sports Physiology and Performance, 14*(6), 841–846.

Giles, B., Kovalchik, S., & Reid, M. (2020). A machine learning approach for automatic detection and classification of changes of direction from player tracking data in professional tennis. *Journal of Sports Sciences, 38*(1), 106–113.

Goes, F. R., Brink, M. S., Elferink-Gemser, M. T., Kempe, M., & Lemmink, K. A. (2021). The tactics of successful attacks in professional association football: Large-scale spatiotemporal analysis of dynamic subgroups using position tracking data. *Journal of Sports Sciences, 39*(5), 523–532.

Goodfellow, I., Pouget-Abadie, J., Mirza, M., Xu, B., Warde-Farley, D., Ozair, S., et al. (2014). Generative adversarial nets. In Z. Ghahramani, M. Welling, C. Cortes, N. Lawrence, & K. Q. Weinberger (Eds.), *Advances in Neural Information Processing Systems* (Vol. 27). Curran Associates, Inc.. https://proceedings.neurips.cc/paper_files/paper/2014/file/5ca3e9b122f61f8f06494c97b1afccf3-Paper.pdf

Goodfellow, I., Bengio, Y., & Courville, A. (2016). *Deep learning*. MIT Press.

Gu, W., Foster, K., Shang, J., & Wei, L. (2019). A game-predicting expert system using big data and machine learning. *Expert Systems with Applications, 130*, 293–305.

Guha, T., & Ward, R. K. (2011). Learning sparse representations for human action recognition. *IEEE Transactions on Pattern Analysis and Machine Intelligence, 34*(8), 1576–1588.

Gupta, A., Mishra, S., Sahu, S. C., Srinivasarao, U., & Naik, K. J. (2023). Application of convolutional neural networks for COVID-19 detection in X-ray images using inceptionV3 and U-Net. *New Generation Computing, 41*(2), 475–502. https://doi.org/10.1007/s00354-023-00217-2

Hagemann, D., Spinath, F. M., & Mueller, E. M. (2022). *Differentielle Psychologie und Persönlichkeitsforschung*. Kohlhammer Verlag.

Haghighat, M., Rastegari, H., Nourafza, N., Branch, N., & Esfahan, I. (2013). A review of data mining techniques for result prediction in sports. *Advances in Computer Science: An International Journal, 2*(5), 7–12.

Hammes, F., Hagg, A., Asteroth, A., & Link, D. (2022). Artificial intelligence in elite sports—A narrative review of success stories and challenges. *Frontiers in Sports and Active Living, 4*, 861466.

Huang, K. Y., & Chang, W. L. (2010, July). A neural network method for prediction of 2006 world cup football game. In *The 2010 International Joint Conference on Neural Networks (IJCNN)* (pp. 1–8). IEEE.

Hubáček, O., Šourek, G., & Železný, F. (2019). Learning to predict soccer results from relational data with gradient boosted trees. *Machine Learning, 108*, 29–47.

Hucaljuk, J., & Rakipović, A. (2011, May). Predicting football scores using machine learning techniques. In *2011 Proceedings of the 34th International Convention MIPRO* (pp. 1623–1627). IEEE.

Inan, T., & Cavas, L. (2021). Estimation of market values of football players through artificial neural network: A model study from the Turkish super league. *Applied Artificial Intelligence, 35*(13), 1022–1042.

Ivanković, Z., Racković, M., Markoski, B., Radosav, D., & Ivković, M. (2010, November). Analysis of basketball games using neural networks. In *2010 11th International Symposium on Computational Intelligence and Informatics (CINTI)* (pp. 251–256). IEEE.

Jamil, M., Phatak, A., Mehta, S., Beato, M., Memmert, D., & Connor, M. (2021). Using multiple machine learning algorithms to classify elite and sub-elite goalkeepers in professional men's football. *Scientific Reports, 11*(1), 22703.

Janiesch, C., Zschech, P., & Heinrich, K. (2021). Machine learning and deep learning. *Electronic Markets, 31*(3), 685–695.

Jaspers, A., De Beéck, T. O., Brink, M. S., Frencken, W. G., Staes, F., Davis, J. J., & Helsen, W. F. (2018). Relationships between the external and internal training load in professional soccer: What can we learn from machine learning? *International Journal of Sports Physiology and Performance, 13*(5), 625–630.

Joseph, A., Fenton, N. E., & Neil, M. (2006). Predicting football results using Bayesian nets and other machine learning techniques. *Knowledge-Based Systems, 19*(7), 544–553.

Kahn, J. (2003). *Neural network prediction of NFL football games* (Vol. 2003, pp. 9–15). World Wide Web Electronic Publication.

Kato, R., Kameda, T., Yamada, E., Fujimoto, H., & Aida, H. (2021, August). Machine learning for handball game analysis using valid statistics linked to victory. In *International Conference on Security, Privacy, and Anonymity in Computation, Communication, and Storage* (pp. 118–121). Springer.

Kelly, A. L., Williams, C. A., Cook, R., Sáiz, S. L. J., & Wilson, M. R. (2022). A multidisciplinary investigation into the talent development processes at an English Football Academy: A machine learning approach. *Sports, 10*(10), 159.

Kempe, M., Grunz, A., & Memmert, D. (2015). Detecting tactical patterns in basketball: Comparison of merge self-organising maps and dynamic controlled neural networks. *European Journal of Sport Science, 15*(4), 249–255.

Kempen, L. (2018). *A fair grade: Assessing difficulty of climbing routes through machine learning. Formal methods and tools*. University of Twente.

Kern, J., Lober, T., Hermsdörfer, J., & Endo, S. (2022). A neural network for the detection of soccer headers from wearable sensor data. *Scientific Reports, 12*(1), 18128.

Keshtkar Langaroudi, M., & Yamaghani, M. (2019). Sports result prediction based on machine learning and computational intelligence approaches: A survey. *Journal of Advances in Computer Engineering and Technology, 5*(1), 27–36.

Kim, M., & Park, S. (2020). Golf swing segmentation from a single IMU using machine learning. *Sensors, 20*(16), 4466.

Knauf, K., Memmert, D., & Brefeld, U. (2016). Spatio-temporal convolution kernels. *Machine Learning, 102*, 247–273.

König, R., Johansson, U., Riveiro, M., & Brattberg, P. (2017). Modeling golf player skill using machine learning. In *Machine Learning and Knowledge Extraction: First IFIP TC 5, WG 8.4, 8.9, 12.9 International Cross-Domain Conference, CD-MAKE 2017, Reggio, Italy, August 29–September 1, 2017, Proceedings 1* (pp. 275–294). Springer.

Koopman, S. J., & Lit, R. (2019). Forecasting football match results in national league competitions using score-driven time series models. *International Journal of Forecasting, 35*(2), 797–809.

Kovalchik, S., & Reid, M. (2018). A shot taxonomy in the era of tracking data in professional tennis. *Journal of Sports Sciences, 36*(18), 2096–2104.

Lang, S., Wild, R., Isenko, A., & Link, D. (2022). Predicting the in-game status in soccer with machine learning using spatiotemporal player tracking data. *Scientific Reports, 12*(1), 16291.

Leser, R., Hoch, T., Tan, X., Moser, B., Kellermayr, G., & Baca, A. (2019). Finding efficient strategies in 3-versus-2 small-sided games of youth soccer players. *Kinesiology, 51*(1), 110–118.

Li, N., & Wu, D. D. (2010). Using text mining and sentiment analysis for online forums hotspot detection and forecast. *Decision Support Systems, 48*(2), 354–368.

Link, D., & Hoernig, M. (2017). Individual ball possession in soccer. *PLoS One, 12*(7), e0179953.

Liu, G., & Schulte, O. (2018). Deep reinforcement learning in ice hockey for context-aware player evaluation. *arXiv preprint arXiv:1805.11088*.

Loeffelholz, B., Bednar, E., & Bauer, K. W. (2009). Predicting NBA games using neural networks. *Journal of Quantitative Analysis in Sports, 5*(1), 7. https://doi.org/10.2202/1559-0410.1156

López-Valenciano, A., Ayala, F., Puerta, J. M., Croix, M. D. S., Vera-García, F., Hernández-Sánchez, S., et al. (2018). A preventive model for muscle injuries: A novel approach based on learning algorithms. *Medicine and Science in Sports and Exercise, 50*(5), 915.

Low, B., Coutinho, D., Gonçalves, B., Rein, R., Memmert, D., & Sampaio, J. (2020). A systematic review of collective tactical behaviours in football using positional data. *Sports Medicine,50*, 343–385.

Lu, W. L., Ting, J. A., Little, J. J., & Murphy, K. P. (2013). Learning to track and identify players from broadcast sports videos. *IEEE Transactions on Pattern Analysis and Machine Intelligence, 35*(7), 1704–1716.

Lu, N., Wu, Y., Feng, L., & Song, J. (2018). Deep learning for fall detection: Three-dimensional CNN combined with LSTM on video kinematic data. *IEEE Journal of Biomedical and Health Informatics, 23*(1), 314–323.

Mahmood, Z., Daud, A., & Abbasi, R. A. (2021). Using machine learning techniques for rising star prediction in basketball. *Knowledge-Based Systems, 211*, 106506.

Maier, T., Meister, D., Trösch, S., & Wehrlin, J. P. (2018). Predicting biathlon shooting performance using machine learning. *Journal of Sports Sciences, 36*(20), 2333–2339.

McCabe, A., & Trevathan, J. (2008, April). Artificial intelligence in sports prediction. In *Fifth International Conference on Information Technology: New Generations (itng 2008)* (pp. 1194–1197). IEEE.

Mehrasa, N., Zhong, Y., Tung, F., Bornn, L., & Mori, G. (2018). Deep learning of player trajectory representations for team activity analysis. In *11th MIT Sloan Sports Analytics Conference* (Vol. 2, p. 3).

Memmert, D. (2023). Reale Datensätze: Positionsdaten. In D. Memmert (Ed.), *Handbuch Sportinformatik*. Springer.
Memmert, D. (2024). Metriken im Fußball. In D. Memmert (Ed.), *Handbuch Sporttechnologie*. Springer.
Memmert, D., & Raabe, D. (2024). *Data analytics in football: Positional data collection, modelling and analysis* (3rd ed.). Routledge.
Memmert, D., Raabe, D., Knyazev, A., Franzen, A., Zekas, L., Rein, R., et al. (2016). Big data im profi-Fußball: Analyse von Positionsdaten der Fußball-Bundesliga mit neuen innovativen key performance Indikatoren. *Leistungssport, 46*(5), 21–26.
Memmert, D., Lemmink, K. A., & Sampaio, J. (2017). Current approaches to tactical performance analyses in soccer using position data. *Sports Medicine, 47*(1), 1–10.
Memmert, D., Raabe, D., Schwab, S., & Rein, R. (2019). A tactical comparison of the 4-2-3-1 and 3-5-2 formation in soccer: A theory-oriented, experimental approach based on positional data in an 11 vs. 11 game set-up. *PLoS One, 14*(1), e0210191.
Memmert, D. (2021). Match Analysis. Abingdon: Routledge.
Miljković, D., Gajić, L., Kovačević, A., & Konjović, Z. (2010, September). The use of data mining for basketball matches outcomes prediction. In *IEEE 8th International Symposium on Intelligent Systems and Informatics* (pp. 309–312). IEEE.
Mohandas, A., Ahsan, M., & Haider, J. (2023). Tactically maximize game advantage by predicting football substitutions using machine learning. *Big Data and Cognitive Computing, 7*(2), 117.
Montoliu, R., Martín-Félez, R., Torres-Sospedra, J., & Martínez-Usó, A. (2015). Team activity recognition in association football using a bag-of-words-based method. *Human Movement Science, 41*, 165–178.
Mukhamediev, R. I., Symagulov, A., Kuchin, Y., Yakunin, K., & Yelis, M. (2021). From classical machine learning to deep neural networks: A simplified scientometric review. *Applied Sciences, 11*(12), 5541.
Mülling, K., Kober, J., Kroemer, O., & Peters, J. (2013). Learning to select and generalize striking movements in robot table tennis. *The International Journal of Robotics Research, 32*(3), 263–279.
Nweke, H. F., Teh, Y. W., Al-Garadi, M. A., & Alo, U. R. (2018). Deep learning algorithms for human activity recognition using mobile and wearable sensor networks: State of the art and research challenges. *Expert Systems with Applications, 105*, 233–261.
O'Donoghue, P., & Williams, J. (2004). An evaluation of human and computer-based predictions of the 2003 rugby union world cup. *International Journal of Computer Science in Sport, 3*(1), 5–22.
O'Donoghue, P., Ball, D., Eustace, J., McFarlan, B., & Nisotaki, M. (2016). Predictive models of the 2015 Rugby World Cup: Accuracy and application. *International Journal of Computer Science in Sport, 15*(1), 37–58.
Odachowski, K., & Grekow, J. (2013). Using bookmaker odds to predict the final result of football matches. In *Knowledge engineering, machine learning and lattice computing with applications: 16th International Conference, KES 2012, San Sebastian, Spain, September 10–12, 2012, Revised Selected Papers 16* (pp. 196–205). Springer.
Oliver, J. L., Ayala, F., Croix, M. B. D. S., Lloyd, R. S., Myer, G. D., & Read, P. J. (2020). Using machine learning to improve our understanding of injury risk and prediction in elite male youth football players. *Journal of Science and Medicine in Sport, 23*(11), 1044–1048.
Pappalardo, L., & Cintia, P. (2018). Quantifying the relation between performance and success in soccer. *Advances in Complex Systems, 21*(03n04), 1750014.
Pathak, N., & Wadhwa, H. (2016). Applications of modern classification techniques to predict the outcome of ODI cricket. *Procedia Computer Science, 87*, 55–60.
Perl, J., Grunz, A., & Memmert, D. (2013). Tactics analysis in soccer: An advanced approach. *International Journal of Computer Science in Sport, 12*, 33–44.
Perri, E., Simonelli, C., Rossi, A., Trecroci, A., Alberti, G., & Iaia, F. M. (2021). Relationship between wellness index and internal training load in soccer: Application of a machine learning model. *International Journal of Sports Physiology and Performance, 16*(5), 695–703.

Peterson, K. D. (2018). Resting heart rate variability can predict track and field sprint performance. *OA Journal-Sports, 1*(1).

Phatak, A. A., Mehta, S., Wieland, F. G., Jamil, M., Connor, M., Bassek, M., & Memmert, D. (2022). Context is key: Normalization as a novel approach to sport specific preprocessing of KPI's for match analysis in soccer. *Scientific Reports, 12*(1), 1117.

Prasetio, D. (2016, August). Predicting football match results with logistic regression. In *2016 International Conference On Advanced Informatics: Concepts, Theory And Application (ICAICTA)* (pp. 1–5). IEEE. https://doi.org/10.1109/ICAICTA.2016.7803111

Purucker, M. C. (1996). Neural network quarterbacking. *IEEE Potentials, 15*(3), 9–15.

Raabe, D., Biermann, H., Bassek, M., Wohlan, M., Komitova, R., Rein, R., et al. (2022). Floodlight—A high-level, data-driven sports analytics framework. *arXiv preprint arXiv:2206.02562.*

Raabe, D., Nabben, R., & Memmert, D. (2023). Graph representations for the analysis of multi-agent spatiotemporal sports data. *Applied Intelligence, 53*(4), 3783–3803.

Ravi, D., Wong, C., Lo, B., & Yang, G. Z. (2016). A deep learning approach to on-node sensor data analytics for mobile or wearable devices. *IEEE Journal of Biomedical and Health Informatics, 21*(1), 56–64.

Razali, N., Mustapha, A., Utama, S., & Din, R. (2018, May). A review on football match outcome prediction using bayesian networks. *Journal of Physics: Conference Series, 1020*(1), 012004. IOP Publishing.

Reed, D., & O'Donoghue, P. (2005). Development and application of computer-based prediction methods. *International Journal of Performance Analysis in Sport, 5*(3), 12–28.

Rein, R., & Memmert, D. (2016). Big data and tactical analysis in elite soccer: Future challenges and opportunities for sports science. *Springerplus, 5*(1), 1–13.

Rein, R., Perl, J., & Memmert, D. (2017). Maybe a tad early for a Grand Unified Theory: Commentary on "Towards a Grand Unified Theory of sports performance". *Human Movement Science, 56*(Part A), 173–175. https://doi.org/10.1016/j.humov.2017.04.011

Rico-González, M., Pino-Ortega, J., Méndez, A., Clemente, F., & Baca, A. (2023). Machine learning application in soccer: A systematic review. *Biology of Sport, 40*(1), 249–263.

Rodas, G., Osaba, L., Arteta, D., Pruna, R., Fernández, D., & Lucia, A. (2019). Genomic prediction of tendinopathy risk in elite team sports. *International Journal of Sports Physiology and Performance, 15*(4), 489–495.

Rodrigues, F., & Pinto, Â. (2022). Prediction of football match results with machine learning. *Procedia Computer Science, 204*, 463–470.

Rojas-Valverde, D., Gómez-Carmona, C. D., Fernández-Fernández, J., García-López, J., García-Tormo, V., Cabello-Manrique, D., & Pino-Ortega, J. (2020). Identification of games and sex-related activity profile in junior international badminton. *International Journal of Performance Analysis in Sport, 20*(3), 323–338.

Rommers, N., Rössler, R., Verhagen, E., Vandecasteele, F., Verstockt, S., Vaeyens, R., et al. (2020). A machine learning approach to assess injury risk in elite youth football players. *Medicine and Science in Sports and Exercise, 52*(8), 1745–1751.

Rossi, A., Pappalardo, L., Cintia, P., Iaia, F. M., Fernández, J., & Medina, D. (2018). Effective injury forecasting in soccer with GPS training data and machine learning. *PLoS One, 13*(7), e0201264.

Schlenger, J., Wunderlich, F., Raabe, D., & Memmert, D. (2023). Systematic analysis of position-data-based key performance indicators. *International Journal of Computer Science in Sport, 22*(1), 80–101.

Sheng, B., Li, P., Zhang, Y., Mao, L., & Chen, C. P. (2020). GreenSea: Visual soccer analysis using broad learning system. *IEEE Transactions on Cybernetics, 51*(3), 1463–1477.

Shi, Z., Moorthy, S., & Zimmermann, A. (2013, September). Predicting NCAAB match outcomes using ML techniques–some results and lessons learned. In *ECML/PKDD 2013 Workshop on Machine Learning and Data Mining for Sports Analytics.*

Stöckl, M., & Morgan, S. (2013). Visualization and analysis of spatial characteristics of attacks in field hockey. *International Journal of Performance Analysis in Sport, 13*(1), 160–178.

Szczepański, Ł., & McHale, I. (2016). Beyond completion rate: Evaluating the passing ability of footballers. *Journal of the Royal Statistical Society. Series A (Statistics in Society), 179*(2), 513–533.

Tabrizi, S. S., Pashazadeh, S., & Javani, V. (2020). Comparative study of table tennis forehand strokes classification using deep learning and SVM. *IEEE Sensors Journal, 20*(22), 13552–13561.

Tax, N., & Joustra, Y. (2015). Predicting the Dutch football competition using public data: A machine learning approach. *Transactions on Knowledge and Data Engineering, 10*(10), 1–13.

Thabtah, F., Zhang, L., & Abdelhamid, N. (2019). NBA game result prediction using feature analysis and machine learning. *Annals of Data Science, 6*(1), 103–116.

Theiner, J., Gritz, W., Müller-Budack, E., Rein, R., Memmert, D., & Ewerth, R. (2022). Extraction of positional player data from broadcast soccer videos. In *Proceedings of the IEEE/CVF Winter Conference on Applications of Computer Vision* (pp. 823–833).

Tixier, A. J. P., Hallowell, M. R., Rajagopalan, B., & Bowman, D. (2016). Application of machine learning to construction injury prediction. *Automation in Construction, 69*, 102–114.

Toosi, A., Bottino, A. G., Saboury, B., Siegel, E., & Rahmim, A. (2021). A brief history of AI: How to prevent another winter (a critical review). *PET clinics, 16*(4), 449–469.

Tsai, T. Y., Lin, Y. Y., Jeng, S. K., & Liao, H. Y. M. (2021). End-to-end key-player-based group activity recognition network applied to basketball offensive tactic identification in limited data scenarios. *IEEE Access, 9*, 104395–104404.

Tümer, A. E., Akyıldız, Z., Güler, A. H., Saka, E. K., Ievoli, R., Palazzo, L., & Clemente, F. M. (2022). Prediction of soccer clubs' league rankings by machine learning methods: The case of Turkish Super League. *Proceedings of the Institution of Mechanical Engineers, Part P: Journal of Sports Engineering and Technology*. https://doi.org/10.1177/17543371221140492

Turing, A. M. (2012). Computing machinery and intelligence 1950. *The Essential Turing: The Ideas That Gave Birth to the Computer Age*, 433–464.

Valero, C. S. (2016). Predicting win-loss outcomes in MLB regular season games—A comparative study using data mining methods. *International Journal of Computer Science in Sport, 15*(2), 91–112.

Valle, X., Mechó, S., Alentorn-Geli, E., Järvinen, T. A., Lempainen, L., Pruna, R., et al. (2022). Return to play prediction accuracy of the MLG-R classification system for hamstring injuries in football players: A machine learning approach. *Sports Medicine, 52*(9), 2271–2282.

van den Tillaar, R., Bhandurge, S., & Stewart, T. (2021). Can machine learning with IMUs be used to detect different throws and estimate ball velocity in team handball? *Sensors, 21*(7), 2288.

van Meurs, E., Buszard, T., Kovalchik, S., Farrow, D., & Reid, M. (2021). Interpersonal coordination in tennis: Assessing the positional advantage index with Australian Open Hawkeye data. *International Journal of Performance Analysis in Sport, 21*(1), 22–32.

Vaswani, A., Shazeer, N., Parmar, N., Uszkoreit, J., Jones, L., Gomez, A. N., et al. (2017). Attention is all you need. *Advances in Neural Information Processing Systems, 30*.

Vijay Fidelis, J., & Karthikeyan, E. (2022). Optimization of artificial neural network parameters in selection of players for soccer match. In *Sustainable Advanced Computing: Select Proceedings of ICSAC 2021* (pp. 275–288). Springer.

Weissbock, J., & Inkpen, D. (2014). Combining textual pre-game reports and statistical data for predicting success in the national hockey league. In *Advances in Artificial Intelligence: 27th Canadian Conference on Artificial Intelligence, Canadian AI 2014, Montréal, QC, Canada, May 6–9, 2014. Proceedings 27* (pp. 251–262). Springer.

Weissbock, J., Viktor, H. L., & Inkpen, D. (2013, September). Use of performance metrics to forecast success in the national hockey league. In *MLSA@ PKDD/ECML* (pp. 39–48).

Wenninger, S., Link, D., & Lames, M. (2020). Performance of machine learning models in application to beach volleyball data. *International Journal of Computer Science in Sport, 19*(1), 24–36.

Werthner, H., Prem, E., Lee, E. A., & Ghezzi, C. (2022). *Perspectives on digital humanism*. Springer.

Whiteside, D., & Reid, M. (2017). Spatial characteristics of professional tennis serves with implications for serving aces: A machine learning approach. *Journal of Sports Sciences, 35*(7), 648–654.

Wolpert, D. H., & Macready, W. G. (1997). No free lunch theorems for optimization. *IEEE Transactions on Evolutionary Computation, 1*(1), 67–82.

Xing, J., Ai, H., Liu, L., & Lao, S. (2010). Multiple player tracking in sports video: A dual-mode two-way bayesian inference approach with progressive observation modeling. *IEEE Transactions on Image Processing, 20*(6), 1652–1667.

Yamashita, R., Nishio, M., Do, R. K. G., & Togashi, K. (2018). Convolutional neural networks: An overview and application in radiology. *Insights Into Imaging, 9*, 611–629.

Zago, M., Sforza, C., Dolci, C., Tarabini, M., & Galli, M. (2019). Use of machine learning and wearable sensors to predict energetics and kinematics of cutting maneuvers. *Sensors, 19*(14), 3094.

Zdravevski, E., & Kulakov, A. (2009, September). System for prediction of the winner in a sports game. In D. Davcev & J. M. Gómez (Eds.), *International Conference on ICT Innovations* (pp. 55–63). Springer.

Zhu, M., Sun, Z., Zhang, Z., Shi, Q., He, T., Liu, H., et al. (2020). Haptic-feedback smart glove as a creative human-machine interface (HMI) for virtual/augmented reality applications. *Science Advances, 6*(19), eaaz8693.

Supervised and Unsupervised Learning 2

Marc Garnica Caparrós

Abstract

This chapter explores the fundamental differences between Supervised and Unsupervised Learning, two important families of algorithms in the field of Machine Learning. The basic methodologies and distinctions between these paradigms are presented and described regarding model optimization and evaluation. In Supervised Learning, classification and regression problems are introduced. Classification problems attempt to learn from known categories in the data, while regression problems involve a continuous numerical variable as the outcome of the prediction. On the other side, Unsupervised Learning is presented as a different paradigm where the discovery and learning from data is performed without prior information or known labels but by exploring the hidden structures of the dataset. Furthermore, this chapter also documents the different evaluation challenges and techniques. The utilization of these different paradigms depends on the business problems and the data requirements at hand. However, having a broad overview of the possible methodologies available and how they interrelate is crucial for sustainable and reliable Machine Learning integration.

2.1 Learning from Data

The use of Machine Learning (ML) for data analysis has become one of the most important drivers of innovation and business growth. Several sectors are profiting from these techniques to leverage large datasets with priceless business value and extract insights to support decision-making. As defined by Kumar et al. (2020), ML

M. G. Caparrós (✉)
German Sport University Cologne, Köln, Germany
e-mail: m.garnica@dshs-koeln.de

D. Memmert (ed.), *Artificial Intelligence and Machine Learning in Sports Science*,
https://doi.org/10.1007/978-3-662-70155-3_2

is a segment of Artificial Intelligence (AI) that provides a system with the ability to learn from data and improve over time without being explicitly programmed for each task. ML originated in academic Computer Science and its path to interdisciplinary research and strategic industries reflected a business turn towards data-driven cultures and decision intelligence approaches. One of the most important high-level benefits of ML integration is the dynamic adaptability and continuous improvement, which facilitates identifying trends, adapting to new strategies, and making informed choices in evolving environments (Jordan & Mitchell, 2015). ML consists of a variety of algorithms that iteratively learn from data. By ingesting training data, these algorithms can produce more accurate models. The term ML model is defined as the result of training an ML algorithm with data. Once the model has been trained, the model is able to yield an output given a specific input. For example, a predictive algorithm will train a predictive model that will deliver a forecast based on the training data when fed with new data.

Before the ML era, human observation could have easily overlooked complex patterns or associations, specifically in large datasets. It is evident that human intelligence and learning are more advanced than current state-of-the-art ML, but computers have a higher capacity to memorize, recall, and process data. In fact, the size of data is one of the most relevant components that provided ML with a fertile terrain to grow and proliferate in science and industry. The accuracy of such models can increase significantly if trained on big data (Günther et al., 2017).

▶ Several concepts sometimes overlap in the Machine Learning community. While not trying to reach a consensus, in this chapter, we refer to *Machine Learning algorithms* as the computer-based technique to train and optimize a learning process from data. The result of the execution of a machine learning algorithm is referred to as a *Machine Learning Model. Training data* refers to the data used in the learning process, and *test data* refers to the part of the data set used to evaluate the performance of the model.

▶ One of the key differences between traditional methods for inferential statistics and Machine Learning models is the shift towards less restrictive, computation-intensive learning of the data. Machine Learning often has a reduced number of requirements in the data, and they provide a robust exploitation method. However, this does not mean that data has to be trusted blindly. Feature engineering is a key component that determines the success of the ML model. Feature engineering techniques might produce transformations on your original dataset that relate to the target variable of study and introduce domain expertise into the process.

Depending on the business use case being addressed, ML algorithms have different approaches to developing, training, and improving the resulting models. These different approaches are diverse in the requirements they expect from the data, which type of algorithms can be explored, and how the models can be employed and evaluated. The first subset of use cases or problems contains a target of interest, a response variable that is the focus of the study (Did the player get injured? What is my team's win probability for the next match?); this field of ML algorithms is referred to as *supervised learning.* Additionally, Supervised Learning involves a set of training data with known and validated values of the target variable (Hastie et al., 2009; Suthaharan, 2016). A visual description of the problem definition in Supervised Learning is illustrated in Fig. 2.1. For instance, we might have in our possession a historical data set of soccer players' load the last month (e.g., how many high-intensity runs they did, how many minutes they played in matches, how many training sessions they performed, etc.) and the information whether the player got injured or not, our target variable or label. In this case, our desire would be that the computer learns from this data and understands the hidden patterns that associate the target variable with the set of input features.

In the second subset of algorithms, imagine we have in possession the same dataset but without the information on whether the player got injured or not. In this case, our initial interest in learning when a player is injured would not be feasible directly. When lacking a target variable, the approaches to learning from data are switched to compile the data for further learning or create new features on which to learn. This set of algorithms is grouped into *unsupervised learning* (see Fig. 2.2). Unsupervised Learning algorithms, such as Clustering, discover natural patterns in the data and allow the automatic dissemination of our data without predefining the differences between these groups (Hastie et al., 2009). Similarly, Dimensionality Reduction algorithms perform transformations on our input data space to explain most of the data's variability, often with fewer features. Additionally, rule mining techniques, such as Association Rule Mining, explore and identify strong rules discovered in datasets based on frequency and interestingness measures, providing

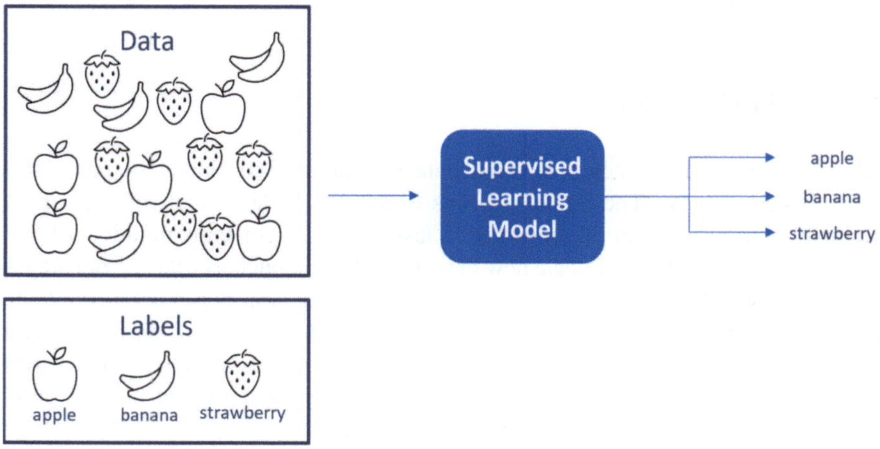

Fig. 2.1 Supervised learning problem definition. Training data include the data samples as well as the target variable, in this case, the type of fruit. The resulting models are able to assign a type of fruit to each data sample

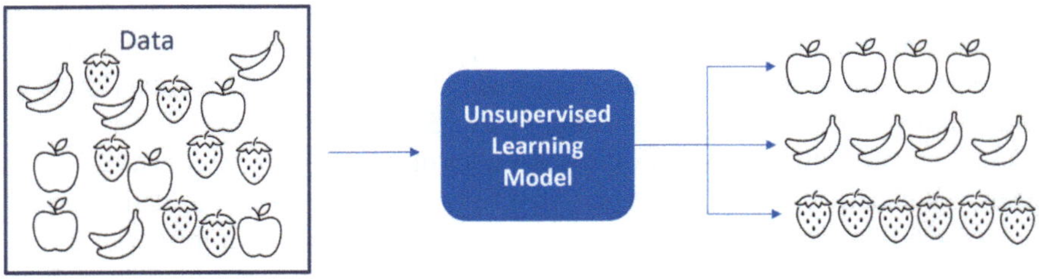

Fig. 2.2 Unsupervised learning problem definition. In this case, the target variable is not known. Therefore, the model cannot learn the types of fruits from the data. However, unsupervised learning algorithms are able to assess the features of the data to, for instance, disseminate the data samples between meaningful groups

insights into hidden relationships between variables without the need for predefined classes or categories.

Both families of ML algorithms play a significant role in solving real-world problems and improving business decisions. The type of model to build depends on your own data and the practical use cases. The following sections will dive deeper into Supervised and Unsupervised Learning details.

2.2 Supervised Learning

Two fundamental methodologies with differentiated algorithms are used within Supervised Learning: Classification and Regression models. Table 2.1 includes the list of the most popular classification algorithms. It is important to mention that although Deep Learning models have not been discussed in detail in this chapter, most Deep Learning algorithms, including Neural Networks, typically perform supervised learning tasks.

2.2.1 Classification

Classification models categorize input data into predefined classes or categories (i.e., discrete values). The model predicts or assigns a class label for each input example from a finite set of categories. Classification algorithms iteratively build classifier objects that can allocate new data into different buckets (classes or categories) (See Fig. 2.1). For example, in the context of sports, injury prediction could be defined as a classification problem where we want to predict if a player will get injured or not depending on a series of factors such as the last month's load metrics and others. In this case, players in our data can be categorized into two classes: players that experienced an injury and those that did not.

Table 2.1 categorizes the classification algorithms between linear and nonlinear decision boundaries. Classification algorithms such as Logistic Regression are linear algorithms (i.e., the decision boundary they are able to learn as the separation

Table 2.1 List of the main supervised learning algorithms categorized by whether they are able to perform classifications or regression tasks and linear or nonlinear decision boundaries (Hastie et al., 2009)

Classification algorithm	Description	Type	Decision boundary
Linear regression	Linear combination of the input features to explain the target variable.	Regression	Linear
Logistic regression	Categorize observations based on quantitative features in binary scenarios. It models the probability that a given input belongs to a particular class.	Classification	Linear
Decision trees	The algorithm splits the training data recursively into subsets based on information gain or other variance indexes. In the end case, a decision tree is built that outputs the predicted class.	Classification and regression	Non-linear
K-nearest Neighbors	A distance formula between instances in the training data allows to compute the "nearest" objects to the given input. The predicted class is chosen based on the classes of these selected neighbors.	Classifications and regression	Non-linear
Random forests/ boosting	Based on decision trees, random forests ensemble several trees with slight modifications to produce a stronger prediction robust to random biases. Boosting algorithms optimize the weight assignment to each tree candidate depending on the misclassified instances.	Classification and regression	Non-linear
Support vector machines	High-dimensional space separation of the different classes to predict the target class.	Classification and regression	Linear (several kernels are available to support nonlinear)
Naïve Bayes	Bayes theorem-based algorithm for classification.	Classification	Non-linear

between classes is a straight line). This might be a limitation in the classification task. Nonlinear algorithms, such as Decision Trees, solve this issue by providing a more flexible decision boundary that learns the data's characteristics. In some other cases, such as with Support Vector Machines, while by definition they are linear algorithms, they include modifications to be able to perform non-linear boundaries. However, nonlinear algorithms are more prone to overfitting.

Classification problems can be distinguished by how many categories the class label contains. Basic classification usually involves binary labels, which involves the optimization of a model to identify whether or not a data object belongs to a class (e.g., player injured or not, goal not goal, etc.). On the other side, when the target variable is disseminated into more than two classes, models must adapt to additional considerations, such as handling class imbalance and extend the evaluation metrics.

▶ For tackling multiclass classification problems (Huang et al., 2012), there are multiple strategies available. The *One-vs-All* (OvA) and *One-vs-One* (OvO) methods simplify multiclass issues into binary classification subproblems. Additionally, alternative techniques incorporate error-correction mechanisms and hierarchical classification structures to improve the precision of multiclass classifications.

2.2.2 Regression

In some other scenarios, discreetly predicting whether a player will be injured might be our model's desired output. For example, coaches might want to predict the number of kilometers run by each player depending on previous match information and training load. Regression algorithms perform a supervised task to predict numerical values.

Linear regression is one of the simplest regression algorithms. This algorithm assumes that the target numerical variable can be explained by a series of additive combinations of the input features as a linear function. The regression model is then obtained by obtaining the best-fitting line on the dataset. Similarly to Classification algorithms, Regression models must adapt to nonlinear data in some use cases. Random Forests or K-nearest neighbors are examples of nonlinear algorithms that can perform in complex, nonlinear data.

Most of the algorithms provide versatility between classification and regression tasks in the case of algorithms based on Decision Trees (including Random Forests and Boosting methods). Predictions are made by passing new data instances through the tree structure. Finally, the predicted value is computed as the mean target value of the training instances in the leaf node (Loh, 2011).

2.2.3 Evaluation

Evaluation techniques offer an objective measure of a model's performance on unseen data. The evaluation and optimization of ML models can be tailored to specific business challenges. In certain scenarios, the primary goal may be to enhance the accuracy of the model's predictions—its ability to identify classes or predict numerical values correctly. Conversely, other scenarios might prioritize avoiding the omission of certain classes. For example, a model designed to predict injuries might focus more on not overlooking actual injury risks rather than misidentifying non-injury situations. In such contexts, stakeholders like coaches or sports scientists might favor a more conservative approach to prevent player injuries.

In evaluating model performance, two main strategies are employed. The first, the holdout method, splits the original dataset into two parts: one split for training and optimizing the model and another, typically smaller, split for evaluating the model's performance on new data. Alternatively, cross-validation offers a more robust but more computationally intensive technique (Hastie et al., 2009). Through k-fold cross-validation, the dataset is split multiple times into training and testing sets. This technique iterates several times, alternating the data splits used for training and testing. The aggregated model at the end of the optimization process has been exposed and validated by the entire dataset.

The approaches to evaluating model performance are similar for both classification and regression techniques (Caruana & Niculescu-Mizil, 2006). In the context of classification, the accuracy of a model is determined by comparing the predicted labels against the actual test labels. Beyond the count-based measure of accuracy, additional metrics like the class-specific confusion matrix, the Receiver Operating Characteristic (ROC) curve, and the area under the ROC curve are also utilized. On the other hand, the accuracy of regression models is assessed by employing a distance function to quantify the discrepancy between the predicted values and the actual numerical outcomes.

2.3 Unsupervised Learning

As introduced at the beginning of this chapter, unsupervised learning involves a series of ML algorithms that learn from data without relying on known categories or variables. In this context, the learning of data differs from actually classifying or predicting a certain value but in exploiting hidden patterns in the data.

2.3.1 Dimensionality Reduction

Unsupervised learning plays a very important role in handling high-dimensional data. In these cases, unsupervised techniques can help us reduce the number of features of the data while not loosing information. These techniques are grouped under Dimensionality Reduction. Dimensionality Reduction techniques perform a sequence of transformations on the data to reduce the feature number while minimizing the loss of variance.

One of the most well-known and effective methods for Dimensionality Reduction is Principle Components Analysis (PCA) (Saporta & Niang, 2009). PCA is a statistical procedure that employs orthogonal transformation to convert potentially correlated variables into a set of linearly uncorrelated variables (i.e., principal components). PCA is a very common procedure for data preprocessing and visualization to provide insights into the structure and distribution of the data that might not be apparent in the high-dimensional space.

2.3.2 Clustering

The main goal of Clustering algorithms is to discover the inherent groups in data. For instance, in sports scenarios, we might want to categorize opponent teams in different categories without specifying which type of categories to look for.

Clustering techniques group data objects in the same group while maximizing the differences between groups. The nature of Clustering techniques makes it a very flexible methodology for exploratory data analysis, anomaly detection or as a pre-processing step for other algorithms. Clustering algorithms can be disseminated into these families and algorithms (Hastie et al., 2009):

- **Hierarchical Clustering** is based in the definition of a dendrogram, a tree that visualized the arrangement of the clusters and the object-wise similarities. In the divisive approaches, the dataset is split recursively until having as many splits as samples in the data. On the other side, agglomerative approaches group pair-wise objects until obtained a single group with all samples. In both approaches, the dendrogram reports all the iterations and allows exploring different grouping possibilities.
- **K-means Clustering** is a partition-based that divides the dataset into non-overlapping subsets. Points that belong to a cluster are as close to each other as possible and as far from points in other clusters as possible.
- **Density-based Clustering** was designed to cope with some of the limitations of other types of clustering (Kriegel et al., 2011). Unlike partitioning-based clustering such as k-means, density-based clustering does not require the number of clusters to be specified in advance and can identify any arbitrary-shaped clusters.

Evaluating Clustering algorithms is not a trivial task because of the lack of known labels or a target attribute to be compared to. Therefore, clustering techniques rely upon relative measures that provide insights into the clustering process and the resulting grouping of the data. On the one hand, internal measures typically assess the compactness or separation of the clusters in the data, such as the Silhouette Score. On the other hand, relative measures compare different clustering models on the same data to determine the optimal clustering configuration.

Evaluating Clustering algorithms is not a trivial task because of the lack of known labels or a target attribute to be compared to. Therefore, clustering techniques rely upon relative measures that provide insights into the clustering process and the resulting grouping of the data. On the one hand, internal measures typically assess the compactness or separation of the clusters in the data, such as the Silhouette Score. On the other hand, relative measures compare different clustering models on the same data to determine the optimal clustering configuration.

2.3.3 Association Rule Mining

Association Rule Mining is a data mining technique that discovers frequent correlations between items organized in sequences or item sets (Agrawal et al., 1993; Han et al., 2000). A classic example of this type of analysis is market basket analysis, where shops and supermarkets want to analyze the transactions their customers usually do and how the different products they sell are linked together.

Example

Have you ever seen beers and diapers located close to each other in a supermarket? Thanks to association rules mining a lot of samples of what a supermarket bill contains, researchers found a strong correlation between the customer buying diapers and buying beers. Therefore, to facilitate the shopping experience of their customers, supermarkets that participated in this study decided to locate these two products close by. ◄

Association rule mining is based on the definition of two key metrics: support and confidence. Support indicates how frequently the itemset (i.e., group of items) appears in the dataset, while confidence measures the likelihood of occurrence of a certain itemset given that another itemset is present in the overall transaction. With this formation, this algorithm identifies the frequent itemsets that meet a minimum support threshold and generates rules from these itemsets that satisfy the minimum confidence threshold. As a result, the dataset is exploited to find which items highly correlate with the appearance of other items or groups of items, and the algorithm learns from these structures.

2.4 Conclusion

In conclusion, this chapter presents the main concepts of Supervised and Unsupervised Learning as a fundamental dissemination of Machine Learning models. Classification and regression modeling are pivotal to Supervised learning, each addressing different types of problems with specific sets of algorithms and evaluation metrics. While classification deals with discrete outcomes and categorizes data into labels, regression focuses on predicting continuous quantities. Unsupervised learning represents a powerful paradigm within the field of ML, characterized by its ability to discern patterns and structures from unlabeled data. Unlike supervised learning, it does not rely on predefined labels or outcomes, allowing for exploring data's intrinsic properties and uncovering hidden relationships. Techniques such as clustering, dimensionality reduction, and association rule learning provide valuable insights into complex datasets, enabling data-driven decision-making and revealing underlying trends that might not be immediately apparent. However, evaluating unsupervised models poses challenges due to the absence of ground truth labels, with the need for metrics that assess the model's ability to identify hidden structures.

References

Agrawal, R., Imieliński, T., & Swami, A. (1993). Mining association rules between sets of items in large databases. *ACM SIGMOD Record, 22*(2), 207–216. https://doi.org/10.1145/170036.170072

Caruana, R., & Niculescu-Mizil, A. (2006). An empirical comparison of supervised learning algorithms. In W. Cohen & A. Moore (Eds.), *Proceedings of the 23rd International Conference on Machine Learning—ICML '06* (pp. 161–168). ACM Press. https://doi.org/10.1145/1143844.1143865

Günther, W. A., Rezazade Mehrizi, M. H., Huysman, M., & Feldberg, F. (2017). Debating big data: A literature review on realizing value from big data. *The Journal of Strategic Information Systems, 26*(3), 191–209. https://doi.org/10.1016/j.jsis.2017.07.003

Han, J., Pei, J., & Yin, Y. (2000). Mining frequent patterns without candidate generation. *ACM SIGMOD Record, 29*(2), 1–12. https://doi.org/10.1145/335191.335372

Hastie, T., Tibshirani, R., & Friedman, J. H. (2009). *The elements of statistical learning: Data mining, inference, and prediction* (2nd ed.). Springer.

Huang, G.-B., Zhou, H., Ding, X., & Zhang, R. (2012). Extreme learning machine for regression and multiclass classification. *IEEE Transactions on Systems, Man, and Cybernetics. Part B, Cybernetics: A Publication of the IEEE Systems, Man, and Cybernetics Society, 42*(2), 513–529. https://doi.org/10.1109/TSMCB.2011.2168604

Jordan, M. I., & Mitchell, T. M. (2015). Machine learning: Trends, perspectives, and prospects. *Science, 349*(6245), 255–260. https://doi.org/10.1126/science.aaa8415

Kriegel, H.-P., Kröger, P., Sander, J., & Zimek, A. (2011). Density-based clustering. *WIREs Data Mining and Knowledge Discovery, 1*(3), 231–240. https://doi.org/10.1002/widm.30

Kumar, Y., Kaur, K., & Singh, G. (2020). Machine learning aspects and its applications towards different research areas. In *2020 International Conference on Computation, Automation and Knowledge Management (ICCAKM)* (pp. 150–156). IEEE. https://doi.org/10.1109/ICCAKM46823.2020.9051502

Loh, W.-Y. (2011). Classification and regression trees. *WIREs Data Mining and Knowledge Discovery, 1*(1), 14–23. https://doi.org/10.1002/widm.8

Saporta, G., & Niang, N. (2009). Principal component analysis: Application to statistical process control. In G. Govaert (Ed.), *Data analysis* (pp. 1–23). Wiley. https://doi.org/10.1002/9780470611777.ch1

Suthaharan, S. (2016). Supervised learning algorithms. In S. Suthaharan (Ed.), *Integrated series in information systems. Machine learning models and algorithms for big data classification* (Vol. 36, pp. 183–206). Springer. https://doi.org/10.1007/978-1-4899-7641-3_8

Model Evaluation in Machine Learning Applications

3

Robert Rein

Abstract

The process of evaluating machine learning models is essential for determining their performance and generalization capabilities beyond training data. Model evaluation assesses how well a model can predict unseen data, ensuring it meets the requirements of the problem at hand. This evaluation involves understanding concepts like overfitting and underfitting, where models either excessively fit to training data or fail to capture underlying patterns, respectively. To measure model performance, various metrics are employed depending on the problem type. For regression models, metrics like Mean Absolute Error, Mean-Squared Error, and R-squared are commonly used. In classification models, metrics such as accuracy, precision, recall, specificity, and F1 score are employed, along with techniques like ROC curves and AUC. Validation techniques like k-fold cross-validation help mitigate biases and variance in performance estimation. Careful consideration of these evaluation methods ensures the development of robust machine learning solutions suitable for specific problems.

Evaluating machine learning models is a crucial step during data analysis (Ding et al., IEEE Signal Processing Magazine 35(6):16–34, 2018). Model evaluation allows to assess whether a chosen model performs well not just on the training data, but also on new, unseen data and is thus able to generalize beyond the given dataset. Through model evaluation the data scientist can assess to what extent the chosen model not only learns the features of the training data, but more importantly, how well the model is able to make predictions on data it has never

R. Rein (✉)
Institute of Exercise Training and Sport Informatics, German Sport University Cologne, Köln, Germany
e-mail: r.rein@dshs-koeln.de

D. Memmert (ed.), *Artificial Intelligence and Machine Learning in Sports Science*,
https://doi.org/10.1007/978-3-662-70155-3_3

seen before, so-called potential observables. The scientist can decide whether the developed model meets the requirements of the specific problem under investigation.

3.1 Understanding Model Evaluation

Closely related to the problem of model evaluation are the concepts of data overfitting and underfitting. In principle, what most machine learning algorithm try to achieve is to find a mapping from predictor variables to a criterion variable. For example, a possible analysis problem could be to find for a given arrangement X of soccer players on the most likely tactical formation of the team Y. The data scientist searches for a function *f* which maps the positions of the players X to a formation Y, or more formally f(X) = Y (James et al., 2013). In a perfect setting this should be fairly straightforward, however every real-world dataset contains some form noise which is typically denoted by the symbol ε. Moreover, the mapping from X to Y may be highly complex and the identification becomes very difficult as the data scientist seldom has sufficient knowledge to specify a specific function. The task now becomes one of finding a function $\check{f}$, an estimator, which approximates the true function *f*. There is allows an infinite number of ways to obtain a mapping from X to Y. Accordingly, a specific metric is needed which to determine a best fitting function $\hat{f}$ under this metric. The problem can thus be stated in the following form: We are searching for a function $\hat{f}$ which is an optimal solution to the problem:

$$y = f(x) + \varepsilon$$

There are two model components which need to be identified from the data. The first term $f(x)$ is the systematic component of the mapping and usually the one the data scientist is most interested in. The second term ε is random or noise term with particular properties like for example an average of 0 or a specific variance σ_{ε}^{2}. When we are trying to find a machine learning algorithm to estimate the true function $f(x)$ we have to make sure that we are not conflating systematic and random components of the data with each other. Therefore, we have to make sure that the algorithm we are choosing to estimate $f(x)$ has the required properties to be able to model the systematic component. If we choose an approach which is too simplistic, we will not be able to model the properties of the data and our results will be erroneous. The algorithm will oversimplify the relationship between X and Y. Similar, if we choose an algorithm which is too complex, the algorithm might model the data too closely and incorporate parts of the noise into the systematic component and we will again obtain erroneous results. A simple example will help to understand these concepts.

Example

Assume we have a dataset from the following data generating process (dgp):

$$y = 2 + x^3 + \varepsilon$$

where ε is a noise term following a normal distribution with mean $\bar{\varepsilon} = 0$ and variance $\sigma_\varepsilon^2 = 1$. The data are depicted in Fig. 3.1a. We will fit three different linear polynomial models to the data.

$$m_0 : y = \beta_0 + \beta_1 x$$
$$m_1 : y = \beta_0 + \beta_1 x^3$$
$$m_2 : y = \beta_0 + \beta_1 x + \cdots + \beta_6 x^6$$

The first model m_0 is just a simple linear equation, the second model m_1 is the actual third degree polynomial model, whereas the third model m_2 is the maximum possible sixth degree polynomial. If we fit these three models to the data using standard linear regression techniques, we obtain the following results (compare Fig. 3.1b)

As can be seen in Figure 3.1b the simple linear model is not able to follow the functional form of the data and therefore is quite far away from the actual dgp. In contrast, m_2 is able to perfectly fit the data including the noise part with zero deviations from the data. However, looking at right border of Fig. 3.1b one can observe that the model starts to deviate from the dgp quite strongly. Similarly, in between the observed data points m_2 also deviates substantially from the actual dgp. Finally, model m_1 is able to follow the actual dgp quite closely and is for most of the visible parts actually covered by the pink line which depicts the shape of the third degree polynomial without the noise term. Thus, m_1 is able to

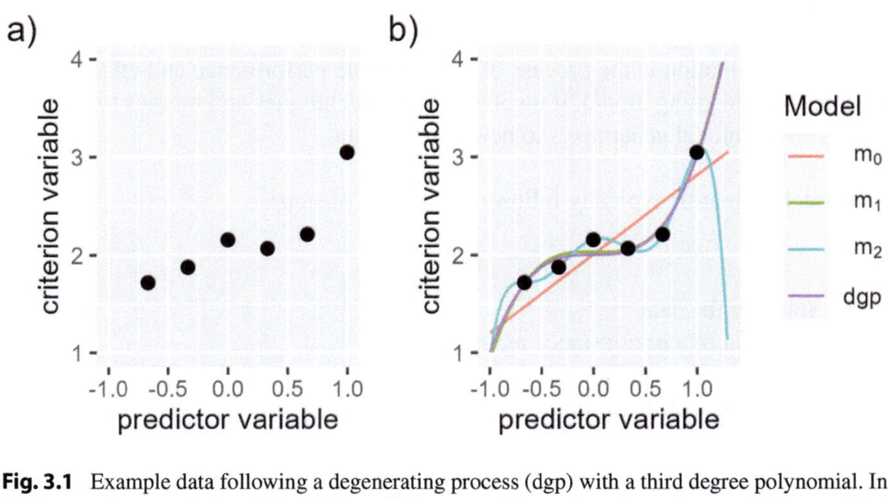

Fig. 3.1 Example data following a degenerating process (dgp) with a third degree polynomial. In (**a**) the raw data are depicted. In (**b**) three different models are fit to data a linear model m_0, a third degree polynomial m_1, and a sixth degree polynomial m_3. The actual dgp is shown in purple

distinguish the systematic and noise components from the raw data whereas m_0 is underfitting the data whilst m_2 is overfitting the data. Accordingly, when we want to prediction Y based on previously unobserved novel data X_{new} not used during the training, we will make rather large errors under both models m_0 and m_1 as both models are not able to generalize beyond the already observed data. Model m_0 is not even able to represent the observed data. Other terms for these concepts are out-of-sample error as opposed to in-sample-error or similarly generalization-error. ◀

This allows us now to provide definitions for the concepts of underfitting and overfitting.

▶ Overfitting occurs when a machine learning model learns the details and noise in the training data to the extent that it negatively impacts its ability to generalize to new, unseen data. The model becomes too complex and fits the training data too closely rather than the underlying patterns. In contrast, underfitting occurs when a machine learning model is too simple to capture the underlying structure of the data. The model fails to learn from the training data effectively, resulting in poor performance both on the training set and on new, unseen data.

The problem lies now therein that a-priori the data analyst rarely has enough information to make conclusive statements about which complexity is needed to model a given dataset. Thus, one needs a mechanism that allows the data analyst to assess to what extent the chosen modelling strategy has underfitted or overfitted the data. As stated above, the data analyst must have a metric which allows her to assess whether the chosen modelling strategy is adequate (James et al., 2013). Model evaluation therefore usually has two goals. First, ensuring that the modelling strategy is adequate and second that the obtained results are of sufficient quality to solve the given problem.

▶ Model evaluation is the process of assessing the performance and effectiveness of a machine learning model. It involves various techniques and metrics to measure how well the model generalizes to new, unseen data.

Model evaluation typically follows a sequence of steps.

- Splitting the data
- Selection of a performance metric
- Selection of a validation technique
- Assessment of the results

3.2 Splitting the Data

As we have seen in the previous section, understanding how a chosen machine learning algorithm performs with unseen data can give us important insights into how well the model performs and is able to generalize. In order to assess our model, it is therefore advantageous if we are able to expose the model to data not used during the training of the algorithm. To this end, the data is split into two parts: a training dataset and a test dataset. The training dataset can be used to train the model, while the test dataset allows to evaluate the performance of model on data not used during training. Sometimes, the training dataset is further split into a third set, called the validation set. The validation set is used during the model tuning phase to optimize model hyperparameters and to validate the model's performance during training optimization (Richter). Again, the approach is the same with using data not seen during training to assess model performance. Once the model is trained and fine-tuned using the training and validation sets, it is evaluated on the test set to get an accurate assessment of its performance before deploying the model.

▶ The key differences between the validation and the validation set are:

- The validation set is used during the training phase to evaluate the performance of a model and fine-tune hyperparameters.
- After training multiple models with different hyperparameters on the training set, the validation set is used to select the best-performing model.
- The test set is used to evaluate the final performance of the trained model and helps to provide an unbiased estimate of the model's performance on new, unseen data.

When splitting the dataset into the three sets, several mistakes can be made leading to erroneous, often over-optimistic performance assessment of the model. One problem can arise when so-called data leakage occurs. Data leakage refers to when information from the validation or test set influences the training. An example could be when data preprocessing steps are applied to the entire dataset before splitting the data into training, validation and testing set. For example, when using positional data from team sports often the positions are scaled or normalized to take into accommodate different pitch sizes. Through this processing step information from the testing and the validation datasets can leak into the training set. Therefore, the data should be split before the data are further processed preprocessing steps should be applied separately to each subset. Similar, when using team data and multiple data points from the team same time or the same players are contained in the dataset. Then, data parts from the same team or the same player can be present in the training and the testing set leading again to overly optimistic performance with respect to applying the model to previously unseen data from a players or teams. Thus, the data analysis needs to make sure that data leakage does not occur and the sets are independent from each other.

Another problem occurs when the dataset shows large class imbalances. Class imbalance occurs when a dataset has significantly more examples for some classes than others. For example, when you are interested in building a machine learning pipeline to predict injuries in athletes during competition from external load data. You might have lots of information about loads from different athletes but only very few instances of the data contain actually injuries. In these cases, just randomly splitting the data into testing, validation and training data set may lead to subsets that do not adequately represent all classes. Again, this may bias the model evaluation towards over-optimistic performance. A possible remedy in these cases is provided through techniques such as stratified sampling.

> Stratified sampling is a technique stemming from statistics to select a sample from a population. Stratified sampling is designed to ensure the final sample accurately reflects the important subgroups, or strata, within the population. First, the population is divided into subgroups. For example, when investigating performance in athletes from a range of different performance levels you would like be the subsets to show the same proportion of performance levels. Thus, you could divide the performance into three sub-categories, like high, intermediate and low performer. Once the dataset is divided into subgroups, objects are randomly select from each subgroup. This ensures that all the subgroups are equally represented in the training, validation and test dataset.

Another problem occurs during the splitting-step when the dataset is of small size. In this case, when the dataset gets divided it into three separate subsets, the data available for training may be too small and model performance suffers, especially if the model is complex and requires a large amount of data to learn effectively. In such cases, techniques like cross-validation may be more appropriate for model evaluation to make the most out of the limited data. There are unfortunately no hard rules when a dataset becomes too small as this will largely depend on the complexity of the problem and the specific algorithm chosen.

Taken together, it is crucial to carefully design the splitting strategy, ensuring randomness, independence, and representativeness in each subset.

3.3 Performance Metrics

The model evaluation requires a specific metric to be chosen in order to be able to assess the performance of the model. Unfortunately, there is no single performance measure which will work in every context but the metric should fit the specific problem. Thus, the data scientist needs to make an informed decision about which metric to choose to evaluation the model. The choice of performance metrics thereby depends also on the type of machine learning model, for example a regression versus a classification problem.

3.3.1 For Regression Models

Popular choice when assessing regression models are:

- Mean Absolute Error (MAE)
- Mean-Squared Error (MSQ) or Root-Mean-Squared (RMSE)
- R-squared (R^2)

Under the Mean Absolute Error (MAE), the average of absolute differences between the predicted and actual values for each data point are calculated. Thus, for each individual data point the difference between the observed data point y_i and the predicted value $\hat{y}_i$ is calculated.

$$|e_i| = |y_i - y_i|$$

Subsequently all MAEs are summed up and the sum is divided by the number n of datapoints. The MAE therefore provides a measure of the average magnitude of errors between predicted and actual values, with lower values indicating better model performance. The sum of the absolute differences is often also called the ℓ_1-Norm of Manhatten norm. The MAE is particularly useful when the dataset contains outliers, as it considers the absolute differences between predictions and actual values without squaring them and is therefore less sensitive to outliers.

Another often used metric in regression modelling is the Mean Squared Error. Here, in contrast to the absolute differences under MAE the squared differences between the observed y_i and the predicted value $\hat{y}_i$ is calculated. The squaring of the differences e_i makes MSE more sensitive to outliers compared to MAE. In linear regression problems, it can be shown that the MSE follows also from a maximum likelihood approach under a conditional normal distribution model (Bishop, 2006). Sometimes it is more convenient to work with the square-root of the sum of squared differences. This is called the Root-Mean-Squared Error (RMSE) which corresponds to the Euclidean norm (Géron, 2022). In linear regression problem often also the R-squared R^2 matric is used to asses the performance of the model. R^2 estimates the proportion of the variance in the dependent variable that is predictable from the independent variable(s) (James et al., 2013). However, this measure is less common in machine learning applications apart from traditional linear regression models.

In summary, for a given regression problem the data scientist needs to choose a particular performance metric to assess the performance of the model. During training the metric is used to tune model parameters or the choose between alternative model formulations. During the testing phase the metric is used to assess the performance of the model on unseen data. The performance metric can also be seen from the perspective of penalizing errors made by the model. The metric estimates the cost of making an error. Therefore, the performance metric is sometimes also called the cost-function. Under this view, using the cost-function during the testing phase aims at finding a model function $\hat{f}$ which minimizes the expected risk (Deisenroth et al., 2020). This approach translates equally to classification problems.

3.3.2 For Classification Models

For classification a different approach to model performance is usually taken as the distance between the observed y_i and the predicted value $\hat{y}_i$ usually does not follow a clear metric. For example, when we want to calculate the outcome of a goal 1 versus a non-goal 0 for a given shot. Therefore, we need a little bit of terminology to understand the typically used performance metrics. The following focusses on the binary case but the concepts easily generalize to multi-class problems.

Staying with the shooting example, we will call a shot resulting in a goal in a positive result whereas a shot resulting in no-goal in a negative result. Accordingly, a model can (1) predict a goal when a goal was observed, (2) predict no-goal when a goal occurred, (3) predict a goal when no goal happened, and (4) predict no-goal when no-goal occurred. In cases (1) and (4) no error was made whereas in cases (2) and (3) an error was made by the model. Case (1) is called a true positive, case (4) a true negative, case (2) a false negative, and case (3) a false positive. Based on this scheme the following term can be defined.

The accuracy of model specifies the proportion of true results (both true positives (TP) and true negatives (TN)) among the total number N of cases examined:

$$\text{accuracy} = \frac{\text{TP} + \text{TN}}{\text{N}}$$

The precision (or confidence or True Positive Accuracy (TPA) (Powers)) of a model is defined as the ratio of true positives (TP) to all predicted positives which is the sum of true positives and false positive (FP):

$$\text{precision} = \frac{\text{TP}}{\text{TP} + \text{FP}}$$

The recall (also called sensitivity or True Positive Rate (tpr) (Powers)) performance of a model is defined as the ratio of true positives to all positives in the data which is the sum of true positives and false negatives (FN):

$$\text{recall} = \frac{\text{TP}}{\text{TP} + \text{FN}}$$

The specificity of a model is specified as the proportion of true negatives to all negative cases consisting of true negatives and false positives.

$$\text{specificity} = \frac{\text{TN}}{\text{TN} + \text{FP}}$$

One needs to be aware that what constitutes a positive case is usually defined by the data scientist. Further, there exist trade-offs between the metrics. For example, precision and recall cannot be modified independently from each other. When I have a classifier there is usually a threshold involved which regulated when a data point is classified as a positive, for example a goal. When the data scientist decreases this threshold, precision will become worse whilst recall will improve. Vice-versa, when increasing the threshold, the classified will miss-classify less negatives as positives which increases precision but at the same time find fewer positives decreasing recall. Using the TP, TN, FP, FN and accordingly accuracy, precision, recall and specificity leads to some of the most commonly classification metrics:

- Confusion matrix
- F_1-Score
- ROC curve and AUC score
- Calibration curve

The confusion matrix in the simple case of only two classes results in a 2 × 2 contingency table showing the TN, TP, FN and FP. Continuing with our goal shot example. Assume we have observed 100 goal shots out of which 20 resulted in a goal. Our model predicted out of the 20 goals only 15 as goals and erroneously predicted 10 goals where actually no goal occurred. This would result in the following confusion matrix (Table 3.1).

From the confusion matrix the data scientist can assess how the model performs with respect to both classes and accordingly tune the model to accommodate specific requirements of the model. Also, from the confusion matrix the values for accuracy, recall, sensitivity and specificity can be easily calculated.

For example, a simple metric used when there is large class imbalance is the balanced accuracy (Erjeng). The balanced accuracy provides a measure of the performance for positive and negative cases and is calculated as the average of recall and specificity:

$$\text{Balanced accuracy}\left(\text{BA}\right) = \frac{\text{recall} + \text{specificity}}{2}$$

Table 3.1 Example of a confusion matrix for a hypothetical goal shot dataset (N = 100) where 20 goals were observed out of which 15 shots were identified by a classifier as goals and 10 shots were misclassified as goals where no goal was scored (*TP* true positive, *FN* false negative, *FP* false positive, *TN* true negative)

		Predicted	
		Goal	Miss
Observed	Goal	15 (TP)	5 (FN)
	Miss	10 (FP)	70 (TN)

For the goal shot example in Table 3.1 the balanced accuracy would be:

$$\mathrm{BA} = \frac{0.75 + 0.875}{2} = 0.33$$

The approach used for the balanced accuracy suggest already that focusing on any single metric may not be sufficient to assess the performance of a model. For example, when the dataset is highly imbalanced and contains 90% positives, then already a naïve classifier which classifies all data points as belonging to the positive class will have a recall score of 0.9 (compare also Chico 2020). Therefore, in order to assess the performance of model always several parameters should be interpreted together.

Another performance measure which combines multiple measures is the F_1 score. The F_1 score is calculates by the harmonic mean of precision and recall which results in a balance between the two metrics (Chinchor & Sundheim, 1993). The F_1 score is defined as:

$$F_1 = \frac{2}{\frac{1}{\text{precision}} + \frac{1}{\text{recall}}}$$

By using the harmonic mean, the F_1 scores tends to give less weight to extreme values compared to the arithmetic mean and works reasonably well with imbalanced datasets. The F_1 score is useful when you want to strike a balance between minimizing false positives and false negatives.

For our goal shot example the F_1 score is:

$$F_1 = \frac{2}{\frac{1}{\frac{15}{25}} + \frac{1}{\frac{15}{20}}} = \frac{2}{3} = 0.\overline{6}$$

One of the great advantages of the F_1 score is based on its easy interpretation. As the F_1 is a single value between $[0, 1]$ one obtains a straightforward summary of the classifier's precision and recall. At the same time, this may also be viewed as a disadvantage as the F_1 score provides only limited insights into why a model performs well or poorly which can also become problematic when used as an optimization metric. Further, the F1 metric is not invariant to class swapping when class labels are changed. For further discussion about the problems and limitations of the F_1 score consult the according literature (Brabec et al., 2020; Chicco & Jurman, 2020; Siblini et al., 2020).

Another widely used approach to assess model performance in binary classification task are Receiver Operator Characteristic curves (ROC) and the related Area Under the Curve (AUC) measure (Fawcett, 2006). Initially, ROC was developed in

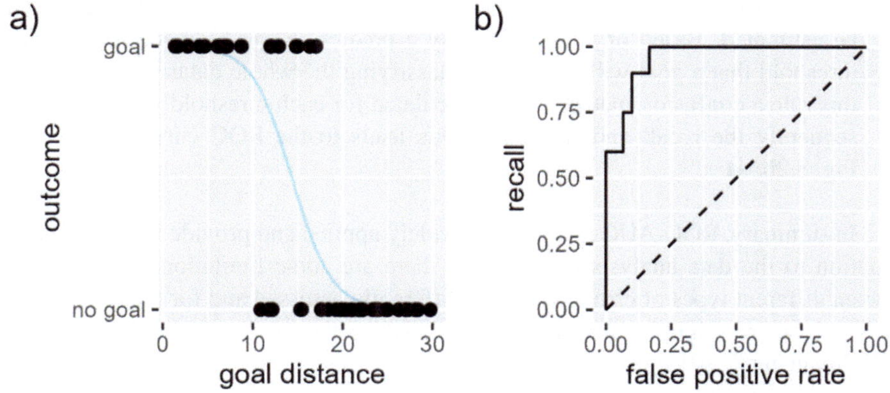

Fig. 3.2 (**a**) Plot of hypothetical goal shot data with respect to shooting distance. The blue line shows a logistic regression fit. (**b**) Receiver operator characteristic curve for the goal shot data

the domain of signal analysis and has been particular popular in medical classification problems. The ROC curve plots the sensitivity (recall) against the false positive rate which equals 1—specificity for different threshold values of the classifier (compare Fig. 3.2b). In our goal shot example, we could for example fit a logistic regression to distance to the goal data in order to predict goal probabilities. The ROC curve shows how well the model distinguishes between the goals and the no-goal shots as the threshold for classification is varied. A perfect classifier will have zero true positive rate and 100% false positive rate whereas a chance classifier will follow the identity line (compare Fig. 3.2b). Essentially, the ROC curve depicts how many correct positive predictions the model makes compared to how many incorrect positive predictions it makes as the decision threshold is adjusted.

Closely related to the ROC curve is the Area Under the Curve (AUC) metric. The AUC is simple the area under the ROC curve. Thus, it is a single number ranging from 0 to 1 that summarizes how well your model performs across all possible thresholds. AUC values close to 1 indicates better performance and an AUC value of 0.5 indicates that the model performs no better than random guessing. Similar, to the F1 score provides a single value which might be impropriate depending on how positive and negative classifications errors are evaluate. Both ROC curves and the AUC score is robust to class imbalance because it measures the model's ability to distinguish between classes across various thresholds. While the AUC score is widely used for binary classification problems, it is not directly applicable to multi-class classification tasks.

Example

A hypothetical example is depicted in Fig. 3.2. Assume we have collected goal shot data in relation to shooting distance (compare Fig. 3.2a). To model the relationship a logistic regression model has been fit to the data resulting in the logit model equation $y = 6.62 - 0.46 \bullet \text{distance}$ which relates the shooting distance to goal probabilities. Thus, for a given distance the probability of scoring a goal can be estimated. Based on this probability a prediction can be made for a given threshold that a goal will be scored. Classifying the whole dataset according to this rule a confusion matrix can be calculated for each threshold value and subsequently the recall and fpr score. This leads to the ROC curve depicted in Fig. 3.2b. ◄

In summary, ROC-AUC methods are widely applied and provide valuable information to the data analysis however but there are some limitations in particular when different types of errors result are differently assessed and for the AUC using a single score might be not sufficient to analyze the performance of the model (Halligan et al., 2015; Hand, 2009).

Another method which is particularly relevant when prediction errors are varying importance is a calibration curve. A calibration curve is a graphical representation of the relationship between predicted probabilities generated by a predictive model and the actual outcomes observed in the data. To generate a calibration curve the chosen model generates probability scores for each observed instance in the dataset. Subsequently, the dataset is divided into groups based on the predicted probabilities. For each group, the average predicted probability and the corresponding empirical probability of the observed proportion of positive instances is calculated. Subsequently, the predicted versus the empirical probabilities are plotted indicating the agreement between predicted probabilities and actual outcomes. In case of perfect agreement, the curve should follow the identity line and deviations accordingly signal possible calibration errors. Based on this approach the data analyst can assess whether the model performs within the intended precision. In contrast to ROC-AUC, which measure discrimination ability, calibration curves directly assess the calibration of the model such as systematic overconfidence, which may not be evident from ROC-AUC measures alone. With respect to the goal shot example, the ROC curve would tell the data analyst how good the model distinguishes between goal and misses whereas the calibration curve shows how accurate the predicted probabilities for goals are (compare Huang et al., 2020; Van Calster et al., 2019).

3.4 Validation Techniques

So far, we have only discussed what is typically referred to as the holdout validation technique (also train-test split). Under holdout validation, the dataset is split into a training set and a testing set and if possible, the training set is further split into a validation set. The hold-out method is straightforward and easy to implement, making it a popular choice for evaluating models, especially when the dataset is large enough to be split into distinct sets. However, it has some limitations, such as potentially high variance in the estimated performance due to the randomness of the data split and the risk of overfitting to the training set, especially if the dataset is small.

To mitigate these issues, techniques like k-fold cross-validation or Leave-One-Out Cross-Validation can be used in conjunction with the hold-out method (Kohavi, 1995).

K-fold cross-validation (CV) is a technique used to evaluate the performance of a predictive model by partitioning the dataset into k equal-sized subsets (folds). The model is trained and evaluated k times, each time using a different fold as the validation set and the remaining folds as the training set. K-fold cross-validation follows therefore the following sequence of steps.

1. Data split: The data is divided into k roughly equal folds.
2. Train-test loop:
   - For each fold:
     - Set aside one fold for testing: The validation set.
     - Train the model on the remaining k − 1 folds: The training set.
     - Evaluate the model's performance on the validation set.
3. Combine results: The performance scores from each fold are averaged to get an overall estimate of the model's generalizability.

K-fold cross-validation helps to provide a more reliable estimate of the model's performance compared to a single train-test split by reducing the variance associated with a single random split of the data. It also helps to mitigate bias in the estimated performance by using multiple partitions of the data for training and testing. K-fold cross-validation is often used for hyperparameter tuning and model selection. By evaluating the model's performance across multiple folds, k-fold cross-validation provides a more robust estimate of the optimal hyperparameters and helps to select the best-performing model.

A alternative to k-fold cross-validation, especially in scenarios with limited datasets, is the Leave-One-Out Cross-Validation (LOOCV) approach. LOOCV can be seen as a special case of cross-validation where K equals the number of data points in the dataset. Thus, a single data point is used for validation and the procedure is repeated over all data points. The performance of the model can subsequently be evaluated by aggregating the results across all iterations. LOOCV provides an unbiased estimate of the model's performance because each sample in the dataset is used as both a training instance and a validation instance. However, LOOCV estimates can have high variance, especially for large datasets in comparison to k-fold CV (Efron & Tibshirani, 1997).

3.5 Conclusion

Evaluating machine learning models is a multifaceted process that involves choosing the right metric, validation techniques, and considering practical issues like overfitting. Using a structured approach allows the data analysis to accurately assess a model's performance and choose between competing model implementations by assess the model's ability to generalize to new, unseen data. By applying this approach, the data analyst can ensure the development of robust and reliable machine learning solutions fitting to the specific problem under investigation. Thereby, no single solution will be equally suitable in different context but the data analysis has to make several informed decisions in order to make sure the specific approach chosen is suitable.

References

Bishop, C. M. (2006). *Pattern recognition and machine learning*. Springer.

Brabec, J., Komárek, T., Franc, V., & Machlica, L. (2020). *On model evaluation under non-constant class imbalance*. Springer.

Chicco, D., & Jurman, G. (2020). The advantages of the Matthews correlation coefficient (MCC) over F1 score and accuracy in binary classification evaluation. *BMC Genomics, 21*(1), 6.

Chinchor, N., & Sundheim, B. M. (1993). *MUC-5 evaluation metrics.* Paper presented at the Fifth Message Understanding Conference (MUC-5): Proceedings of a Conference Held in Baltimore, Maryland, August 25–27, 1993.

Deisenroth, M. P., Faisal, A. A., & Ong, C. S. (2020). *Mathematics for machine learning*. Cambridge University Press.

Ding, J., Tarokh, V., & Yang, Y. (2018). Model selection techniques: An overview. *IEEE Signal Processing Magazine, 35*(6), 16–34.

Efron, B., & Tibshirani, R. (1997). Improvements on cross-validation: The 632+ bootstrap method. *Journal of the American Statistical Association, 92*(438), 548–560.

Fawcett, T. (2006). An introduction to ROC analysis. *Pattern Recognition Letters, 27*(8), 861–874.

Géron, A. (2022). *Hands-on machine learning with Scikit-Learn, Keras, and TensorFlow*. O'Reilly Media, Inc.

Halligan, S., Altman, D. G., & Mallett, S. (2015). Disadvantages of using the area under the receiver operating characteristic curve to assess imaging tests: A discussion and proposal for an alternative approach. *European Radiology, 25*, 932–939.

Hand, D. J. (2009). Measuring classifier performance: A coherent alternative to the area under the ROC curve. *Machine Learning, 77*(1), 103–123.

Huang, Y., Li, W., Macheret, F., Gabriel, R. A., & Ohno-Machado, L. (2020). A tutorial on calibration measurements and calibration models for clinical prediction models. *Journal of the American Medical Informatics Association, 27*(4), 621–633.

James, G., Witten, D., Hastie, T., & Tibshirani, R. (2013). *An introduction to statistical learning* (Vol. 112). Springer.

Kohavi, R. (1995). *A study of cross-validation and bootstrap for accuracy estimation and model selection.* Paper presented at the Ijcai.

Siblini, W., Fréry, J., He-Guelton, L., Oblé, F., & Wang, Y.-Q. (2020). Master your metrics with calibration. In M. Berthold, A. Feelders, & G. Krempl (Eds.), *Advances in intelligent data analysis XVIII. IDA 2020. Lecture notes in computer science* (Vol. 12080). Springer. https://doi.org/10.1007/978-3-030-44584-3_36

Van Calster, B., McLernon, D. J., van Smeden, M., Wynants, L., Steyerberg, E. W., & On behalf of Topic Group 'Evaluating diagnostic tests and prediction models' of the STRATOS initiative. (2019). Calibration: The Achilles heel of predictive analytics. *BMC Medicine, 17*(1), 230.

Predictive Bayesian Modeling in Sport Science

4

Alexander Asteroth

Abstract

Predictive, data-driven models are of central importance for the application of artificial intelligence in sport. The data situation in sport is characterized by small numbers of cases, sometimes high noise and incomplete data, which make the use of suitable machine learning methods necessary. Bayesian inference is considered a suitable approach to deal with this problem. With normally distributed parameters and data, exact inference is possible with Gaussian processes, for example. Marcov chain Monte Carlo methods allow the use of arbitrary distributions for prior and likelihood. By using suitable priors and selecting models that are not unnecessarily complex, meaningful predictions can be made even with small and incomplete data sets.

4.1 Challenges in Data-Driven Inference in Sport

Artificially intelligent applications (AI) make informed decisions in their environment on the basis of models (Russell & Norvig, 2020; Brooks, 1986). For this to work well, these models must represent the known environment with sufficient accuracy, but at the same time they must generalize well, even in unknown situations. Such predictive models are usually learned from data about the AI's environment using machine learning methods (inference). In sport, this environment is generally represented by athletes, teams or something comparable (Hammes et al., 2022) (see Fig. 4.1).

A. Asteroth (✉)
Hochschule Bonn-Rhein-Sieg, University of Applied Science, Sankt Augustin, Germany
e-mail: alexander.asteroth@h-brs.de

D. Memmert (ed.), *Artificial Intelligence and Machine Learning in Sports Science*,
https://doi.org/10.1007/978-3-662-70155-3_4

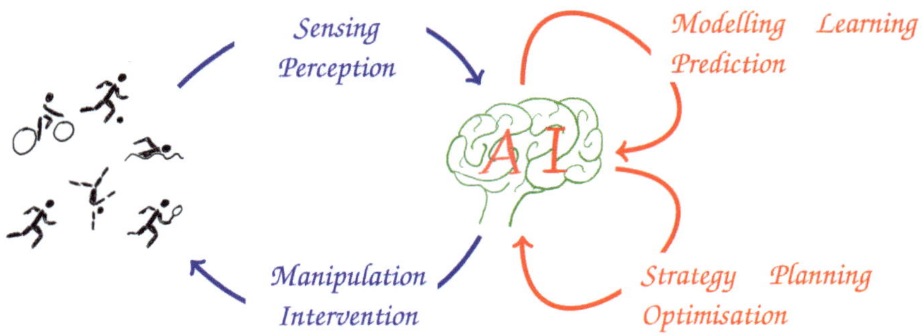

Fig. 4.1 Illustration of the embedding of an AI in the context of sport

Due to low acceptance and an already low number of athletes, the number of cases in elite sport is small (Hecksteden & Kellner, 2021, p. 21 ff). But even in amateur sport, the amount of data is limited due to complex test procedures (e.g. laboratory diagnostics) or data of questionable quality as the tests lack control. In addition, the oftentimes poor signal-to-noise ratio of physiological measurement data makes modeling in sport difficult in many cases.

4.2 Predictive Data-Driven Modeling in Sport

Suitable models and machine learning methods consider the data situation in sport by making optimal use of the limited data available and explicitly taking into account the possibly poor quality of the data. The latter is possible through approaches that consider how confident the model is in making a prediction. Models can be very simple representations of reality, such as linear functions, or very complex, such as deep neural networks with millions of free parameters (Goodfellow et al., 2016). Depending on the choice of model class and machine learning method for parameters and hyperparameters, there are different advantages and disadvantages, particularly regarding accuracy, generalization capability and the quantity and quality of the required data.

When a predictive model is learnt by machine learning, it is important to avoid both overfitting and underfitting and thus simultaneously enable good generalization and the highest possible accuracy (cf. Goodfellow et al., 2016, p. 128). Overfitting occurs when the model reproduces the data used to learn the model very well, but the accuracy of the model is low on unknown test data (generalization). In small data situations, as is the case in sport science, the risk of overfitting is particularly high.

Overfitting of the model usually occurs when the model has too many degrees of freedom for too little data or, vice versa, if there is too little data for the degrees of freedom of the model. To prevent overfitting, the principle of Ockham's razor (William of Ockham (1288–1347)) is applied, which states that "of various theories, the simplest is to be preferred". In machine learning, "simple" refers to model complexity, i.e. the number of model parameters, the size of the hypothesis space or the VC dimension (Kearns & Vazirani, 1994).

In machine learning practice, models are often used, such as artificial neural networks, whose complexity is theoretically too high for the given data situation or which are even able to represent arbitrary functions of a certain functional class (universal function approximators (Pinkus, 1999)). In order to avoid overfitting, so-called regularization methods are used (cf. Goodfellow et al., 2016), which are intended to minimize generalization errors. The most common and simplest technique is to terminate the learning process prematurely. By splitting the data into training and validation data, two model errors E_{train} and E_{val} are determined during the learning process, with the latter providing an estimate for the generalization error of the model. If E_{val} increases persistently during the learning process while E_{train} decreases, the boundary from underfitting to overfitting zone is reached and the learning process is aborted. However, this assumes that there is sufficient data to train an appropriately complex model that generalizes well and that sufficient data remains to be used as a validation set.

Conversely, by choosing a particularly simple model, overfitting can be avoided and the number of samples required can be reduced. However, such simple models considerably restrict the distributions that can be represented, so that the true distribution of the data may no longer be among the representable distributions and therefore the error on the training data will inevitably remain high.

A comprehensive overview of the topic of machine learning methods from a probabilistic perspective is beyond the scope of this chapter, so interested readers are referred to the comprehensive book by Murphy (2012).

4.3 Bayesian Inference

Bayesian approaches are considered promising in small data situations (cf. Van de Schoot & Miocevic, 2020; Hecksteden et al., 2022; Hagg et al., 2021). Bayes' rule is a formulation of the definition of conditional probability. Applied to the unknown parameters θ of a model M_θ, it states that the a posteriori distribution of the model parameters is the product of likelihood and prior divided by the evidence of the data:

$$p(\theta|D) = \frac{p(D|\theta)p(\theta)}{p(D)} = \frac{\text{likelihood} \cdot \text{prior}}{\text{evidence}} \tag{4.1}$$

When learning a concrete model M_θ one is interested in the most probable value of θ for the given data D. The probability distribution of $\theta \mid D$ can (at least theoretically) be determined according to (4.1). The likelihood can be determined by predicting D for a given θ. The prior offers the possibility of integrating information about θ and thus controlling which parameter values are more or less probable. The marginal likelihood is

$$p(D) = \int p(D|\theta)\, d\theta \tag{4.2}$$

It serves as a normalization constant and contains an integral whose value is often difficult to determine. For classical statistics, several points are critical in this approach:

1. the parameters θ of the model M_θ are unknowns and not accessible to sampling, and thus do not have a distribution in the frequentist sense
2. a priori assumptions made about the distribution of θ must be evidence-based, there is a risk of subjectivity
3. the product of likelihood and prior is generally not accessible to an analytical solution and must be determined numerically

Even if the subjectivity of the results inherent in the Bayesian approach will always be a cause for dispute (see Excursus 4.1), the validity of the Bayesian rule is beyond question and, moreover, always leads to admissible decisions in the sense of Wald's decision theory (Wald, 1939; Cornfield, 1967).

Excursus 4.1 Bayesian and Frequentist Concepts of Probability
For centuries there has been a dispute between Bayesians and frequentists. After inference methods were initially based on the Bayesian approach (Bayes, 1763), frequentist methods became established for a long time. Bayesian methods have only been gaining ground again since the middle of the last century. In particular, the foundations for today's subjectivist interpretation of the Bayesian approach were laid in the 1950–70s (De Finetti, 1974; Cornfield, 1967). Frequentist statistics is based on the interpretation of probability as the limit value of relative frequencies. Model parameters are regarded as fixed unknowns, for the determination of which the likelihood is maximized. Bayesian statistics is based on the interpretation of probability as the degree of belief. This also makes it possible to consider values that are not accessible to sampling as random variables.

4.4 Bayesian Modeling in Sport

By selecting suitable informative priors, background knowledge, for example from previous analyses or expert knowledge, can be used in sport. For example, the results of a meta-analysis on the effectiveness of cold water immersions on sprint performance in soccer were used as a prior in a Bayesian approach with small numbers of cases (Hecksteden et al., 2022). Another approach uses the approximate linear relationship between power and heart rate as a prior in noisy, incomplete data to determine the maximum average power in endurance sport (Asteroth et al., 2020). In addition to a literature search, Van de Schoot et al. use expert knowledge collected via interviews with 22 experts to create a weakly informative prior. The results show that this does not restrict the parameter space too much and at the same time makes it possible to deal with very small numbers of cases (van de Schoot et al., 2018).

Since the 1990s, two main approaches have been established for determining the a posteriori distribution (4.1). If distributions can be used whose product can be determined analytically, a closed-form solution to the inference problem is possible. This is the case, for example, with normally distributed prior and likelihood and is used in so-called Kriging, which today is also referred to as Gaussian process (see Sect. 4.6) (Krige, 1951; Williams & Rasmussen, 2006). Otherwise, Marcov chain Monte Carlo methods (MCMC) can be used to numerically approximate the a posteriori distribution (Metropolis et al., 1953; Geyer, 1992).

An important aspect of modeling with small numbers of cases and noisy data is the choice of optimal model complexity. According to the principle of parsimony, the model should be as simple as possible, but not simpler. The Bayesian approach offers the possibility of integrating this type of model selection into the inference process (MacKay, 1992; Williams & Rasmussen, 2006). For this purpose, we consider a class of models $\{M_i\}$. Each of the models depends on a parameter vector θ. For a model M_i and a given data set, the density of θ is defined as (cf. (4.1)):

$$p(\theta|D,M_i) = \frac{p(D|\theta,M_i)\,p(\theta|M_i)}{p(D|M_i)} \tag{4.3}$$

The evidence of the data for the M_i model appears here as marginal likelihood:

$$p(D|M_i) = \int p(D|\theta,M_i)\,p(\theta|M_i)\,d\theta \tag{4.4}$$

This is the likelihood for the a posteriori probability of the model M_i.

$$P(M_i|D) = \frac{p(D|M_i)\,P(M_i)}{p(D)} \tag{4.5}$$

Since the Bayesian approach, in contrast to the frequentist approach, does not make point estimates for the unknowns θ and M_i, but instead determines their entire a posteriori distribution, this automatically takes into account the principle of parsimony for the following reason: simple models can only explain a limited amount of data sets. Since (4.5) is a probability distribution, the values must sum to 1. Therefore, models in which the probability mass is distributed over a larger range of explainable data sets also have lower probabilities, even if they may explain the data somewhat better than simpler models. This is expressed by the so-called Ockham factor (MacKay, 1992).

4.5 Bayesian Modeling Using Gaussian Processes

A Gaussian process is a generalization of the multidimensional normal distribution. Formally, it is a stochastic process or a family of random variables with the characteristic that each finite subset is jointly normally distributed (Williams & Rasmussen, 2006). A Gaussian process on the set X is defined by a mean function $\mu : X \rightarrow R$ and a covariance function $k : X \times X \rightarrow R$. By selecting the mean function and the covariance function, prior knowledge about the expected function curve (e.g. linear) and properties of the function (e.g. smoothness or periodicity) can be incorporated into the prior. For a set of points $x_1, \ldots, x_n \in X$ the distribution of the values $y = (y_1, \ldots, y_n)$ is defined as

$$y \sim N\left(y|,\bar{y}|,\Sigma\right) \quad with \quad \bar{y} = \left(\mu\left(x_1\right),\ldots,\mu\left(x_n\right)\right),\ \Sigma = \left(k\left(x_i,x_j\right)\right)_{i,j=1\ldots n} \tag{4.6}$$

In this simplest case, the likelihood is also normally distributed and the same applies to the a posteriori distribution:

$$p\left(y|D,x\right) \sim N\left(K_D^T K^{-1} y_D, K - K_D^T K_{D,D}^{-1} K_D\right) \tag{4.7}$$

in which case (x, y) is the prediction, $D = (x_D, y_D)$ is the observed data and K, K_D and $K_{D,D}$ are the covariance matrices between x and x, x and x_D as well as x_D and x_D.

4.6 Gaussian Processes for Predicting Maximum Mean Power in Endurance Disciplines

Since the introduction of the critical power concept (CP) by Monod and Scherrer (1965), various extensions and modifications of this model have been proposed, see Morton (2006). Here, the maximum average power P that an athlete can maintain for a certain duration T is estimated. In order to obtain a good estimate of P, maximal workloads must be available for different durations. Their absence is the main limiting factor for the use of these models. Since training software and portals have become increasingly popular in professional sport in recent years, such models are now widely used. However, as training data is usually used to fit the model, maximal loads are not always available, and even when they are, it is difficult to distinguish maximal loads from submaximal loads.

Asteroth et al. (2020) use a Gaussian process with a linear mean function to enable the prediction of P even from submaximal efforts. The idea is to model the joint distribution $p(P, hr| T)$ where the average power P serves as a measure of the effort over the interval duration T and the average heart rate hr as a measure of the stress. Assuming that both measures are simultaneously maximized at maximum average power, the Gaussian process can be used to determine the distribution of power values predicted at the expected maximum heart rate for the interval length T in question. An athlete's historical training data is used to determine the frequency

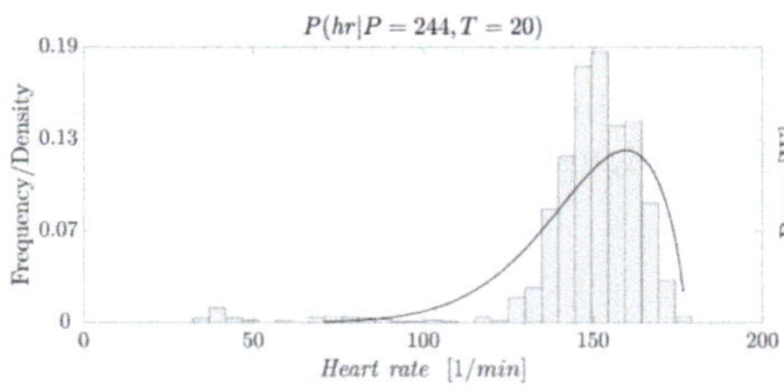

(a) Density estimation

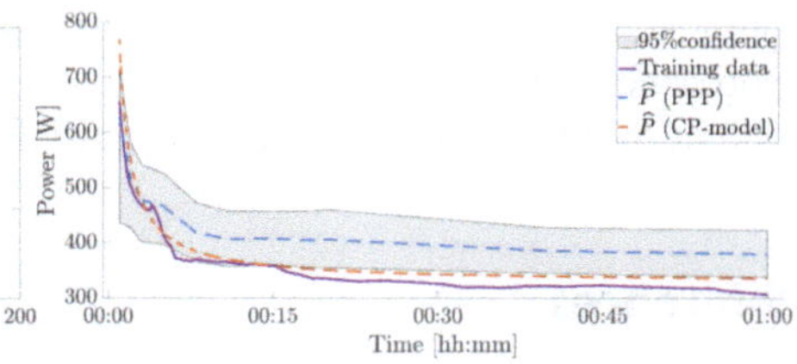

(b) Prediction $\widehat{P}$ and CP-model

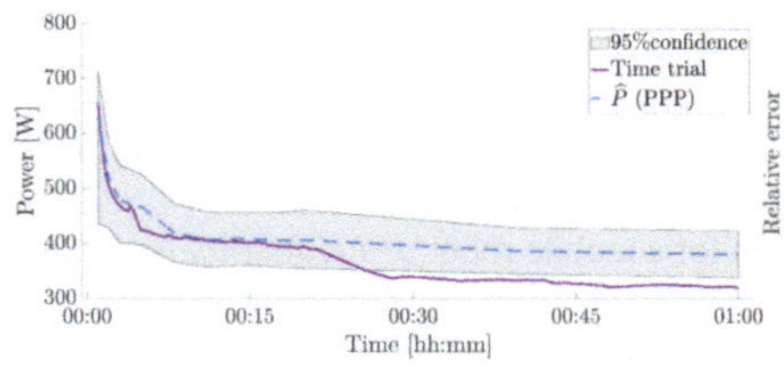

(c) Evaluation in 20 min. time trial

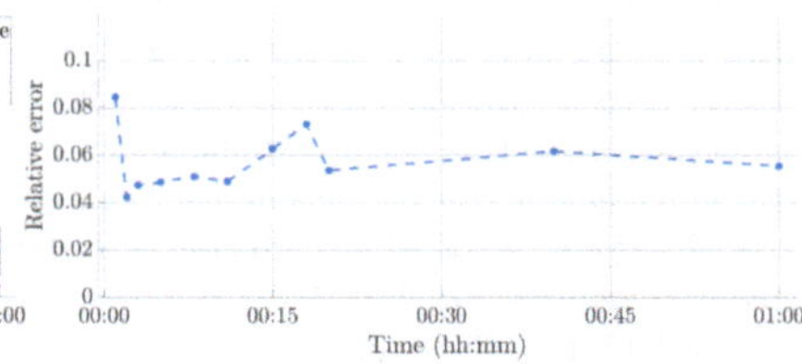

(d) Relative predictive error

Fig. 4.2 (**a**) Estimation of the density from frequencies; (**b**) distribution of the predicted P based on submaximal training data. The prediction of the classical CP model is shown as a comparison. (**c**) Maximum mean power output of a cycling training session including a 20-min time trial; (**d**) Average relative prediction error of the probabilistic model for 6 athletes (cf. Asteroth et al., 2020)

of pairs (P, hr) for all durations 3 min $\leq T \leq$ 60 min; Fig. 4.2a shows example frequencies and the estimated distribution.

Figure 4.2b shows an example of the resulting distribution of P. As can be seen, a distribution of maximum power outputs is predicted for each interval duration. The prediction is based only on submaximal training and not on maximal competition loads. The actual performance, on the other hand, was determined using a 20-min time trial of this athlete in the context of a competition and is nevertheless well predicted by the model (see Fig. 4.2c).

Training data from six professional cyclists was used to evaluate the approach and a random subset of this data was used to predict maximum performance. Additional Competition data was used for the evaluation. Overall, the mean deviation of the prediction from the observed values was on average around 5% (Fig. 4.2d).

4.7 Conclusion

Bayesian modeling methods are particularly suitable for dealing with the challenging data situation in sport. If prior and likelihood can be modeled as normally distributed, an analytical, exact solution is available with Gaussian processes. Otherwise, MCMC methods provide a proven and efficient numerical method for determining the a posteriori distribution. In recent years, various applications of Bayesian methods have been successfully carried out and published in the sport sciences. It was possible to form a prior based on traditional knowledge and thus combine the advantages of the classical approach in sport with modern data-driven approaches. This opens up the possibility for future projects to use all available sources of information in order to master the challenging data situation in sport.

References

Asteroth, A., Ludwig, M., & Bach, K. (2020). Probabilistic performance profiling: A model of the power duration relation in endurance sports. In *Proceedings of the 12th International Symposium on Computer Science in Sport (IACSS 2019)* (pp. 123–131). Springer.

Bayes, T. (1763). LII. An essay towards solving a problem in the doctrine of chances. By the late Rev. Mr. Bayes, FRS communicated by Mr. Price, in a letter to John Canton, AMFR S. *Philosophical Transactions of the Royal Society of London, 53*, 370–418.

Brooks, R. (1986). A robust layered control system for a mobile robot. *IEEE Journal on Robotics and Automation, 2*(1), 14–23.

Cornfield, J. (1967). Bayes theorem. *Revue de l'Institut International de Statistique*, 34–49.

De Finetti, B. (1974). Bayesianism: Its unifying role for both the foundations and applications of statistics. *International Statistical Review/Revue Internationale de Statistique*, 117–130.

Geyer, C. J. (1992). Practical Markov chain Monte Carlo. *Statistical Science, 7*(4), 473–483.

Goodfellow, I., Bengio, Y., & Courville, A. (2016). *Deep learning*. MIT Press. http://www.deeplearningbook.org

Hagg, A., Asteroth, A., Rasche, C., Bach, K., & Pfeiffer, M. (2021). *Künstliche Intelligenz für den Spitzensport im Spannungsfeld zwischen Big und Small Data: (KISs-BiS)* (No. 2021| 03). Bundesinstitut für Sportwissenschaft (BISp).

Hammes, F., Hagg, A., Asteroth, A., & Link, D. (2022). Artificial intelligence in elite sports—A narrative review of success stories and challenges. *Frontiers in Sports and Active Living, 4*, 861466.

Hecksteden, A., Forster, S., Egger, F., Buder, F., Kellner, R., & Meyer, T. (2022). Dwarfs on the shoulders of giants: Bayesian analysis with informative priors in elite sports research and decision making. *Frontiers in Sports and Active Living, 4*, 793603.

Hecksteden, A., & Kellner, R. (2021). *Wissenschaftlicher Umgang mit den kleinen Fallzahlen in der Spitzensportforschung*. Sportverlag Strauß.

Kearns, M. J., & Vazirani, U. V. (1994). *An introduction to computational learning theory*. MIT Press.

Krige, D. G. (1951). A statistical approach to some basic mine valuation problems on the Witwatersrand. *Journal of the Southern African Institute of Mining and Metallurgy, 52*(6), 119–139.

MacKay, D. J. (1992). Bayesian interpolation. *Neural Computation, 4*(3), 415–447.

Metropolis, N., Rosenbluth, A. W., Rosenbluth, M. N., Teller, A. H., & Teller, E. (1953). Equation of state calculations by fast computing machines. *The Journal of Chemical Physics, 21*(6), 1087–1092.

Monod, H., & Scherrer, J. (1965). The work capacity of synergic muscle group. *Ergonomics, 8*, 329–338.

Morton, R. H. (2006). The critical power and related whole-body bioenergetic models. *European Journal of Applied Physiology, 96*(4), 339–354.

Murphy, K. P. (2012). *Machine learning: A probabilistic perspective*. MIT Press.

Pinkus, A. (1999). Approximation theory of the MLP model in neural networks. *Acta Numerica, 8*, 143–195.

Russell, S., & Norvig, P. (2020). *Artificial intelligence—a modern approach* (4th ed.). Prentice-Hall.

Van de Schoot, R., & Miocevic, M. (2020). *Small sample size solutions: A guide for applied researchers and practitioners*. Taylor & Francis.

van de Schoot, R., Sijbrandij, M., Depaoli, S., Winter, S. D., Olff, M., & Van Loey, N. E. (2018). Bayesian PTSD-trajectory analysis with informed priors based on a systematic literature search and expert elicitation. *Multivariate Behavioral Research, 53*(2), 267–291.

Wald, A. (1939). Contributions to the theory of statistical estimation and testing hypotheses. *The Annals of Mathematical Statistics, 10*(4), 299–326.

Williams, C. K., & Rasmussen, C. E. (2006). *Gaussian processes for machine learning* (Vol. 2). MIT Press. http://www.gaussianprocess.org

Computer Vision and Deep Learning for Sports Analytics

5

Eric Müller-Budack, Jonas Theiner, and Ralph Ewerth

Abstract

Cutting-edge technologies from the field of artificial intelligence (AI), in particular deep learning, have revolutionized the field of sports science. Computer vision approaches play a pivotal role in converting visual data (e.g., video recordings of competitions or training sessions) into actionable insights, delivering valuable information to researchers, coaches, analysts, and athletes. The applications of computer vision in sports are diverse and range from player detection, tracking of movements and action recognition to the extraction of key performance indicators (KPIs) and tactical patterns via position data estimation. Hence, keeping pace with the latest developments in deep learning is crucial in order to understand the evolving landscape of computer vision in sports analytics. In this chapter, we provide an overview of both foundations and recent research advancements of computer vision in sports science. We present state-of-the-art algorithms and offer practical insights for key applications in sports analytics, including object detection and tracking, action recognition, and sports field registration. Finally, we address existing challenges and discuss potential future research directions such as generative AI models for sports analytics.

E. Müller-Budack (✉) · J. Theiner
TIB—Leibniz Information Centre for Science and Technology, Hannover, Germany
e-mail: eric.mueller@tib.eu; jonas.theiner@tib.eu

R. Ewerth
TIB—Leibniz Information Centre for Science and Technology, Hannover, Germany

L3S Research Center, Leibniz University Hannover, Hannover, Germany

University of Marburg & hessian.AI—Hessian Center for Artificial Intelligence, Marburg, Germany
e-mail: ralph.ewerth@tib.eu

D. Memmert (ed.), *Artificial Intelligence and Machine Learning in Sports Science*, https://doi.org/10.1007/978-3-662-70155-3_5

5.1 Computer Vision for Sports Analytics: An Overview

The field of sports science has experienced a substantial evolution, primarily driven by technologies such as sensors, tracking devices, and calibrated camera systems that led to the availability of big data in many sports disciplines. Combined with the extensive potential offered by AI, this has opened up numerous opportunities and novel applications for obtaining valuable information for researchers, coaches, analysts, and athletes, e.g., regarding athlete's performances and tactical patterns (Chap. 1). While many approaches in sports science rely on positional or other sensor data (Chaps. 8 and 10), video recordings of competitions and training sessions can offer a more comprehensive view, as they capture details of athletes' actions and movements, such as postures, biomechanics, and tactical nuances. The level of detail with which these actions are depicted often exceeds what can be determined from position or sensor data alone. Thus, computer vision approaches can offer valuable insights into the internal dynamics of sports.

Recent advancements in computer vision and particularly deep learning (Dosovitskiy et al., 2021; He et al., 2016), have led to significant improvements in many sports analytics tasks. Network architectures tailored to general video understanding including action recognition (Fan et al., 2021; Liu et al., 2022) have been fine-tuned to detect specific actions in sports, e.g., shots, goals (Deliège et al., 2021). Approaches for object detection (Redmon & Farhadi, 2017; Ren et al., 2015; Zhu et al., 2021) and tracking (Zhang et al., 2021, 2022) allow us to locate, identify, and track relevant objects in a video sequence and have been adopted to detect players, balls, and other game equipment, e.g., in soccer (Cioppa et al., 2022; Theiner et al., 2022). In combination with sports field registration (Chen & Little, 2019; Theiner & Ewerth, 2023), these detected objects can be mapped to a two-dimensional model of the sports field to extract position data from a single camera (Theiner et al., 2022). This allows for the analysis of position data without data from expensive sensor devices that are rarely available outside professional sports. Approaches for pose estimation identify key joints of the human body to represent the spatial arrangement of body parts (Cao et al., 2021; Kreiss et al., 2019) and can help analyze the movement of athletes, e.g., to evaluate potential pass options (Sangüesa et al., 2020) or evaluate penalty kicks in soccer (de Sousa Pinheiro et al., 2022). These examples demonstrate how the recent progress in computer vision can support research in sports science. For a more comprehensive review of computer vision in sports, we refer to Naik et al. (2022).

The progress in computer vision has been made possible by the availability of large-scale datasets (e.g., Lin et al., 2014; Russakovsky et al., 2015) and fast computational resources which led to the availability of general-purpose models for many tasks such as object recognition (e.g., Dosovitskiy et al., 2021; He et al., 2016). In recent years, a couple of large-scale datasets and challenges (e.g., Deliège et al., 2021; Ingwersen et al., 2023; Jiang et al., 2020) have also been introduced to optimize these models for sports-related tasks (e.g., player detection) and domains (e.g., basketball, soccer). For instance, the *SoccerNet*[1] challenge (Deliège et al., 2021) has released an extensive dataset of 550 soccer game broadcasts from major European leagues, covering various tasks such as action spotting, camera calibration, player re-identification, and tracking. Furthermore, top-tier conferences in computer vision (e.g., *IEEE/CFV Conference on Computer Vision and Pattern Recognition*) and multimedia (e.g., *ACM International Conference on Multimedia*) have initiated dedicated workshops on sports analytics (e.g., *CVSports—IEEE International Workshop on Computer Vision in Sports, MMSports—International ACM Workshop on Multimedia Content Analysis in Sports*) to encourage research in the field of sports science.

In the remainder of this chapter, we provide the foundations of the most crucial steps involved in a deep learning pipeline (Sect. 5.2). Subsequently, we present state-of-the-art approaches along with practical tips for three selected computer vision tasks with high relevance to sports analytics (Sect. 5.3). Finally, we discuss open challenges and future research directions including new opportunities that arise from novel generative AI approaches.

5.2 Deep Learning Pipeline

▶ Deep learning is a subfield of machine learning that focuses on artificial neural networks. It involves training algorithms that automatically learn and extract features from data to understand complex patterns and representations. The term "deep" refers to the use of many hidden layers in a neural network.

While the exact structure of a deep learning pipeline can vary depending on the task and its specific requirements, Fig. 5.1 illustrates a general overview of the key components. These steps can be iteratively repeated to improve the performance of a deep learning model. Please note that this pipeline is not different to the processing pipeline of other supervised machine learning methods.

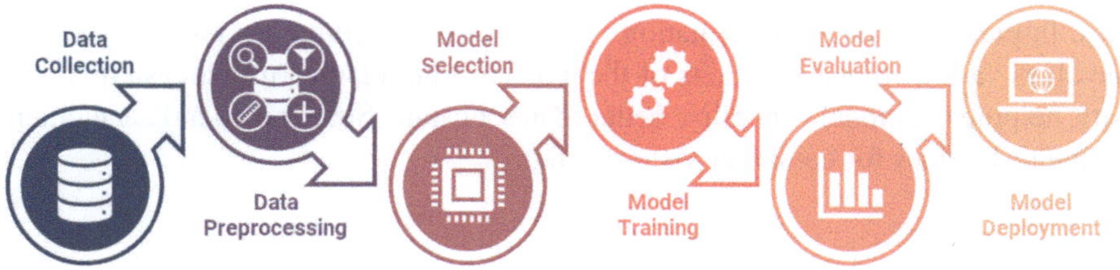

Fig. 5.1 Illustration of a deep learning pipeline

[1] https://www.soccer-net.org/

5.2.1 Data Collection

Data is crucial since deep learning models learn patterns and representations directly from the data. Thus, it is necessary to gather a large and diverse dataset that is representative of the problem to be solved. In computer vision, data typically comprises images or videos that can be gathered, e.g., from the web, existing benchmark datasets, or own data collections. There are many open-source tools for image or video annotation (e.g., *VGG Image Annotator,*[2] *Label Studio,*[3] *labelme*[4]) that can be used to create ground truth labels required to train models via supervised learning (Chap. 9).

The quality, quantity, and diversity of the training data play a significant role for the performance and generalization ability of the deep learning model and depend on several factors such as the complexity of the task as well as intra and extra-class variations. Also, data bias needs to be considered for a responsible use of deep learning models (Chap. 21). In broad terms, the model's ability to predict and generalize to unseen examples improves with more diverse and representative training data. Smaller, task-specific datasets can be used to fine-tune models that have been pre-trained on large and diverse datasets for similar tasks. This transfer learning approach leverages the knowledge gained from the larger dataset to improve performance on a more narrow task.

As mentioned in Sect. 5.1, there are many large-scale datasets (and resulting models) for various computer vision tasks. Prominent examples are the *ImageNet Large Scale Visual Recognition Challenge* (*ILSVRC*, Russakovsky et al., 2015) for image classification and the *Microsoft Common Objects in Context* (*MS COCO*; Lin et al., 2014) for object detection. Datasets specific to selected computer vision tasks in sports are covered in Sect. 5.3.

5.2.2 Data Preprocessing

The data collected in the first step can be further preprocessed, e.g., to clean or augment the data, making it more suitable for training. This may involve steps such as data filtering, normalization, resizing, data augmentation, and handling missing values. In particular, data augmentation is crucial to train computer vision models. It involves various transformations of the existing training data to increase its diversity, helping the model generalize better to unseen data. Some popular data augmentation techniques for computer vision include:

[2] https://www.robots.ox.ac.uk/~vgg/software/via/

[3] https://github.com/HumanSignal/label-studio

[4] https://github.com/labelmeai/labelme

- Randomly **cropping** the image to rectangular regions with different aspect ratios to add variations regarding scale, zoom, and translation;
- **Rotation** of the image by a certain angle to expose the model to variations in orientation;
- **Flipping** the image horizontally or vertically to account for different perspectives;
- **Changing the brightness and contrast** levels to account for variations in lighting conditions;
- Random **color adjustment** of the image to introduce diversity in color distribution;
- **Inject random noise** (e.g., Gaussian blur) to the image to simulate noisy inputs.

▶ Widely used deep learning libraries such as *PyTorch*[5] or *TensorFlow*[6] include many data augmentation techniques for images. *PyTorch* offers image augmentation approaches from the literature such as *RandAugment*[7] (Cubuk et al., 2020) or *TrivialAugment*[8] (Müller & Hutter, 2021).

5.2.3 Model Selection and Training

In the past decade, various deep learning architectures have been introduced. The choice of a suitable architecture depends on various factors such as the task itself (e.g., image classification, object detection), its complexity (e.g., the number of classes, intra- and extra-class variations), the computational resources, and the availability of models pre-trained for similar tasks that can save time and reduce the need for training data. Common architectures used in computer vision include *convolutional neural networks* (*CNNs*; e.g., He et al., 2016) and *vision transformers* (*ViT*; Dosovitskiy et al., 2021). Table 5.1 lists a selection of widely applied backbone architectures in computer vision. As the results show, there is typically a trade-off between the model complexity (e.g., number of parameters and operations) and performance depending on the application.

During training, the training data is passed through the model and a loss function measures the difference between the model's predictions and the ground-truth data. An optimization algorithm, e.g., *Stochastic Gradient Descent* (*SGD*; Bottou, 2010)

[5] https://pytorch.org/vision/stable/transforms.html

[6] https://www.tensorflow.org/tutorials/images/data_augmentation

[7] https://pytorch.org/vision/stable/generated/torchvision.transforms.RandAugment.html

[8] https://pytorch.org/vision/main/generated/torchvision.transforms.TrivialAugmentWide.html

Table 5.1 Top-1 (Acc@1) and top-5 accuracy (Acc@5) on the ILSVRC dataset (Russakovsky et al., 2015) for object classification for a selection of popular deep learning architectures along with their number of parameters (Params) and number of floating point operations per second (GFLOPS)

Architecture	Acc@1	Acc@5	Params	GFLOPS
MobileNetV3 (Howard et al., 2019)	75.27	92.57	5.5M	0.22
ResNet50 (He et al., 2016)	80.86	95.43	25.6M	4.09
EfficientNet-B5 (Tan & Le, 2019)	83.44	96.63	30.4M	10.27
ViT-B/16 (Dosovitskiy et al., 2021)	85.30	97.65	86.9M	55.48

A full list of models and results can be found on: https://pytorch.org/vision/stable/models.html

or *Adaptive Momentum* (*Adam*; Kingma & Ba, 2015) is applied to minimize this loss during training via *backpropagation*. This process is iteratively repeated until the model converges to a satisfactory performance level, ideally on an independent validation dataset to avoid overfitting or underfitting (Chap. 3). For theoretical and mathematical details, we refer to the deep learning book from Goodfellow et al. (2016).

▶ *Python* libraries such as *PyTorch* and *TensorFlow* offer a wide range of architectures and pre-trained models[9,10] along with extensive tutorials[11,12] for training and fine-tuning.

5.2.4 Model Evaluation & Deployment

After training the model can be evaluated using unseen test data and appropriate evaluation metrics, e.g., accuracy, precision, recall, F1 score for classification tasks (Chap. 3). As mentioned in Sect. 5.1, a lot of public benchmark datasets have been introduced for computer vision tasks that include training, validation, and test data (Lin et al., 2014; Russakovsky et al., 2015). Table 5.1 shows the results of widely applied architectures for visual concept recognition. In Sect. 5.3, we cover benchmark datasets for selected sports analytics tasks.

After evaluation, the model can be deployed in a target system or application. To monitor the model performance in real-world scenarios, feedback from users can be gathered. The model can also be periodically updated using new data, if available, to adapt to changing patterns and improve performance. Based on the evaluation, feedback, and new data, the steps in the proposed pipeline can be iteratively repeated to improve performance.

---

[9] https://pytorch.org/vision/stable/models.html

[10] https://www.tensorflow.org/resources/models-datasets

[11] https://pytorch.org/tutorials/beginner/introyt/introyt1_tutorial.html

[12] https://www.tensorflow.org/tutorials/quickstart/beginner

5.3 Computer Vision Applications for Sports Analytics

In this section, we present a selection of state-of-the-art approaches and tools crucial for many sports domains and analyses. This includes approaches for action recognition, object detection, multi-object tracking, and sports field registration.

5.3.1 Action Spotting

Video recordings of sports matches or competitions depict a large number of actions, e.g., goals and shots in soccer. The detection of such actions can facilitate efficient searches in individual videos and large video collections. For instance, action recognition in videos enables the retrieval of standard situations or scenes that led to goals can empower game analysts to, e.g., seek particular tactical patterns.

Numerous tasks involve the identification and detection of actions in videos, spanning from action recognition, where actions are classified in short video clips, to action spotting, which also entails the prediction of the time (interval) when an action is performed making it particularly important for analyzing salient actions, e.g., shots, in team sports (Giancola et al., 2018). Action spotting approaches frequently use popular image recognition backbones (Table 5.1) combined with methods for temporal pooling, for example, using *Network Vector of Locally Aggregated Descriptors* (*NetVLAD*; Arandjelovic et al., 2016), to predict actions for video sequences (e.g., Giancola et al., 2023). More recently, vision transformers for videos (Fan et al., 2021; Liu et al., 2022) trained on large datasets for action recognition such as "*something-something*" (Goyal et al., 2017) or *Kinetics* (Kay et al., 2017) have been proposed that directly predict actions given a video sequence as input. Given that these models use video sequences as input, which are typically much shorter than the entire video, a set of overlapping (or non-overlapping) sequences is used to predict actions across the entire video. To eliminate multiple predictions of the same action, post-processing steps like non-maximum suppression are applied to spot actions. Recent work has optimized these models for sports data with promising results, e.g., in action spotting within soccer videos (e.g., Giancola et al., 2023; Xarles et al., 2023). State-of-the-art models for video understanding and action recognition can be implemented using Python libraries such as *MMAction2*[13] and *PySlowFast.*[14]

To apply action spotting to the soccer domain, Deliège et al. (2021) have introduced the *SoccerNet* challenge that comprises 500 videos annotated for 17 crucial soccer actions, including goals, fouls, and shots both on and off target. Giancola et al. (2023) have trained action spotting using several datasets including *SoccerNet* based on features extracted from a *ResNet* (He et al., 2016) model and *NetVLAD++* (Giancola & Ghanem, 2021) for temporal pooling. They employed an active

[13] https://github.com/open-mmlab/mmaction2

[14] https://github.com/facebookresearch/SlowFast

learning approach that uses few labeled training data at the beginning and iteratively optimizes the model based on new data annotated by humans (Sect. 5.2.4).

A notable drawback of existing approaches and the *SoccerNet* dataset is the lack of details about actions like passes and attributes (e.g., flat or high passes). Biermann et al. (2021) have introduced a taxonomy in order to comprehensively cover actions in team sports. Through appropriately annotated training data, models can be fine-tuned to encompass a wider and more fine-grained range of actions. However, the differentiation between fine-grained actions is much more difficult since they are more ambiguous, which poses additional challenges to action spotting approaches.

5.3.2 Object Detection and Tracking

The identification and tracking of key entities, such as players, balls, or equipment, is crucial for many applications in sports analytics since they provide insights into player positions and movements, team formations, (inter)actions, and game dynamics. It is also an important preprocessing step for other tasks, e.g., to analyze actions or position data for individual players.

Approaches for object detection use convolutional neural networks or transformers such as *Faster d Region-based Convolutional Neural Network* (*R-CNN;* Ren et al., 2015), *You Only Look Once* (*YOLO*; Redmon & Farhadi, 2017), and *DEtection TRansformer* (*DETR*; Carion et al., 2020) to solve the tasks of object localization and classification. As explained in Sect. 5.2.3, these models are trained on large datasets such *MS COCO* labeled with bounding boxes for a wide range of objects (e.g., cars, animals, objects, humans) (Lin et al., 2014). For videos, object detection approaches are often combined with multi-object tracking (MOT) algorithms to simultaneously track the detected objects over time. Popular tracking algorithms include *CenterTrack* (Zhou et al., 2020), *FairMOT* (Zhang et al., 2021), and *ByteTrack* (Zhang et al., 2022). General-purpose models for object detection and tracking have been fine-tuned using labeled training data (Sect. 5.2.1) to adapt to key entities, such as players, ball, etc., for specific sports domains such as soccer (e.g., Cioppa et al., 2022). There are several *Python* libraries that allow for the application and fine-tuning of state-of-the-art models for object detection and tracking such as *Detectron2,*[15] *Hugging Face,*[16] *PaddleDetection,*[17] *MMDetection,*[18] *MMTracking,*[19] and *Norfair.*[20]

An example for a joint object detection and tracking approach in team sports has been proposed by Cioppa et al. (2022). They have released the *SoccerNet-Tracking*

[15] https://ai.meta.com/tools/detectron2/

[16] https://huggingface.co/

[17] https://github.com/PaddlePaddle/PaddleDetection

[18] https://github.com/open-mmlab/mmdetection

[19] https://github.com/open-mmlab/mmtracking

[20] https://github.com/tryolabs/norfair

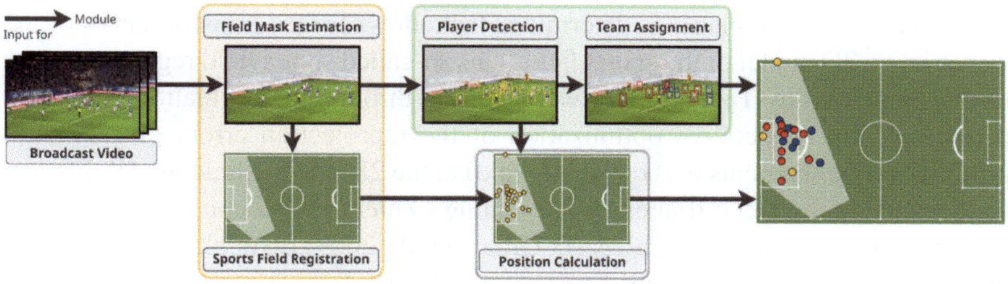

Fig. 5.2 Pipeline for position data extraction from a broadcast soccer video (Theiner et al., 2022). The broadcast image is taken from Wikimedia Commons (©Markus Unger, CC BY 2.0, link: https://commons.wikimedia.org/wiki/File:2012-11-17_Hamburg3_(8208561891).jpg)

challenge along with a dataset of 200 video sequences from 12 soccer matches, each of 30 s length. With this dataset, they optimized various state-of-the-art tracking methods including *FairMOT* and *ByteTrack* using the *PaddleDetection* library for simultaneous player and ball tracking. The *ByteTrack* approach achieved the best results which were overall promising but also revealed limitations and challenges.

In particular, the detection of the ball remains challenging due to its small size, occlusions, and high speed. Also, the re-identification of players that have not been detected in previous frames because they were, for example, moving fast, outside the visible field captured by the camera, or occluded by other players can be challenging. Moreover, the detection of persons outside the pitch can have a negative impact on the results and might require additional preprocessing steps to, for example, separate the pitch from the stands (Fig. 5.2).

5.3.3 Position Data Estimation from Videos Using Sports Field Registration

Position data is used for a wide variety of tasks in sports science (Chaps. 8 and 10). Modern tracking technologies enable the collection of millions of data points, depicting both player and ball positions throughout a match. However, the acquisition of position data can be challenging, often due to factors like licensing, financial constraints, or competitive considerations. Clubs typically refrain from sharing or disclosing their teams' data. On the contrary, accessing broadcast video recordings of soccer matches is comparatively easier.

Theiner et al. (2022) introduced a system (Fig. 5.2) that combines several computer vision tasks to automatically estimate position data from television broadcasts and scouting feed recordings of soccer matches. First, the *pix2pix* model (Isola et al., 2017), fine-tuned for soccer pitches according to Chen and Little (2019), is applied to a video image to separate the playing field from the stands and to detect the field markings. Then, a deep learning approach is used to compare the resulting image to a reference database of synthetic images with known camera parameters (Chen & Little, 2019). Based on the camera parameters of the most similar reference image and an image alignment to the video frame (Evangelidis & Psarakis,

2008), a homography matrix is determined that allows for the transformation of the image into a 2D model of the sports field. This so-called sports field registration was further improved by Theiner and Ewerth (2023) through the estimation of camera parameters via an iterative optimization of the reprojection errors of geometric primitives (line segments of the playing field) to the 2D playing field model. Players are detected in the video frames by fine-tuning *CenterTrack* (Zhou et al., 2020; see Sect. 5.3.2) on two soccer datasets (D'Orazio et al., 2009; Lu et al., 2017). Finally, the detected players are mapped to the 2D playing field using the homography matrix to determine their position data. Team assignment is conducted using clustering with *DBScan* (Ester et al., 1996) based on simple HSV color features from the detected player bounding boxes. The system has achieved promising results and provides an initial foundation for various analyses based on position data (Chap. 8).

However, Theiner et al. (2022) observed limitations that may affect the performance of subsequent steps. Most notably, the challenges discussed in Sect. 5.3.2 limit the performance for ball and player tracking. Also, the team assignment of players can be challenging, especially when they occlude each other or if goalkeepers, referees, or uninvolved persons (e.g., substitutes) are visible. In addition, sports field registration is typically optimized for a specific camera setting and lacks generalization when applied to other settings with significantly different camera position or pitch dimensions as, for example, in other sports domains. To adapt to other settings, fine-tuning with appropriate training data (Sect. 5.2.1) or modifications to the architecture may be necessary.

5.4 Conclusion

Significant advancements in computer vision have been made for many tasks, due to the availability of large-scale datasets and deep learning. In this chapter, we have described how this progress has paved the way for numerous sports science applications, including action recognition, player detection, and sports field registration. However, many challenges remain as well that may limit real-world applications. For example, fine-grained actions like passes are often disregarded by action recognition approaches, object detectors can fail to detect small, fast moving objects such as the ball, and player tracking remains challenging since players are difficult to identify when they were not visible in previous frames. To address these issues, recently, large-scale datasets and competitions like *SoccerNet* have been introduced to foster research for computer vision in sports. Moreover, novel generative AI approaches such as *GPT-4* (Generative Pretrained Transformer) have revolutionized research in many computer science domains and have demonstrated to perform well in computer vision tasks, such as video understanding (e.g., Song et al., 2023), that are relevant for sports science. Multimodal approaches that combine computer vision with other approaches that use sensor data (e.g., position data; see Chaps. 8 and 10) are another promising research direction.

References

Arandjelovic, R., Gronat, P., Torii, A., Pajdla, T., & Sivic, J. (2016). NetVLAD: CNN architecture for weakly supervised place recognition. In *IEEE Conference on Computer Vision and Pattern Recognition*, CVPR 2016 (pp. 5297–5307).

Biermann, H., Theiner, J., Bassek, M., Raabe, D., Memmert, D., & Ewerth, R. (2021, October). A unified taxonomy and multimodal dataset for events in invasion games. In *Proceedings of the 4th International Workshop on Multimedia Content Analysis in Sports* (pp. 1–10).

Bottou, L. (2010). Large-scale machine learning with stochastic gradient descent. In *International Conference on Computational Statistics*, COMPSTAT 2010 (pp. 177–186).

Cao, Z., Hidalgo, G., Simon, T., Wei, S.-E., & Sheikh, Y. (2021). OpenPose: Realtime multi-person 2D pose estimation using part affinity fields. *IEEE Transactions on Pattern Analysis and Machine Intelligence, 43*, 172–186.

Carion, N., Massa, F., Synnaeve, G., Usunier, N., Kirillov, A., & Zagoruyko, S. (2020). End-to-end object detection with transformers. In *European Conference on Computer Vision*, ECCV 2020 (pp. 213–229).

Chen, J., & Little, J. J. (2019). Sports camera calibration via synthetic data. In *IEEE Conference on Computer Vision and Pattern Recognition Workshops,* CVPRW 2019 (pp. 2497–2504).

Cioppa, A., Giancola, S., Deliège, A., Kang, L., Zhou, X., Cheng, Z., Ghanem, B., & Droogenbroeck, M. V. (2022). SoccerNet-tracking: Multiple object tracking dataset and benchmark in soccer videos. In *IEEE/CVF Conference on Computer Vision and Pattern Recognition Workshops*, CVPRW 2022 (pp. 3490–3501).

Cubuk, E. D., Zoph, B., Shlens, J., & Le, Q. V. (2020). Randaugment: Practical automated data augmentation with a reduced search space. In *IEEE/CVF Conference on Computer Vision and Pattern Recognition Workshops*, CVPRW 2020 (pp. 702–703).

D'Orazio, T., Leo, M., Mosca, N., Spagnolo, P., & Mazzeo, P. L. (2009). A semi-automatic system for ground truth generation of soccer video sequences. In *IEEE International Conference on Advanced Video and Signal Based Surveillance,* AVSS 2009 (pp. 559–564).

de Sousa Pinheiro, G., Jin, X., Da Costa, V. T., & Lames, M. (2022). Body pose estimation integrated with notational analysis: A new approach to analyze penalty kicks strategy in elite football. *Frontiers in Sports and Active Living, 4*, 1–10.

Deliège, A., Cioppa, A., Giancola, S., Seikavandi, M. J., Dueholm, J. V., Nasrollahi, K., Ghanem, B., Moeslund, T. B., & Droogenbroeck, M. V. (2021). SoccerNet-v2: A dataset and benchmarks for holistic understanding of broadcast soccer videos. In *IEEE Conference on Computer Vision and Pattern Recognition Workshops*, CVPRW 2021 (pp. 4508–4519).

Dosovitskiy, A., Beyer, L., Kolesnikov, A., Weissenborn, D., Zhai, X., Unterthiner, T., Dehghani, M., Minderer, M., Heigold, G., Gelly, S., Uszkoreit, J., & Houlsby, N. (2021). An image is worth 16×16 words: Transformers for image recognition at scale. In *International Conference on Learning Representations*, ICLR 2021.

Ester, M., Kriegel, H. P., Sander, J., & Xu, X. (1996). A density-based algorithm for discovering clusters in large spatial databases with noise. In *International Conference on Knowledge Discovery and Data Mining*, KDD 1996 (pp. 226–231).

Evangelidis, G. D., & Psarakis, E. Z. (2008). Parametric image alignment using enhanced correlation coefficient maximization. *IEEE Transactions on Pattern Analysis and Machine Intelligence, 30*(10), 1858–1865.

Fan, H., Xiong, B., Mangalam, K., Li, Y., Yan, Z., Malik, J., & Feichtenhofer, C. (2021). Multiscale vision transformers. In *IEEE/CVF International Conference on Computer Vision*, CVPR 2021 (pp. 6824–6835).

Giancola, S., & Ghanem, B. (2021). Temporally-aware feature pooling for action spotting in soccer broadcasts. In *IEEE/CVF Conference on Computer Vision and Pattern Recognition*, CVPR 2021 (pp. 4490–4499).

Giancola, S., Amine, M., Dghaily, T., & Ghanem, B. (2018). SoccerNet: A scalable dataset for action spotting in soccer videos. In *IEEE Conference on Computer Vision and Pattern Recognition Workshops*, CVPRW 2018 (pp. 1711–1721).

Giancola, S., Cioppa, A., Georgieva, J., Billingham, J., Serner, A., Peek, K., et al. (2023). Towards active learning for action spotting in association football videos. In *IEEE/CVF Conference on Computer Vision and Pattern Recognition*, CVPR 2023 (pp. 5097–5107).

Goodfellow, I., Bengio, Y., & Courville, A. (2016). *Deep learning*. MIT Press. http://www.deeplearningbook.org

Goyal, R., Ebrahimi Kahou, S., Michalski, V., Materzynska, J., Westphal, S., Kim, H., et al. (2017). The "something something" video database for learning and evaluating visual common sense. In *IEEE International Conference on Computer Vision*, ICCV 2017 (pp. 5842–5850).

He, K., Zhang, X., Ren, S., & Sun, J. (2016). Deep residual learning for image recognition. In *IEEE Conference on Computer Vision and Pattern Recognition*, CVPR 2016 (pp. 770–778).

Howard, A., Sandler, M., Chu, G., Chen, L. C., Chen, B., Tan, M., et al. (2019). Searching for MobileNetV3. In *IEEE/CVF International Conference on Computer Vision*, ICCV 2019 (pp. 1314–1324).

Ingwersen, C. K., Mikkelstrup, C. M., Jensen, J. N., Hannemose, M. R., & Dahl, A. B. (2023). SportsPose—A dynamic 3D sports pose dataset. In *IEEE/CVF Conference on Computer Vision and Pattern Recognition*, CVPR 2023 (pp. 5218–5227).

Isola, P., Zhu, J. Y., Zhou, T., & Efros, A. A. (2017). Image-to-image translation with conditional adversarial networks. In *IEEE Conference on Computer Vision and Pattern Recognition*, CVPR 2017 (pp. 1125–1134).

Jiang, Y., Cui, K., Chen, L., Wang, C., & Xu, C. (2020). SoccerDB: A large-scale database for comprehensive video understanding. In *International Workshop on Multimedia Content Analysis in Sports*, MMSports 2020 (pp. 1–8).

Kay, W., Carreira, J., Simonyan, K., Zhang, B., Hillier, C., Vijayanarasimhan, S., et al. (2017). The kinetics human action video dataset. *arXiv preprint arXiv:1705.06950.*

Kingma, D. P., & Ba, J. (2015). Adam: A method for stochastic optimization. In *International Conference on Learning Representations*, ICLR 2015.

Kreiss, S., Bertoni, L., & Alahi, A. (2019). PifPaf: Composite fields for human pose estimation. In *IEEE Conference on Computer Vision and Pattern Recognition*, CVPR 2019, pp. 11977–11986.

Lin, T.-Y., Maire, M., Belongie, S. J., Hays, J., Perona, P., Ramanan, D., Dollár, P., & Zitnick, C. L. (2014). Microsoft COCO: Common objects in context. In *European Conference on Computer Vision*, ECCV 2014 (pp. 740–755).

Liu, Z., Ning, J., Cao, Y., Wei, Y., Zhang, Z., Lin, S., & Hu, H. (2022). Video swin transformer. In *IEEE/CVF Conference on Computer Vision and Pattern Recognition*, CVPR 2022 (pp. 3202–3211).

Lu, K., Chen, J., Little, J. J., & He, H. (2017). Light cascaded convolutional neural networks for accurate player detection. In *British Machine Vision Conference 2017*, BMVC 2017.

Müller, S. G., & Hutter, F. (2021). TrivialAugment: Tuning-free yet state-of-the-art data augmentation. In *IEEE/CVF International Conference on Computer Vision*, ICCV 2021 (pp. 774–782).

Naik, B. T., Hashmi, M. F., & Bokde, N. D. (2022). A comprehensive review of computer vision in sports: Open issues, future trends and research directions. *Applied Sciences, 12*(9), 4429.

Redmon, J., & Farhadi, A. (2017). YOLO9000: Better, faster, stronger. In *IEEE Conference on Computer Vision and Pattern Recognition*, CVPR 2017 (pp. 6517–6525).

Ren, S., He, K., Girshick, R. B., & Sun, J. (2015). Faster R-CNN: Towards real-time object detection with region proposal networks. In *Annual Conference on Neural Information Processing Systems,* NIPS 2015 (pp. 91–99).

Russakovsky, O., Deng, J., Su, H., Krause, J., Satheesh, S., Ma, S., Huang, Z., Karpathy, A., Khosla, A., Bernstein, M. S., Berg, A. C., & Li, F.-F. (2015). ImageNet large scale visual recognition challenge. *International Journal on Computer Vision, 115*(3), 211–252.

Sangüesa, A. A., Martín, A., Fernández, J., Ballester, C., & Haro, G. (2020). Using player's body-orientation to model pass feasibility in soccer. In *IEEE/CVF Conference on Computer Vision and Pattern Recognition*, CVPRW 2020 (pp. 3875–3884).

Song, E., Chai, W., Wang, G., Zhang, Y., Zhou, H., Wu, F., Guo, X., Ye, T., Lu, Y., Hwang, J.-N., & Wang, G. (2023). MovieChat: From dense token to sparse memory for long video understanding. *arXiv preprint, abs/2307.16449.*

Tan, M., & Le, Q. V. (2019). EfficientNet: Rethinking model scaling for convolutional neural networks. In *International Conference on Machine Learning*, ICML 2019 (pp. 6105–6114).

Theiner, J. & Ewerth, R. (2023). Keypoint-less camera calibration for sports field registration in soccer. In *IEEE/CVF Winter Conference on Applications of Computer Vision*, WACV 2023 (pp. 1166–1175).

Theiner, J., Gritz, W., Müller-Budack, E., Rein, R., Memmert, D., & Ewerth, R. (2022). Extraction of positional player data from broadcast soccer videos. In *IEEE/CVF Winter Conference on Applications of Computer Vision*, WACV 2022 (pp. 1463–1473).

Xarles, A., Escalera, S., Moeslund, T. B., & Clapés, A. (2023). ASTRA: An action spotting transformer for soccer videos. In *International Workshop on Multimedia Content Analysis in Sports*, MMSports 2023 (pp. 93–102).

Zhang, Y., Wang, C., Wang, X., Zeng, W., & Liu, W. (2021). FairMOT: On the fairness of detection and re-identification in multiple object tracking. *International Journal of Computer Vision, 129*(11), 3069–3087.

Zhang, Y., Sun, P., Jiang, Y., Yu, D., Weng, F., Yuan, Z., et al. (2022). Bytetrack: Multi-object tracking by associating every detection box. In *European Conference on Computer Vision*, ECCV 2022 (pp. 1–21).

Zhou, X., Koltun, V., & Krähenbühl, P. (2020). Tracking objects as points. In *European Conference on Computer Vision*, ECCV 2020 (pp. 474–490).

Zhu, X., Su, W., Lu, L., Li, B., Wang, X., & Dai, J. (2021). Deformable DETR: Deformable transformers for end-to-end object detection. In *International Conference on Learning Representations*, ICLR 2021.

6 Human and Artificial Intelligence Integration for Decision-Making in Sports Performance

Sam Robertson

Abstract

As in most areas of human life, technology and artificial intelligence (AI) continue to permeate many functions and decisions pertaining to performance sport. Although these developments are predicated on providing faster, more accurate and increasingly efficient solutions, little is known about how best to integrate technology and AI with the current functions and structures common to high performance sport. Using an eight-stage model of decision-making, this chapter provides an overview of how technology and AI can be integrated into the deliberate and structured decision-making of individuals and groups. Providing an optimistic view of the future and providing various practical tips, it expands on the key benefits presented by technology and AI, whilst emphasizing the continued importance of the role of the human.

6.1 Introduction

For the human of 2024, technology and AI influences just about every part of their life, from ordering food to monitoring their health and wellness, to work and social interactions. For the purposes of this chapter, AI is defined as the development of methods and computer systems which can perform tasks and functions that typically required human intelligence. These tasks include but are not limited to problem-solving, learning, recognizing complex patterns, and ultimately using these tasks to make decisions (Russell & Norvig, 2016). Throughout the chapter, AI is discussed in tandem with the various technologies that produce the data without which, such systems could not exist.

S. Robertson (✉)
Institute for Health and Sport, Victoria University Melbourne, Footscray, VIC, Australia
e-mail: Sam.robertson@vu.edu.au

D. Memmert (ed.), *Artificial Intelligence and Machine Learning in Sports Science*,
https://doi.org/10.1007/978-3-662-70155-3_6

Despite the rapid and immense developments to technology and AI, presently very few of their processes and functions run completely autonomously without any need whatsoever for human input. Consequently, whilst a further increased role of technology and AI in our lives may feel inevitable, for the moment at least there is a gap—some may say a chasm—between the present and a world of full AI automation.

Consequently, understanding how the interaction between AI, technology, humans and organisations can be optimized has never been more important. Despite the doomsayers consistently pitting humans against AI, the reality is that this present situation should be viewed through the lens of humans AND AI—ideally working together. Without this approach, it is likely that humans lose control of it and let it dictate their lives. Some may argue that has already occurred. For the first time in history, technology is being developed that is not borne out of necessity. We are not developing a hammer to help us build, or a machine to manufacture at scale; rather, a considerable portion of the technology of today is being driven solely by commercial incentives; to get us to click, share or purchase. Instead of merely being a tool to assist, these new technologies actually often demand something back from us, most notably our attention (Kelly, 2016). And because this new way of interacting with technology is all so new, we know very little about how it affects us, both positively and negatively, particularly over the long term.

To help navigate this changing space, this chapter seeks to outline key principles for integration and collaboration between human and AI. Using the example of sports performance, the chapter does so through a lens of eight key steps of decision-making:

1. Questioning
2. Collection
3. Storage
4. Access
5. Insights
6. Decisions
7. Actions, and
8. Evaluation.

Although each organization (and sometimes the individuals within it) will employ different processes to structured decision making, these eight steps are likely interchangeable with any system employed, from one that is highly organized and analytical (i.e., in large businesses and similar to the model presented here), through to approaches grounded predominantly in intuition and experience (i.e., those often seen in sports coaching).

For all its uniqueness as an industry, sport is not exempt from any of the challenges mentioned above. Sports performance in particular faces some of the most complex and challenging questions of all. Figure 6.1 shows an example of the types of problems faced by various role types within a typical sports performance

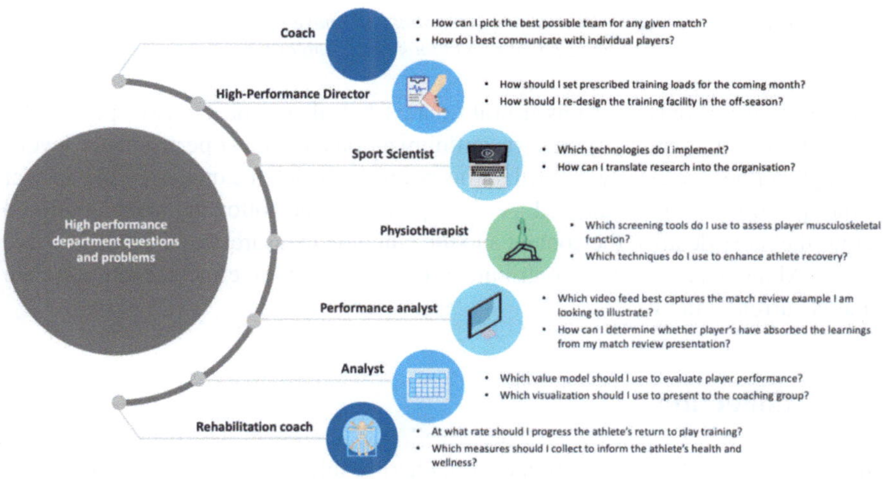

Fig. 6.1 Example decisions and problems faced by high performance sport staff

department. Throughout the chapter a range of these decisions will be used to illustrate these eight stages of decision-making.

6.2 Questioning

The first stage of decision-making should be to specifically define the question or problem. Finding the right solution to a problem may not necessarily be contingent on having expressed the specific question clearly beforehand, but it certainly helps. Of course, solutions to problems can arise from many sources: trial and error, long-term systematic analysis, and sometimes pure luck. However, when a decision needs to be made more than once, having these steps formalized helps to ensure appropriate allocation of resources and ideally refinement of the solution over time through recording the key elements along the way.

Good questioning consists of framing a problem in a way that it can be satisfactorily solved within a specific context (Lin et al., 2003). For the moment at least, it remains an exercise requiring human leadership, as opposed to being generated by AI. Despite the apparent limited need for AI at this point, it warrants mention as the first of these eight stages of decision-making because all of the remain seven should be shaped by the nature of the question. To bring this eight-stage framework to life practically, consider the following example question:

> *I am a high-performance director of a professional soccerl team. My team trains or plays four times a week and I want to know their injury risk prior to each time of training or competition. I want to adopt an accurate decision-making solution that optimizes risk minimization whilst not meaningfully affecting player performance levels. The regular decision I am faced with is whether to clear a player for full training or competition, modify their activity (i.e., different or lower levels of training, reduced match minutes) or to restrict them*

altogether. Financial resources are not a constraint and I have recently signed a three-year contract and thus any solution I adopt should look to the longer term.

The question not only contains useful context but also indicates two, potentially contradictory, outcome measures (risk minimization and player performance levels, although neither are clearly defined to this point) by which the quality of the answer to the question can be evaluated. As is explored in the following stages outlined below, the more detailed the context and the outcome measure(s), the more possible it is for AI and technology to contribute to providing a more effective solution than what is currently in place.

6.3 Collection

The second stage, Collection, relates to gathering information with the purpose of finding this satisfactory solution to the question stated in Stage 1. Some of the key considerations for each question at this stage of the process are outlined below.

- Which data is needed?
- Until relatively recently, the question of what data is needed to answer a given problem required much more pragmatism than it does today. Whilst many organisations certainly always sought the best possible data, historically there was nonetheless an element of 'taking what one could get'. Without the use of dynamic, scalable computer systems, there was a somewhat limited ceiling of data that could be organised and comprehended by the human user (Miller, 1956). With exponential improvements in computing and more recently AI, this is now less of a consideration. AI also has built-in abilities to inform just how much information to use, thus balancing the offset between perfect accuracy and processing time. Examples include information gain, goodness of fit statistics and cross-validation (Christian & Griffiths, 2016). All of these serve to help find a parsimonious model; one that only values information that meaningfully adds to a further understanding of the problem at hand. Figure 6.1 provides a simple representation between the human and AI when it comes to systematic information use on structured decision-making tasks. This not only shows the capabilities of AI but illustrates that often when examining a problem, humans and AI are often working from different information sources and thus can be said to be examining different things (see McIntosh et al., 2019 for an example)! (Fig. 6.2).
- Which data is currently accessible, but could be improved upon?
- The explosion of technology and AI has predictably led to commercial incentives and ultimately, competition between companies. This is largely good news for the end user; not only do they have new types of information potentially accessible (i.e., biomarkers during sports competition), but also enhanced versions of existing data (i.e., marker-less skeletal data on athletes, rather than solely tracking centre of mass). I refer to the process of examining a problem for new,

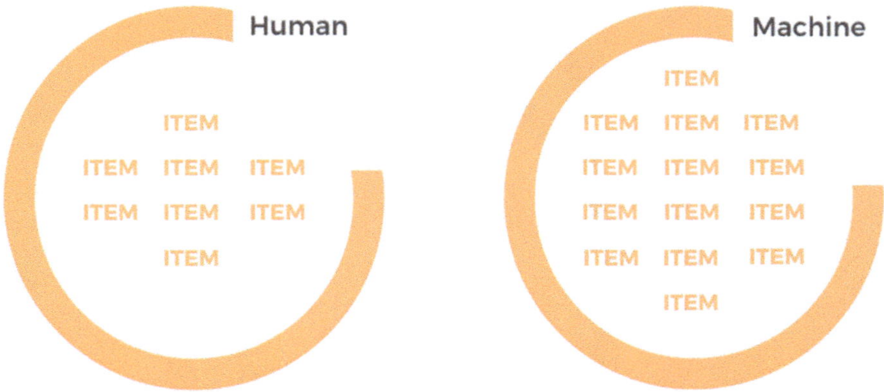

Fig. 6.2 Differences between human and AI in data use

enhanced or faster data sources to help solve it as an 'Information Stocktake' and in scenarios where technology is developing rapidly, it may require undertaking somewhat regularly. It also helps to identify technology-derived information sources that traditionally may have come from a human. Despite this, even in a situation whereby technology has revealed more detailed information than a human and done so in a faster manner, it may not necessarily mean that it should form part of a data collection process moving forwards. Sometimes irrespective of its clear value, the technology may be prohibitively expensive, or unreliable. In such circumstances where it is ultimately determined that the technology should not be used to replace the existing human judgement, it could nonetheless be used to refine it. For instance, a scout may be asked to rate the speed of an athlete before approaching to draft or sign them. Consulting the athlete's sprint times from timing gates or a global positioning system (GPS) device, may help to refine their judgement moving forwards. For example, an athlete viewed as 'quick' by the scout may develop an ability to visually determine that athlete as having a top speed of 32 km/h. The decision around whether to 'refine or replace' in various measures used for decision-making is one of the most pervasive problems currently faced for humans and organisations as they grapple with the emergence of AI.

- What data is not available?
- The old saying 'Not everything that can be counted counts, and not everything that counts can be counted' has never been truer than it is today. Being aware of what is not being measured or used to inform a decision is almost as important as being aware of what is. Figure 6.3 shows a conceptual example of this exercise in practice. The inner circle relates to information that is used by an organisation or individual to address a given problem, with the outer circle displaying the hypothetical 'optimal' solution. The closer the inner circle reflects the other circle, the more complete the data set is to fully informing the problem. The middle blue circle includes information that an organisation is aware of, but chooses to

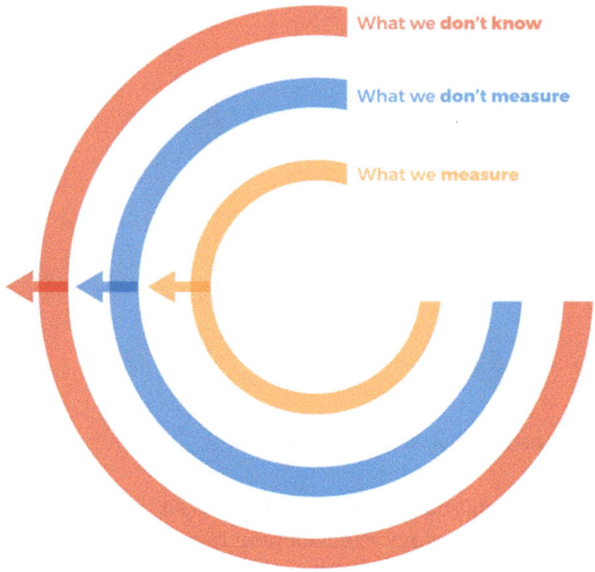

Fig. 6.3 What is missing in a model?

not utilise to inform the problem. A range of reasons exist as to why certain information is not considered; it could be unreliable, or too costly in terms of time or finances. However again as technology continues to improve, it is likely that many forms become cheaper and more accessible, this expanding the outer edges of the circle shown in the figure.

- How much data is enough in order to make an 'informed' decision?
- Determining a sufficient volume of information to help solve a question will likely differ between organisations, based on culture, resourcing and importance. Some will use a brainstorm with existing staff, others will refer to the literature. Some will merely base it on current practice. As mentioned above, computer science has well established tools available to define the completeness of a solution to a given problem or question. In organisations already overrun with technology and data, these tools could do well to be more widely adopted by organisations. The conundrum can be viewed through various metaphors but can simply be viewed as a gauge such as in Fig. 6.4. The red section of the gauge indicates insufficient information, yellow may be a suitable range, whereas green represents the range that could be achieved with appropriate developments to technology or resourcing within the organisation.
- Good organisations are implementing technology strategies to ensure they have the most impactful information and to avoid being overrun by data, much of which is either useless or adding limited value to enhancing decision-making.

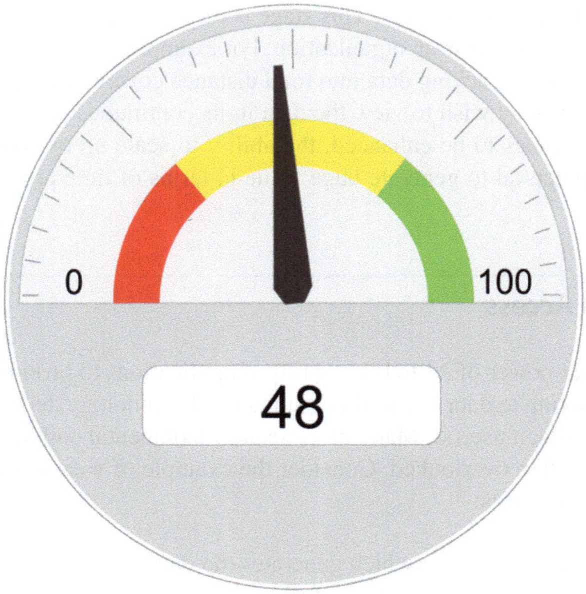

Fig. 6.4 Model completeness gauge, signifying how well the data collection has informed the given question or problem of interest

6.4 Storage

Soon, almost no processes will remain that use manual data storage means. The filing cabinet is an entirely foreign object to the new generation of sports performance staff and even the handwritten notes infamously preferred by members of the medical profession are slowly becoming obsolete. The complete shift to digital is inevitable, however it comes with its own set of challenges. As outlined in Sect. 6.3, as the quality of tech improves, so too does the size of data. Consequently, decisions need to be made when, if at all, to upgrade data due to the implication this has for local or cloud- or remote-based storage. This decision will at least partially depend on how complete the data collection is for the specific problem (also discussed in Sect. 6.3). For instance, if the solution to a question would be helped by access to in-game 50 Hertz (Hz) skeletal tracking data or a higher-resolution medical imaging. When the previous stage is not undertaken, it is not uncommon for sports organisations to enter into a constant upgrade cycle where there really isn't a strong rationale to do so. This is often predicated on the idea of keeping up with competitors rather than specifically enhancing decision-making.

Consequently, an increasingly key advantage of AI and technology is its ability to process and reformat data to make it more manageable for the human user. Even when this wrangling, filtering and summarizing is being done well by humans presently, it still has a manual component to the exercise. Again, the intended use for the data (i.e., which question or problem it is intending to solve) should dictate the nature of this stage of the process. This stage is of particularly value when the data has various use cases across an organization. For example, a live soccer broadcaster may aggregate 25 Hz tracking data into total distance covered by a player, whereas performance staff may wish to view the data in its continuous form. As the resolution of data continues to be enhanced, the ability to scale up or down in terms of detail has the potential to generate huge value in terms of time efficiency and also commercially.

6.5 Data Access

Despite the clear power of AI to bring a raft of applications to prominence, a range of questions relating to data access (both human and technology derived) remain the domain of the human user(s). Many of these are fundamental and have large implications, yet are often overlooked. Consider the example of wearable technology in professional team sports:

- Who owns the technology (which generates the data) for the specific use case?
  - Is it the athlete, who wore the wearable technology to generate the data?
  - Is it the professional team who acquired the device?
  - Is it the technology company, who stores the data on their cloud network?
  - Is it the league or governing body, who permits the technology to be used in-game? Is it the broadcaster, who purchased the data for use in television coverage or fan engagement?
  - Is it the fan themselves, who purchased a license to access the data as part of a betting company subscription?
  - Is it all of these stakeholders?
- Did they provide consent for that specific use?
  - Examples are everywhere of companies using data on athletes without their consent for commercial gain. Vision is perhaps the most common example of this.
- Whose responsibility?
  - In many performance environments this is simply left to whoever is interacting most with the data, from the analyst to sport scientist. Responsibility lines to senior staff may be loose at best and even when they are present, governance protocols are typically non-existent.
- How is their privacy protected?
  - How does an athlete have recourse in case of a breach?
  - Is the athlete even aware of their rights with respect to privacy?

Designing and implementing solutions to these questions remains largely a human exercise and requires considerable time and investment. After all, AI will not govern itself.

6.6 Insights

No other aspect of AI has been so widely publicized and hyped as its purported ability to generate precise predictive insights which are more accurate than humans (for examples of such comparisons in sports see Pappalardo et al., 2017; Blythe & Király, 2016; Pretorius & Parry, 2016). This hype is not entirely unwarranted. AI does have an ability to systematically trawl through large data sets, finding meaningful patterns indiscernible by the human. It can also do this at great speed.

Evidence of machine learning and more recently deep learning's ability to perform well on predictive tasks is well established (Grace et al., 2018). However, its effectiveness in sports performance environments (and anywhere else for that matter) is still highly dependent on the human. To be specific, if a question is poorly posed in Stage 1, then the AI model, regardless of how accurate it may be, is largely useless. It may be answering the question using data that is not available to the user, has quality issues or is simply answering a question in a manner that nobody was asking in the first place. Second, AI is subject to the same biases that exist with human decision making. For example, if information is missing or incomplete (Sect. 6.3) this can lead to bias or overweighting types of data that is present, also leading to racial or demographic discrimination (Raghavan et al., 2020; Lee, 2018).

Despite its undoubted power to automate and improve efficiency, AI's (or at least, machine learning's) greatest impact on sports performance to date has arguably been more it providing a framework for the human user to consider multiple ways in which a problem could be addressed (Ofoghi et al., 2013). Relating back to the initial example, Fig. 6.5 shows how by considering the problem through the lens of four different algorithm types it is transformed into a question that could lead to very different decisions and actions. This can be useful in challenging dogma or the status quo and lead to change with respect to how an organisation decides to address a process or question they experience regularly (as well as potentially provide a more accurate solution to what is currently in place).

Whilst on tasks whereby AI shows predictive power above those displayed by the human decision-maker there may be a mad rush to adopt these techniques, caution should be observed. Many questions or problems in sport require a knowledge of how an insight was arrived at, which is not always possible with AI. 'Black box' approaches, those for which the weightings of the information leading to the decision cannot be directly interpreted, will often not be suitable for use in the sports setting. How often could the performance practitioner follow the advice of an algorithm, regardless of how accurate, but yet not be able to explain the rationale behind the decision? Whilst in some circumstances all that matters is the predictive performance, more often than not a level of trust needs to be developed with the human end user and this is difficult to do without transparency and communication. In all of this it is worth noting that the generation of Insights with the intention of influencing Decisions and Actions (the next two stages) is often still a human exercise.

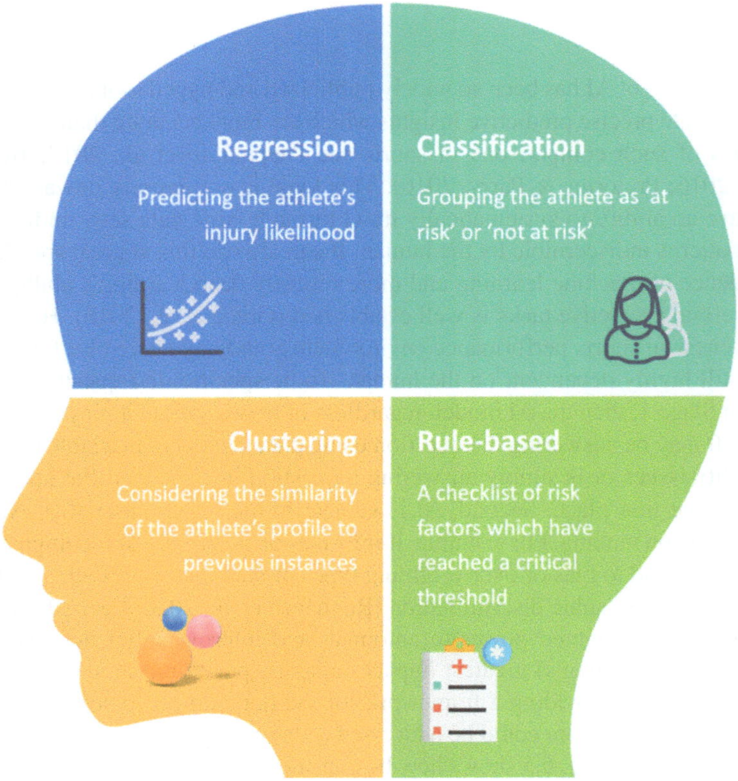

Fig. 6.5 Example of how different machine learning techniques can provide different options for a problem to be solved

6.7 Decisions

It sounds obvious, but the insights obtained from technology and AI need to match the nature of a decision. As outlined in Sect. 6.6, classification algorithms may be useful at assigning items into groups, but will be less suitable when the intention is to predict a number of interest. Conversely, predictions algorithms are often misemployed for Boolean or categorical decisions. In the example used at the start of this chapter, this might relate to an injury risk being outputted for a certain player, whereas the actual decision to be made by the end user is whether the player is available for the match or not. As discussed in Sect. 6.6, sometimes this mismatch in algorithm output and the decision can be positive and encourage a different view of a problem.

6.8 Actions

The above section discussed the importance of Insights matching a Decision. Invariably Decisions also often do not match the Action. Considering the example scenario, many AI-generated commercial offerings in sport injury prediction focus on predicting or classifying whether an athlete will be injured or not. However, the action to remove an athlete from competition or modify their training is only part of the action required. The end user also needs to decide what the athlete does instead of training or playing. This requires considerable more detail than the AI can currently provide. Does the athlete undertake 25 min of massage treatment? Do they undertake cross-training instead on a bike? What do they eat? This represents one of many examples whereby the AI is only providing part of the solution to a problem. Because by their very design AI does not account for this level of nuance in its output, the role of the human continues to be crucial when faced with complex questions. I refer to the extent to which output of an AI matches the way in which it is to be used as its 'operational compatibility' and it is likely to be a fervent area for AI research in the coming years.

6.9 Evaluation

The final stage of decision-making is (along with the first) perhaps the most overlooked. Evaluation of the entire decision-making process is much more than simply how accurate the solution was, or whether it was right or wrong. Each stage of the process should be evaluated to identify inefficiencies and areas for improvement. Increasingly, these refinements will be filled by technology. Further, evaluation should be even broader than the process itself: what was the opportunity cost of the decision? Even when a decision and following action(s) clearly appear to be correct, does following the AI decision lead to a reliance on it and potentially even a loss of expertise from the end user? Such questions will only increase in prominence as humans increasingly come to terms with a world of further automation.

6.10 Conclusion

Some of sports most important questions and challenges also remain those most unresolved. Whilst technology and AI will play a role in moving towards answering these the role of the human in sports decision-making is as important as ever—in fact, one could argue more so. In order to best move forward, humility and openness are needed. AI clearly has the ability to rapidly model the complex. In turn, humans have the creativity to identify and develop new ways of measuring information missing from these models to begin with. The interaction between humans and AI is not a competition, despite what media dialogue is out there. In order for real progress to be made, humans and organisations need to honestly and systematically identify areas in each of the eight stages illustrated in this chapter whereby the technology and AI can help, whilst where relevant retaining the human characteristics we all know and love in sport.

References

Blythe, D. A., & Király, F. J. (2016). Prediction and quantification of individual athletic performance of runners. *PLoS One, 11*(6), e0157257.

Christian, B., & Griffiths, T. (2016). *Algorithms to live by: The computer science of human decisions*. Macmillan.

Grace, K., Salvatier, J., Dafoe, A., Zhang, B., & Evans, O. (2018). When will AI exceed human performance? Evidence from AI experts. *Journal of Artificial Intelligence Research, 62*, 729–754.

Kelly, K. (2016). *The inevitable: Understanding the 12 technological forces that will shape our future*. Penguin.

Lee, N. T. (2018). Detecting racial bias in algorithms and machine learning. *Journal of Information, Communication and Ethics in Society, 16*(3), 252–260.

Lin, J., Quan, D., Sinha, V., Bakshi, K., Huynh, D., Katz, B., & Karger, D. R. (2003). What makes a good answer? The role of context in question answering. In *INTERACT*.

McIntosh, S., Kovalchik, S., & Robertson, S. (2019). Comparing subjective and objective evaluations of player performance in Australian Rules football. *PLoS One, 14*(8), e0220901.

Miller, G. A. (1956). The magical number seven, plus or minus two: Some limits on our capacity for processing information. *Psychological Review, 63*(2), 81.

Ofoghi, B., Zeleznikow, J., MacMahon, C., & Raab, M. (2013). Data mining in elite sports: A review and a framework. *Measurement in Physical Education and Exercise Science, 17*(3), 171–186.

Pappalardo, L., Cintia, P., Pedreschi, D., Giannotti, F., & Barabasi, A. L. (2017). Human perception of performance. *arXiv preprint arXiv:1712.02224*.

Pretorius, A., & Parry, D. A. (2016). Human decision making and artificial intelligence: A comparison in the domain of sports prediction. In *Proceedings of the Annual Conference of the South African Institute of Computer Scientists and Information Technologists* (pp. 1–10).

Raghavan, M., Barocas, S., Kleinberg, J., & Levy, K. (2020, January). Mitigating bias in algorithmic hiring: Evaluating claims and practices. In *Proceedings of the 2020 Conference on Fairness, Accountability, and Transparency* (pp. 469–481).

Russell, S. J., & Norvig, P. (2016). *Artificial intelligence: A modern approach*. Pearson.

Match Predictions in Soccer: Machine Learning vs. Poisson Approaches

7

Mirko Fischer and Andreas Heuer

Abstract

Predicting the results of soccer matches is of great interest. This is not only due to the popularity of the sport and the joy of private "betting rounds", but also due to the large sports betting market. Where previously expert knowledge and intuition were used, today there are models that analyze large amounts of data and make predictions based on them. In addition to Poisson models, approaches that belong to the machine learning (ML) category are increasingly being used. These include, for example, neural network or random forest models, which are compared in this article with each other as well as with Poisson models with regard to single-match prediction. In each case, the match results of a season are used as the data basis. The analysis is carried out for five European top leagues. A statistical analysis shows that the performance levels of the teams do not change systematically during a season. In order to characterize performance levels as accurately as possible, all match results, except from the match to be predicted, can be used as features with equal weighting. It can be seen that both, the exact choice of features and the choice of model, have only a minor influence on the prediction quality. Possible improvements in match prediction are discussed.

7.1 Introduction and Motivation

7.1.1 Relevance of the Match Predictions

Since the 1960s, attempts have been made to systematically predict the results of soccer matches using statistical models (Dubitzky et al., 2019), while previous

M. Fischer · A. Heuer (✉)
Institute of Physical Chemistry, University of Münster, Münster, Germany
e-mail: mirko.fischer@uni-muenster.de; andheuer@uni-muenster.de

D. Memmert (ed.), *Artificial Intelligence and Machine Learning in Sports Science*,
https://doi.org/10.1007/978-3-662-70155-3_7

predictions were essentially based on "expert knowledge". One of the main reasons for the high level of interest in such models and predictions is the growing international sports betting market, which is worth several billion US dollars per year (Etuk et al., 2022).

A match prediction essentially consists of two steps. First, the most meaningful information possible must be collected about the two teams. Matches are now analyzed and quantified in detail, especially using automated techniques (Goes et al., 2020). Teams can use this data to make important tactical decisions or identify suitable players for potential future transfers. In the context of match prediction, the performance strengths of the teams in particular can be estimated. Due to the limited information available before the match, the assessment of the performance of the two teams will always be subject to a certain degree of inaccuracy, but this can at least be minimized by optimizing the data used.

In the second step, a match prediction model is required, which is "fed" with the available data. This shows that the result is not completely predictable. Reep and Benjamin (1968) already recognized that chance plays a significant role in the outcome of a game. For example, the teams with the higher market value do not always win—this is a major part of the appeal of soccer. Therefore, the actual prediction consists of determining probabilities for all possible outcomes. There are various fundamental reasons why exact predictions are not possible (Heuer, 2012; Heuer & Rubner, 2014). (1) Relatively few goals are scored (on average between 2 and 3 goals in European professional leagues), so that the match result is generally a consequence of relatively few relevant actions and referee decisions, for example, have a major influence. This effect would disappear if (hypothetically) matches lasted much longer with the same level of performance. (2) There are irreversible match-immanent effects (red cards, in-game injuries) that are unpredictable but can have a significant impact on the final result. (3) There are match day-specific effects such as the prior absence of players due to injury or specific match tactics whose impact on the expected match result can hardly be quantified. All of these reasons can be summarized as random effects.

We can speak of a *perfect* tip if both steps are optimal, i.e. the estimation of the previous performance level and the model are optimal. Differences in the quality of the two points explain, for example, the slight differences that were observed in the "Soccer Prediction Challenge" (see below).

7.1.2 Machine Learning in Match Prediction

Poisson models imply that the random effects relating to the goals scored can essentially be described according to Poisson statistics. In contrast, data-driven approaches, used here synonymously with machine learning (ML), do not make these assumptions. Another advantage is the simple integration of the ever-growing volumes of data in professional soccer. The first ML models were developed in the early 2010s. Kumar (2013) compared various ML models for predicting soccer results. In the following years, further models were presented and tested. The aim is often to predict the outcome of a match based on previously known data. However, some models also use data collected during a match (e.g. strength of tackles, ball possession, shots on goal, etc.) to increase the predictive power.

Here is a brief overview of some models:

- Poisson (Angelini & De Angelis, 2017; Goddard, 2005; Koopman & Lit, 2015)
- Bayesian Networks (Constantinou, 2019; Constantinou et al., 2012; Joseph et al., 2006)
- Neural networks (Jain et al., 2021; Mendes-Neves & Mendes-Moreira, 2020)
- Random Forest (Stübinger et al., 2020)
- K-Nearest-Neighbor (Berrar et al., 2019; Esme & Kiran, 2018)
- Boosting (XGBoost, CatBoost) (Hubáček et al., 2019; Ren & Susnjak, 2022; Yeung et al., 2024)

In 2018, Dubnitzky et al. launched the "Soccer Prediction Challenge", in which various teams were able to submit models that were supposed to predict the outcome of future matches based on previous results from 52 different leagues (Dubitzky et al., 2019). Only information on the times and results of previous matches could be used as input for the models. When comparing all the models described in the literature (Berrar et al., 2019; Constantinou, 2019; Hubáček et al., 2019; Tsokos et al., 2019), it is noticeable that they show only slight differences in predictive power (RPS score 0.2054–0.2087, accuracy: 51.46–53.88), although different features were used in some cases. A similar observation is described in the review article by Hubáček et al. (2022). The challenge was repeated in a similar form in 2023. The results can be viewed at https://sites.google.com/view/2023soccerpredictionchallenge/home. The aim of our article is to analyze the second step of the match prediction, i.e. the model selection, whereby different features are also considered with the same data set.

7.2 Data Set

In the following, a data set is considered, which consists of over 14,000 match results from the Bundesliga (Germany), Premier League (England), La Liga (Spain), Serie A (Italy) and Ligue 1 (France). The data was collected over a period of eight seasons (2014/2015–2021/2022). The 2019/2020 season was canceled in France and was therefore not included in the analysis.

7.3 Team Performance Over the Course of the Season

The mean goal difference that a specific team would score against an average team after averaging over all random effects can be chosen as a useful measure of the performance strength of a team within a league (Heuer, 2012). To find out whether the performance of a team changes within a season, the league-averaged autocorrelation function of the goal difference ΔG

$$\mathrm{corr}_{\Delta G}\left(\Delta n\right)=\Delta G\left(n\right)\Delta G\left(n+\Delta n\right)_{t,\mathrm{teams},\mathrm{leagues}}$$

$$=\frac{1}{\#\,\mathrm{terms}}\sum_{\mathrm{leagues}}\sum_{m=1}^{N_{\mathrm{teams}}}\sum_{n=1}^{\mathrm{N_{match\ days}}-\Delta n}\Delta G_m\left(n\right)\cdot\Delta G_m\left(n+\Delta n\right)$$

can be used, whereby N_{teams} describes the number of teams and $N_{\mathrm{match\ days}}$ describes the number of match days within a league. The argument Δn describes the time difference and thus records whether and to what extent the performance level changes after matches. $\Delta G_m(n)$ describes the goal difference from the point of view of team m on match day n. Some technical aspects are still important: (1) Terms are omitted from the sum if it is the same opponent of team m on match days n and $n + \Delta n$. (2) $\Delta G_m(n)$ is corrected by the home advantage available for the corresponding season. (3) Especially in the Corona years, the sequence of games is often different than originally planned. Therefore, the temporal argument (e.g. $n = 10$) generally describes the tenth game of team m, which does not necessarily have to take place on the tenth match day.

The autocorrelation function has a finite positive value if the performance of a team is still positively correlated after a number of match days. If both are completely uncorrelated after a time Δn the autocorrelation function disappears.

As can be seen in Fig. 7.1a, the autocorrelation function shows no significant dependence on Δn but fluctuates around a constant value. This results in the surprising observation that the performance of a team does not change during a season within the statistical errors. This observation is consistent with preliminary earlier studies (Heuer, 2012; Heuer & Rubner, 2009). It should be noted that this does not rule out fluctuations in performance from game to game—but without any systematic time dependency. On the other hand, systematic changes in performance levels do occur when comparing successive seasons (Heuer & Rubner, 2009).

7.4 Single Match Predictions

7.4.1 Prediction of Match Outcome, Goal Difference and Exact Match Result

There are different ways of predicting the outcome of a game. In the simplest variant, only the result in the form win/draw/loss is considered. The prediction of the goal difference is somewhat more specific. In the most informative variant, the exact match result is predicted, as already mentioned, by specifying the probabilities for all possible outcomes. To predict the exact match result using data-driven models, more data is naturally required than when restricting to goal differences, as the number of possible exact match results is significantly larger than the number of possible goal differences. In this paper, we focus on the prediction of goal differences and will also discuss the prediction of total number of goals for comparison.

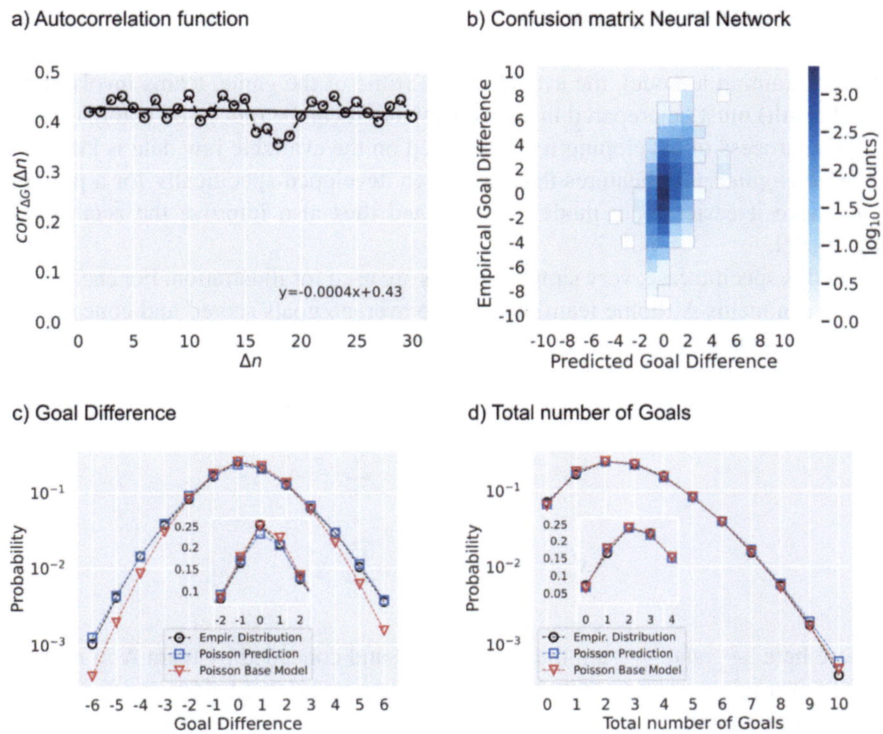

Fig. 7.1 (**a**) Autocorrelation function of the goal differences. (**b**) Confusion matrix for prediction and true empirical distribution of the goal differences for the neural network model trained with the features $x^i_{+,A}$, $x^i_{-,A}$, $x^i_{+,B}$, $x^i_{-,B}$ for all 33 match days that are used for cross-validation. (**c**) and (**d**) Probability distribution of goal differences and total number of goals for the prediction of the Poison model and comparison with the Poisson base model. The inset shows a selected range on a linear scale

7.4.2 Neglecting the Chronological Order

When predicting the results of the last game, all previous games of the teams are available. Since (see above) the performance of a team does not change systematically during a season (within the statistical inaccuracies), the order of the matches is not statistically relevant. Consistency means that a stronger consideration of the last matches does not significantly improve the prediction (Heuer, 2012). For our model investigations, we use the information about all other matches of the involved teams, regardless of whether the match is before or after the match to be predicted. Of course, the results can be directly generalized to include only the previous results, see also (Berrar et al., 2019; Hubáček et al., 2019).

7.4.3 Feature Engineering

Before training a model, the available data (time of the game, teams involved and final result) must be prepared in such a way that it can be processed automatically.

The process of developing features based on the available raw data is known as *feature engineering*. Features that have been developed specifically for a problem can make it easier for a model to learn and thus also improve the accuracy of the model.

In this specific case, very simple features are used for illustration. For each match i between teams A (home team) and B, the average goals scored and conceded by the two teams ($x^i_{+,A}$, $x^i_{-,A}$, $x^i_{+,B}$, $x^i_{-,B}$) during the other matches of the season are calculated, e.g.

$$x^i_{+,A} = \frac{1}{N_{\text{match days}} - 1} \sum_{j \neq i}^{N_{\text{match days}}} g^j_{+,\mathrm{A}}$$

$$x^i_{-,A} = \frac{1}{N_{\text{match days}} - 1} \sum_{j \neq i}^{N_{\text{match days}}} g^j_{-,\mathrm{A}}$$

where here $g^j_{+,\mathrm{A}}$ and $g^j_{-,\mathrm{A}}$ are the goals scored and conceded by team A in match j. When averaging, it does not matter whether the matches are before or after match i in terms of time.

The features obtained in this way can be further combined. The difference between $x^i_{+,A}$ and $x^i_{-,A}$ results in a new feature $x^i_{\Delta G,A} = x^i_{+,A} - x^i_{-,A}$. Equivalent also $x^i_{\Delta G,AB} = x^i_{\Delta G,A} - x^i_{\Delta G,B}$ can be obtained. In the hypothetical limiting case of an infinitely long season, i.e. the determination of random effects, $x^i_{\Delta G,AB}$ corresponds to the expected goal difference of the match i when averaged over a large number of realizations of the match (Heuer, 2012). Similarly, the feature associated with the total number of goals of a team $x^i_{\Sigma G,A} = x^i_{+,A} + x^i_{-,A}$ can be defined analogously. As shown in Heuer (2012), in the hypothetical limiting case outlined above, the variable $x^i_{\Sigma G,AB} = x^i_{\Sigma G,A} + x^i_{\Sigma G,B} - G$ is a consistent estimate of the expected goals in the match i under consideration, where $\langle G \rangle$ is the mean number of goals during the season under consideration. In the approach chosen here, the same amount of information is available for each game. Here, all teams are considered equally, which increases the transferability of the model to data from other leagues and seasons. Furthermore, the introduction of a team index does not lead to any improvement in the prediction.

For some applications, the goals are adjusted for home advantage (obtained as the average of all games except game i). This removes the effect of how often a team has played at home. This analysis is based on the finding that home advantage is not team-specific (but certainly season-specific) (Heuer, 2012; Heuer et al., 2010). The corresponding variables are abbreviated with the symbol $\tilde{x}$ (instead of x). The home advantage indicates how many more goals the home team scores on average than the away team.

At the transition from $x^i_{+,A}$ and $x^i_{-,A}$ to $x^i_{\Delta G,A}$ the information is reduced. On the one hand, this makes modeling easier and requires less data; on the other hand, it is associated with a loss of information, which could potentially reduce the prediction quality in the edge case of a very large amount of available data. The actual modeling can now be carried out for different selections of features. By comparing the respective prediction qualities, the extent to which the individual features influence the match result and are therefore relevant for a soccer match can be derived directly.

Many other features are discussed in the literature, which often explicitly include time dependency. Some of them are listed here:

- Historical form
- Importance of a game
- League
- Home advantage
- Page rank (Hubáček et al., 2019)
- ELO rating (Robberechts & Davis, 2019)
- Pi rating (Constantinou et al., 2012; Hubáček et al., 2019)
- Berrar rating → Characterization of team performance through dynamic ratings that are adjusted after each game (Berrar et al., 2019)

The paper by Yeung et al. (2024) provides a good overview of the features used in literature.

7.4.4 Models for Match Prediction

Three different models are compared for the match prediction: Neural networks (1 intermediate layer with 8 neurons, activation function: logistic, solver: ADAM), Random Forest (100 decision trees, maximum depth: 4, min. 32 data points for a split) and the statistical Poisson model.

The two ML models are used for classification in this context. This means that each goal difference or total number of goals is regarded as a class. For each class, the models predict the probability with which it corresponds to the true class based on the input features.

In machine learning, classification refers to a task in which a model has to find the correct class for an associated input among several classes. The model calculates a probability for each class that it is the true class. If the model is very certain, for example, a class has a very high probability.

To train the models, the error in the prediction needs to be minimized. More precisely, a so-called loss function needs to be minimized in which the prediction and the true value are compared with each other. In our example, we use cross entropy, which compares the predicted probabilities with the true probabilities (here 1 for the true result and 0 otherwise) for each class.

The cross entropy is given as

$$L(p,q) = -\sum_i p_i \log q_i$$

where p_i is the true and q_i is the predicted probability for each class i. If j corresponds to the actual match result, then $L(p, q) = - \log q_j$ is chosen.

In contrast to ML models, which have only been increasingly used for match prediction in recent years, Poisson models have been used successfully for some time (Angelini & De Angelis, 2017; Goddard, 2005; Heuer et al., 2010; Koopman & Lit, 2015). For the prediction of game k in the Poisson model, first the home-advantage adjusted variables $\tilde{x}^i_{+,A}$ and $\tilde{x}^i_{-,A}$ for team A and analogously for team B for each game $i \neq k$ are calculated. Game k is excluded from the averaging process. For all games i (between teams C and D), the linear superposition $a_{\Delta,\text{home}}\, \tilde{x}^i_{\Delta G,CD} + a_{\Sigma,\text{home}}\, \tilde{x}^i_{\Sigma G,CD} + a_{0,\text{home}}$ is compared with the actual number of goals scored by team C and, after taking all games i into account, the three coefficients $a_{\Delta,\text{ home}}$, $a_{\Sigma,\text{ home}}$ and $a_{0,\text{ home}}$ are determined. The same procedure is used for the goals scored by the away teams. The expected values of the goals of team A are then determined for match k (based on $\tilde{x}^k_{\Delta G,AB}, \tilde{x}^k_{\Sigma G,AB}$) and similarly for team B. Then using the Poisson distribution with the expectation value λ

$$P_\lambda(G) = \frac{\lambda^G}{G!} e^{-\lambda}$$

the probability distribution of the goals G scored by teams A and B is calculated. The simple assumption that the goals of the two teams are uncorrelated is used here. This gives a probability for each result and therefore in particular for each goal difference and each total number of goals.

7.4.5 Metrics

To compare the different models, we calculate two different metrics. The cross entropy is already known from the loss function. In addition, we calculate the RPS score (ranked probability score), which takes into account the probability distribution of the prediction and the order of the classes:

$$RPS = \frac{1}{n_c - 1} \sum_{i=1}^{n_c} \left(\sum_{j=1}^{i} q_j - \sum_{j=1}^{i} p_j \right)^2$$

n_c indicates the number of classes, p_j the true and q_j the predicted probability for each class j.

7.4.6 Comparison of Models and Features

The cross-validation method is used to compare the models based on the metrics. It allows a statistically more accurate determination of the metrics including standard errors compared to a simple calculation of the metrics using a single test set.

To do this, the data set is divided into $N_{\text{match days}}$ subsets, each representing one match day. Based on this, models are now trained on $N_{\text{match days}} - 1$ match days (training set) and the metrics for a match day (test set) are calculated. This is repeated so that each match day acts as a test set once. The final metrics are averaged over all match days and compared including standard errors in Table 7.1 for different models and features. Five randomly selected seasons are not included in the calculation of these metrics and are instead used to optimize the hyperparameters. The results are shown in Table 7.1.

For the prediction of goal differences (−10 to +10), all models are fundamentally better than a simple base model in which the same probabilities $p(\Delta G)$ are used for each game. $p(\Delta G)$ is the normalized frequency with which ΔG occurs in the data set.

It is also observed that the use of several features does not lead to any significant improvement in the prediction. In general, the neural network model provides

Table 7.1 Comparison of metrics for different models and features

		Goal difference		Total # of goals	
Model	Features	Cross entropy	RPS score	Cross entropy	RPS score
Poisson	$\tilde{x}^i_{\Delta G,AB}$	1.902 ± 0.006	–	1.884 ± 0.005	–
Poisson	$\tilde{x}^i_{\Delta G,\exp}$, $x^i_{\Sigma G,\exp}$	1.901 ± 0.006	–	1.884 ± 0.005	–
Neural network	$x^i_{\Delta G,\exp}$	1.907 ± 0.011	0.0445 ± 0.0004	1.894 ± 0.007	0.0614 ± 0.0005
Neural network	$x^i_{\Delta G,A}$, $x^i_{\Delta G,B}$	1.906 ± 0.011	0.0444 ± 0.0004	1.893 ± 0.007	0.0613 ± 0.0005
Neural network	$x^i_{\Delta G,A}$, $x^i_{\Delta G,B}$, $x^i_{\Sigma G,A}$, $x^i_{\Sigma G,B}$	1.905 ± 0.011	0.0444 ± 0.0004	1.883 ± 0.007	0.0606 ± 0.0004
Neural network	$x^i_{\Delta G,A}$, $x^i_{\Delta G,B}$, $x^i_{\Sigma G,A}$, $x^i_{\Sigma G,B}$, $x_{ID,A}$, $x_{ID,B}$, $x_{ID,\text{Season}}$	1.906 ± 0.011	0.0444 ± 0.0004	1.884 ± 0.007	0.0606 ± 0.0004
Random Forest	$x^i_{\Delta G,\exp}$	1.917 ± 0.012	0.0445 ± 0.0005	1.896 ± 0.008	0.0611 ± 0.0005
Random Forest	$x^i_{\Delta G,A}$, $x^i_{\Delta G,A}$	1.916 ± 0.010	0.0448 ± 0.0004	1.891 ± 0.007	0.0611 ± 0.0005
Random Forest	$x^i_{\Delta G,A}$, $x^i_{\Delta G,B}$, $x^i_{\Sigma G,A}$, $x^i_{\Sigma G,B}$	1.918 ± 0.010	0.0445 ± 0.0004	1.888 ± 0.007	0.0608 ± 0.0005
Base model	–	2.08	0.0533	1.895	0.0616

slightly better results than the random forest and performs similarly compared to the Poisson model.

Instead of a single model for all leagues, several models can also be trained for each league individually. The metrics differ slightly depending on the league (the more balanced the league, the more difficult the prediction) but are generally worse due to the smaller training set and have a higher standard error (cross entropy of 1.90 ± 0.03 to 2.03 ± 0.03) compared to the model trained on all leagues.

Similar results can be observed for the prediction of the total number of goals (taking into account values from 0 to 16 goals). However, the models are only slightly better in their predictions than the base model defined analogously to above. This observation implies that the total number of goals in a match is less team-specific than the goal difference. Betting on the number of goals in a match is therefore similar in nature to simply rolling the dice.

Consideration of specific season and team IDs ($x_{ID,\,\text{Season}}$, $x_{ID,\,A}$, $x_{ID,\,B}$) as a feature does not significantly improve the metrics.

7.4.7 Predicted Distributions

The models output the goal difference or total number of goals with the highest probability as the final result. As goal differences with a value greater than 2 are largely due to random events, these are not predicted. This is shown in the confusion matrix in Fig. 7.1b.

Even though the metrics for the Poisson model and the neural network are very similar, there are still differences in the exact predictions of the models. To illustrate this, we look at $P(\Delta G)$ averaged over all games (Fig. 7.1c). In the Poisson model, draws ($\Delta G = 0$) are predicted with a lower probability than expected (Heuer, 2012). Otherwise, the distribution is predicted very accurately. The neural network learns the distribution function directly from the data, so that $P(\Delta G)$ naturally matches the empirical distribution of the data set. For the total number of goals, both the prediction of the Poisson model and the neural network agree well with the empirical distribution of probabilities (Fig. 7.1d).

7.5 Conclusion

This chapter illustrated how ML models can be used to predict the goal difference or total number of goals of a soccer match. Feature engineering was explained and the application of different features was discussed. The neural network is comparable to a Poisson model in terms of cross entropy and does not require any model assumptions. Compared to a base model, both models deliver significantly better results for the goal difference, while the results for the total number of goals are comparable to the base model. Conversely, the predictive power of goal chances as shown in Heuer (2012) and Heuer & Rubner (2014) is significantly increased compared to the predictive power of goals. Therefore, the choice of model is less important than the quality of the data and the associated features in order to successfully predict match results.

Acknowledgement We would like to thank Dr. Fabian Wunderlich for providing the list of match results.

References

Angelini, G., & De Angelis, L. (2017). PARX model for football match predictions. *Journal of Forecasting, 36*(7), 795–807. https://doi.org/10.1002/for.2471

Berrar, D., Lopes, P., & Dubitzky, W. (2019). Incorporating domain knowledge in machine learning for soccer outcome prediction. *Machine Learning, 1*(108), 97–126. https://doi.org/10.1007/s10994-018-5747-8

Constantinou, A. C. (2019). Dolores: A model that predicts football match outcomes from all over the world. *Machine Learning, 108*, 49–75. https://doi.org/10.1007/s10994-018-5703-7

Constantinou, A. C., Fenton, N. E., & Neil, M. (2012). pi-football: A Bayesian network model for forecasting association football match outcomes. *Knowledge-Based Systems, 36*, 322–339. https://doi.org/10.1016/j.knosys.2012.07.008

Dubitzky, W., Lopes, P., Davis, J., et al. (2019). The Open International Soccer Database for machine learning. *Machine Learning, 108*, 9–28. https://doi.org/10.1007/s10994-018-5726-0

Esme, E., & Kiran, M. S. (2018). Prediction of football match outcomes based on bookmaker odds by using k-nearest neighbor algorithm. *International Journal of Machine Learning and Computing, 8*(1), 26–32. https://doi.org/10.18178/ijmlc.2018.8.1.658

Etuk, R., Xu, T., Abarbanel, B., Potenza, M. N., & Kraus, S. W. (2022). Sports betting around the world: A systematic review. *Journal of Behavioral Addictions, 11*(3), 689–715. https://doi.org/10.1556/2006.2022.00064

Goddard, J. (2005). Regression models for forecasting goals and match results in association football. *International Journal of Forecasting, 21*(2), 331–340. https://doi.org/10.1016/j.ijforecast.2004.08.002

Goes, F. R., Meerhoff, L. A., Bueno, M. J. O., Rodrigues, D. M., Moura, F. A., Brink, M. S., Elferink-Gemser, M. T., Knobbe, A. J., Cunha, S. A., Torres, R. S., & Lemmink, K. A. P. M. (2020). Unlocking the potential of big data to support tactical performance analysis in professional soccer: A systematic review. *European Journal of Sport Science, 21*(4), 481–496. https://doi.org/10.1080/17461391.2020.1747552

Heuer, A. (2012). *Der perfekte Tipp: Statistik des Fussballspiels*. Wiley-VCH.

Heuer, A., & Rubner, O. (2009). Fitness, chance, and myths: An objective view on soccer results. *The European Physical Journal B, 67*, 445–458.

Heuer, A., & Rubner, O. (2014). Optimizing the prediction process: From statistical concepts to the case study of soccer. *PLoS One, 9*(9), e104647. https://doi.org/10.1371/journal.pone.0104647

Heuer, A., Müller, C., & Rubner, O. (2010). Soccer: Is scoring goals a predictable Poissonian process? *EPL, 89*, 38007. https://doi.org/10.1209/0295-5075/89/38007

Hubáček, O., Šourek, G., & Železný, F. (2019). Learning to predict soccer results from relational data with gradient boosted trees. *Machine Learning, 108*, 29–47. https://doi.org/10.1007/s10994-018-5704-6

Hubáček, O., Šourek, G., & železný, F. (2022). Forty years of score-based soccer match outcome prediction: An experimental review. *IMA Journal of Management Mathematics, 33*(1), 1–18. https://doi.org/10.1093/imaman/dpab029

Jain, S., Tiwari, E., & Sardar, P. (2021). Soccer result prediction using deep learning and neural networks. In J. Hemanth, R. Bestak, & J. I. Z. Chen (Eds.), *Intelligent data communication technologies and internet of things. Lecture notes on data engineering and communications technologies* (Vol. 57). Springer. https://doi.org/10.1007/978-981-15-9509-7_57

Joseph, A., Fenton, N. E., & Neil, M. (2006). Predicting football results using Bayesian nets and other machine learning techniques. *Knowledge-Based Systems, 19*(7), 544–553. https://doi.org/10.1016/j.knosys.2006.04.011

Koopman, S. J., & Lit, R. (2015). A dynamic bivariate Poisson model for analyzing and forecasting match results in the English Premier League. *Journal of the Royal Statistical Society Series A: Statistics in Society, 178*(1), 167–186. https://doi.org/10.1111/rssa.12042

Kumar, G. (2013). *Machine learning for soccer analytics.* PhD Thesis. https://doi.org/10.13140/RG.2.1.4628.3761

Mendes-Neves, T., & Mendes-Moreira, J. (2020, September). Comparing state-of-the-art neural network ensemble methods in soccer predictions. In *International symposium on methodologies for intelligent systems* (pp. 139–149). Springer.

Reep, C., & Benjamin, B. (1968). Skill and chance in association football. *Journal of the Royal Statistical Society, Series A (General), 131*(4), 581–585. https://doi.org/10.2307/2343726

Ren, Y., & Susnjak, T. (2022). Predicting football match outcomes with explainable machine learning and the kelly index. arXiv. https://doi.org/10.48550/arXiv.2211.15734

Robberechts, P., & Davis, J. (2019). Forecasting the FIFA World Cup—combining result-and goal-based team ability parameters. In *Machine learning and data mining for sports analytics: 5th International Workshop, MLSA 2018, Co-located with ECML/PKDD 2018, Dublin, Ireland, September 10, 2018, Proceedings 5* (pp. 16–30). Springer.

Stübinger, J., Mangold, B., & Knoll, J. (2020). Machine learning in football betting: Prediction of match results based on player characteristics. *Applied Sciences, 10*, 46. https://doi.org/10.3390/app10010046

Tsokos, A., Narayanan, S., Kosmidis, I., et al. (2019). Modeling outcomes of soccer matches. *Machine Learning, 108*, 77–95. https://doi.org/10.1007/s10994-018-5741-1

Yeung, C., Bunker, R., Umemoto, R., & Fujii, K. (2024). Evaluating soccer match prediction models: A deep learning approach and feature optimization for gradient-boosted trees. *Machine Learning, 113*, 7541–7564.

Part II

Metrics in Team Sports

Machine Learning in Soccer

8

Ashwin Phatak

Abstract

The rise of big data has transformed soccer analysis, providing rich insights through high-volume, high-velocity, and high-variety information. This data's potential benefits are immense, offering a deeper understanding of team tactics, player performance, and more. However, challenges in data governance and processing must be addressed. Machine learning has emerged as a potent tool in soccer analytics. It's used to analyze events, evaluate player performance, predict injuries, and estimate team tactics. Its applications span from predicting match outcomes to measuring individual and collective performance. Open datasets fuel ongoing research in this area. The knowledge discovery pipeline in soccer involves data collection, preprocessing, mining, modelling, and evaluation. This iterative process supports performance measurement, success determinants identification, and outcome prediction. Libraries like Floodlight, Codeball, soccermatics, and optapro aid soccer data analysis. They provide streamlined tools for handling data, creating visualizations, and extracting insights. Big data and machine learning revolutionize soccer analysis, while the knowledge discovery process and dedicated libraries facilitate deeper insights and strategic decision-making.

A. Phatak (✉)
Institute of Exercise Training and Sport Informatics, German Sport University Cologne, Köln, Germany
e-mail: a.phatak@dshs-koeln.de

D. Memmert (ed.), *Artificial Intelligence and Machine Learning in Sports Science*,
https://doi.org/10.1007/978-3-662-70155-3_8

8.1 Introduction

8.1.1 Rise of Big Data

► "**Big data** is defined as high-volume, high-velocity, high-variety and high veracity information assets (4Vs) that demand cost-effective, innovative forms of information processing for enhanced insight and decision making." In this definition, volume refers to the magnitude or size of the data, variety refers to structural heterogeneity in the dataset, velocity refers to the rate at which data are generated and veracity refers to the truthfulness or reliability of the data (Phatak et al., 2021; Roy et al., 2020).

The rise of big data in soccer has revolutionized the way the sport is analyzed, strategized, and played (Biermann et al. 2023; Memmert, 2021, 2023; Memmert et al., 2019). With the advent of next-generation tracking technologies and miniature sensor technologies, detailed game logs and physiological training data have become available for research (Rein & Memmert, 2016). These technologies capture detailed spatiotemporal events such as passes, shots, fouls, and more, providing a rich source of information for analysis (Pappalardo et al., 2019). The analysis and management of big data in soccer can provide new ideas and methods for data management and analysis commonly used in sports (Zhang et al., 2022).

This abundance of data presents both opportunities and challenges for soccer research. One of the main challenges in utilizing big data in soccer research is data governance and access to technologies. Standardized open privacy protection mechanisms, such as encrypting individual data items, can help ensure the privacy and security of player data. Additionally, the complexity of data processing in soccer research requires the development of specialized processing pipelines to manage and analyze large volumes of data (Rein & Memmert, 2016). Machine Learning seems to be the tool which may help us explore the opportunities and mitigate challenges offered by big data in the domain of soccer.

8.1.2 Use of ML Within Soccer Analytics

Machine learning has emerged as a powerful tool in soccer analytics, enabling researchers and practitioners to gain deeper insights into various aspects of the game. The availability of high-fidelity data streams from sensing technologies has facilitated the application of machine-learning techniques in soccer analysis (Pappalardo et al., 2019).

By leveraging the large volumes of data available, machine learning algorithms can measure and evaluate individual and collective performance, as well as identify the determinants of success and failure (Berrar et al., 2018). These analyses can provide valuable insights for coaches, players, and teams, helping them make data-driven decisions to improve their performance on the field.

ML techniques in combination with big data have been used in the fields of self and opponent analysis, recruiting, coaching, talent identification, injury prediction, workload management, betting etc. not only in soccer but across sports (Phatak, 2023). In the following subchapters, we will go into the details of what kind of data sources exist, applications in the soccer industry to solve specific problems and programming libraries which exist to perform some of the analysis mentioned. We explore this through the knowledge discovery process, which is the basis of problem-solving in data science.

8.2 The Knowledge Discovery Process

Knowledge discovery pipelines involve the systematic extraction of insights from various data sources to enhance team performance and strategic decision-making. These pipelines typically encompass data collection, pre-processing, feature engineering (data transformation), data mining, modelling, and evaluation stages (Fayyad et al., 1996). In soccer, raw data such as player statistics, match outcomes, event data, spatiotemporal tracking data and gameplay footage are collected and transformed into a structured format for analysis. ML and data mining techniques are then applied to identify patterns, predict player performance, and optimize tactical strategies (Rico-González et al., 2023). The pipeline's iterative nature ensures continuous improvement as new data is integrated and models are refined, ultimately offering valuable insights to coaches and analysts in soccer teams. Figure 8.1 below illustrates the general knowledge discovery process.

The insights gained from the model can be used for various purposes in soccer analytics. This includes measuring and evaluating performance at both individual and collective levels, studying the determinants of success and failure, and predicting future match outcomes (Berrar et al., 2018; Lee & Jung, 2022; Pappalardo et al., 2019). The application of machine learning in soccer has the potential to

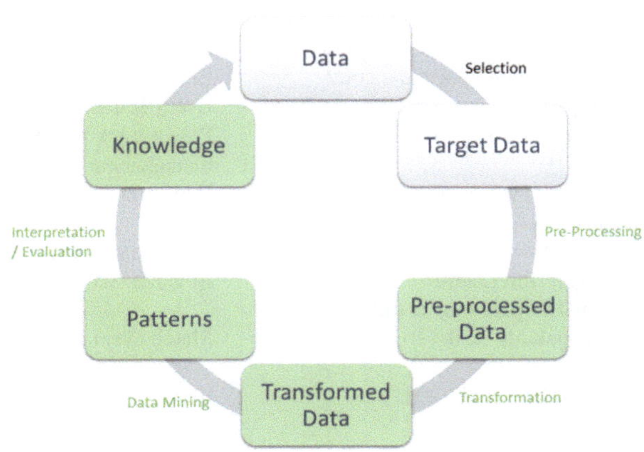

Fig. 8.1 Data science pipeline (Fayyad et al., 1996; Phatak, 2023)

revolutionize the evaluation of tactical performance and provide valuable insights into the game (Goes et al., 2019).

8.3 Components of the Knowledge Discovery Pipeline

Primarily there are two important components to the knowledge discovery pipeline viz. '(Big) Data', modelling and interpretation (may or may not be machine learning). Below we shall take a look at applications of big data and ML for the domain of soccer analytics.

8.3.1 'Big Data' in Soccer

In the realm of soccer, the diversity of disciplines contributes to substantial data variability, encompassing physiological, positional tracking, psychological, scouting, and video data (Rein & Memmert, 2016). A comprehensive array of data sources for soccer is presented in Table 8.1, including commercial data providers and open-source platforms offering data under creative commons licenses. The surge in available data has notably expanded research prospects in sports analytics, necessitating a coherent framework for harnessing data in knowledge discovery processes across various sports and domains (Bai & Bai, 2021; Rein & Memmert, 2016).

Table 8.1 A non-exhaustive list of Soccer data sources (Phatak, 2023)

Data source	Reference	Data type
Data providing companies		
OPTA sports	https://www.optasports.com/	In-game event & tracking data
Hudl	https://www.hudl.com/	Video data & analysis
Instat	https://instatsport.com/	Event & tracking data
Statsbomb	https://statsbomb.com/	Event & tracking data
Stats perform	https://www.statsperform.com/	Event & tracking data
Sportradar	https://www.sportradar.com/	End-to-end services
Wyscout	https://wyscout.com/	Event & tracking data
Kitmanlabs	https://www.kitmanlabs.com/	Player tracking and load management
Catapult sports	https://www.catapultsports.com/	End-to-end services
Open source websites		
Whosoored	https://whoscored.com/	In-game statistics for Soccer
FBref	https://fbref.com/	In-game statistics for Soccer
Scientific literature with open-source data sets		
Published data set	Pappalardo et al. (2019)	Match event data
Published data set	Biermann et al. (2021)	Event data

Data from the above sources is collected and stored in various formats, encompassing game videos, spatial-temporal position tracking data, biometric information, event records, coaching insights, scouting details, and psychological assessments. This range of data varies from highly structured to unstructured forms. The predominant methods for collecting tracking data involve Real-Time Location Systems (RTLS) and computer vision. Manual event tagging through video analysis is subsequently employed to generate notational data. Challenges arise across different methods, including tracking in densely populated scenarios, the absence of physiological measurements, issues in tracking high-speed objects, and potential human errors linked to the tracking technology used (Bialkowski et al., 2014).

Moreover, the abundance of data poses challenges in distinguishing pertinent information from the noise. To address these hurdles, advancements in spatial-temporal data collection through technologies like computer vision, wearable sensors, and RTLS offer prospects for acquiring substantial real-time data. AI/ML emerges as a promising approach to surmount these challenges and derive insights from the data derived from these technological enhancements (Phatak et al., 2021). Below we take a deeper look at what kind of problems AI/ML methods have been used to solve in the field of soccer analytics (Table 8.2).

8.3.2 Machine Learning Algorithms in Soccer (Modelling and Interpretation)

Over the last two decades, advancements in technology for data collection, machine learning methodologies, increased computational capabilities, and the availability of tools like built-in libraries in Python and R have significantly propelled the adoption of machine learning and data science techniques in the sports domain. This surge has permeated both the professional sports industry and the realm of academic research in sports science (Hao & Ho, 2019; Phatak et al., 2021; Rein & Memmert, 2016; Swartz, 2020). Notably, dedicated conferences such as MIT SLOAN have emerged, solely focused on data analysis in sports, with a considerable portion of presented works revolving around applications of artificial intelligence and machine learning in sports contexts. This trend is further reflected in numerous articles published across diverse multidisciplinary journals, delving into data analytics within sports (Swartz, 2020).

A PubMed search using the term 'sports analytics' reveals a total of 4348 hits as of August 3, 2023, with 3539 of them being published after 2010. An up-to-date publication specifically reviewing machine learning applications in soccer highlighted 32 studies investigating aspects like injury prediction, performance analysis, and talent forecasting utilizing machine learning algorithms (Rico-González et al., 2023). The review emphasizes the significance of exploring the necessary data volume required to achieve relevant and accurate predictions across these areas. In line with this, Table 8.2 below showcases a selection of these studies that employ machine learning methods in the domain of soccer tactics.

Table 8.2 A non-exhaustive list of publications which used Machine Learning for performance analysis in soccer (Phatak, 2023)

Topic	Central concept	References
VAEP & XT	Ranking threat of teams and players based on their on-the-ball actions, based on if they increase or decrease the chances of goal scoring using gradient boosting classification model	Decroos et al. (2019, 2020a), Roy et al. (2020)
Which pass is better	Assess the effectiveness of a pass by evaluating its impact on majority situations and space control in front of the goal using Voronoi diagrams	Rein et al. (2017)
Not all passes are created equal	Estimating the risk and reward of a pass on a continuous scale using tracking data and outlining multiple used cases for evaluating player performances.	Power et al. (2017)
Incorporating domain knowledge in ML models for outcome prediction	Feature extraction and learning by incorporating domain-specific knowledge to improve model performance	Berrar et al. (2018)
DCNN to Predict Goal-Scoring Opportunities in Soccer	Using GoogLeNet and three layered CNN architecture to classify enhanced images of ball possession phases to find which one of them ends up on a shot on goal	Wagenaar et al. (2017)
Predicting match outcome according to the quality of the opponent in the English premier league using situational variables and team performance indicators	Decision trees performed the best while predicting the result of matches when an input feature of opponent team strength was provided along with 22 situational variables and performance in-game	Bilek and Ulas (2019)
Quantifying the relation between performance and success in soccer	Using ML algorithms and technical data the study simulated outcomes of an entire season based on 6396 games and ten million events from top six European soccer leagues	Pappalardo et al. (2019)
Not every pass can be an assist	Quantifying the effectiveness of passes based on measures evaluated from tracking data which can be used for analysing complex dynamics in build-up play and space creation in soccer	Goes et al. (2019)
Discovering team structures in soccer from spatiotemporal data	Classified teams and ranked players based on heat maps and pass locations with an accuracy of 87%. The study also predicted occurrence shots for each possession phase with AU-CROC of 0.785	Brooks et al. (2016)
Using ML to draw inferences from pass location data in soccer	Applying unsupervised learning on 21.5 million frames of player tracking data the study presents a method for automatic formation template (team structure) detection	Bialkowski et al. (2016)

(continued)

Table 8.2 (continued)

Topic	Central concept	References
Learning to rate player positioning in soccer	The study describes a purely data-driven approach using deep reinforcement learning to rate multiplayer positions based on tracking data. It validates this by comparing the automated ratings with actual dangerous situations in the game	Dick and Brefeld (2019)
Spatiotemporal convolution kernels	The study proposes a novel class of Spatio-temporal convolution kernels which capture similarities in multi-object scenarios. Furthermore, it compares it to baseline techniques for clustering real and simulated team sport data	Knauf et al. (2016)
Finding efficient strategies in 3-versus-2 small-sided games of youth soccer players	The study used unsupervised learning to find the efficiency of tactical patterns using 30 technical and tactical parameters for a shot and an event right before the shot occurred	Leser et al. (2019)
Individual ball possession in soccer	The study automatically classified ball possession phases into six different categories using Bayesian Network based on features pertaining to the position and mechanics of the ball and the layers. The F-score for all phases were between 0.77 and 0.88	Link and Hoernig (2017)
Team activity recognition in soccer	Automatic recognition from match video data was performed using the Bag of Words technique. NN-classifier, KNN, SVM and RF were used to classify video clips as Ball Possession, Quick Attack and Set Piece. The RF performed the best in one of the experiments with an accuracy of 92.89%	Montoliu et al. (2015)
ML and wearable sensors to predict energetics and kinematics	Four supervised ML techniques were used to predict turn direction, speed, and mechanical work based on 18 features derived from IMU data. The classification of turning direction yielded a score of >98.4% while the regression task showed R^2 of 0.42–0.43 for predicting mechanical work and 0.66–0.69 for running speed before and after the turn	Zago et al. (2019)
The tactics of successful attacks in professional Soccer	The study analyses the behaviour of dynamic subgroups in relation to successful attacks. The unsupervised classification was used to classify attacks as successful or unsuccessful. Successful attacks strongly depend on defenders creating space for the attackers along	Goes et al. (2021)

8.4 Libraries for Soccer Data Analysis

A number of libraries are available in both Python and R which are available open source and can be used to perform multiple steps in the knowledge discovery pipeline, viz. preprocessing, data mining, modelling and visualization. Below we mention a few and their primary utility.

8.4.1 Python

Floodlight is a Python package tailored for efficient sports data analysis, with a strong emphasis on scientific computing. Leveraging well-known libraries like numpy and pandas, Floodlight simplifies loading, merging, and refining tracking, event data, codes, and match details from prominent data sources. It supplies standardized data structures for organizing sports data and offers a range of standard processing tasks like transformations and data manipulation functions (Raabe et al., 2022), (https://floodlight.readthedocs.io/en/latest/).

Codeball is a centralized Python package designed for tactical analysis of soccer matches using data-driven methods. Unlike scattered code repositories, Codeball offers a centralized hub for various data-driven tactical analysis tools. It facilitates the connection of these analyses with match footage, making it easy to correlate insights with actual game video. The package allows you to work extensively with tracking and event data, employing Pandas Data Frames and specialized methods. It enables the creation and utilization of tactical models, such as Zones, for filtering events based on specific game areas. Additionally, Codeball supports the creation and application of Patterns, units of analysis that identify specific moments in matches. The package also integrates annotations for Metrica Play, enabling seamless event visualization and analysis using various tools provided by the platform (https://codeball.metrica-sports.com/).

Kloppy is a Python package that tackles the issue of varying soccer data formats used by different vendors for game analysis. This diversity typically demands customized software for each vendor, hindering compatibility with others. Kloppy aims to be a foundational tool for processing soccer tracking and event data. It offers tools like (de)serializers, standardized data models, filters, and transformers. This facilitates the seamless handling of tracking and event data from multiple vendors, streamlining the analysis process (https://kloppy.pysport.org).

MplSoccer is a Python library that expedites the creation of informative football/soccer visualizations. With features like customizable pitch types, radar charts, Nightingale/pizza charts, bumpy charts for time-based changes, arrows, heatmaps, hexbins, scatter, and comet lines, the library simplifies data representation. It integrates StatsBomb data, offers standardized pitch coordinates, and ultimately empowers users to craft compelling graphics efficiently, reducing the need to start from scratch (https://mplsoccer.readthedocs.io/en/latest/).

8.4.2 R

soccermatics: This package focuses on soccer analytics and provides functions to analyze match events, player actions, and team performance metrics. It offers tools for examining various aspects of the game, such as pass networks, shot locations, and player movements (https://www.rdocumentation.org/packages/soccermatics/versions/0.9.5).

WorldfootballR package is designed to enable users to easily extract diverse global football (soccer) outcomes and player metrics from prominent soccer data platforms such as FBref, Transfermarkt, Understat, and Fotmob. This package facilitates the retrieval of comprehensive soccer-related information, including match results and player statistics (https://cran.r-project.org/web/packages/worldfootballR/readme/README.html).

ggsoccer package complements this by furnishing a collection of functions that enable the elegant presentation and examination of soccer event data using 'ggplot2'. Designed with adaptable layers and themes, it seamlessly integrates with diverse leading sports data platforms, enhancing compatibility and facilitating the display and analysis of soccer-related information (https://cran.r-project.org/web/packages/ggsoccer/index.html).

8.5 Conclusion

Machine learning's applications have extended to various facets of soccer, from predicting match outcomes to assessing player behaviour and optimizing strategies. Its impact is evident in performance analysis, injury risk assessment, and even estimating team tactics. Open datasets and advanced methodologies have fueled the rapid evolution of this field, contributing to a more data-driven approach to soccer. Looking ahead, the fusion of big data and machine learning is poised to further revolutionize soccer analysis. The continued development of tracking technologies and sensor advancements will provide even richer and more detailed data streams. This will enable researchers to unearth even deeper insights into multiple facets of performance in the soccer industry.

Moreover, the integration of artificial intelligence and machine learning algorithms into real-time match analysis holds great promise. This could lead to predictive tactical suggestions during games, enhancing coaches' decision-making on the sidelines. The use of wearable technology and real-time physiological data could also contribute to player load management, optimizing performance and minimizing the risk of injuries.

In conclusion, the journey of soccer analysis through big data and machine learning has only just begun. The future holds exciting possibilities for uncovering the intricate dynamics of the sport, optimizing strategies, and redefining how teams approach matches. The ongoing evolution of technology and methodology will continue to shape the way soccer is played, understood, and enjoyed.

References

Bai, Z., & Bai, X. (2021). Sports big data: Management, analysis, applications, and challenges. *Complexity, 2021*, 11. https://doi.org/10.1155/2021/6676297

Berrar, D., Lopes, P., & Dubitzky, W. (2018). Incorporating domain knowledge in machine learning for soccer outcome prediction. *Machine Learning, 1*(108), 97–126. https://doi.org/10.1007/s10994-018-5747-8

Bialkowski, A., Lucey, P., Carr, P., Yue, Y., Sridharan, S., & Matthews, I. (2014). Large-scale analysis of soccer matches using spatiotemporal tracking data. In *2014 IEEE International Conference on Data Mining* (pp. 725–730). https://doi.org/10.1109/ICDM.2014.133

Bialkowski, A., Lucey, P., Carr, P., Matthews, I., Sridharan, S., & Fookes, C. (2016). Discovering team structures in soccer from spatiotemporal data. *IEEE Transactions on Knowledge and Data Engineering, 28*(10), 2596–2605. https://doi.org/10.1109/TKDE.2016.2581158

Biermann, H., Theiner, J., Bassek, M., Raabe, D., Memmert, D., & Ewerth, R. (2021). A unified taxonomy and multimodal dataset for events in invasion games. In *Proceedings of the 4th International Workshop on Multimedia Content Analysis in Sports* (pp. 1–10). https://doi.org/10.1145/3475722.3482792

Biermann, H., Wieland, F. G., Timmer, J., Memmert, D., & Phatak, A. (2023). Towards expected counter-using comprehensible features to predict counterattacks. In *International Workshop on Machine Learning and Data mining for sports analytics* (pp. 3–13). Springer. https://doi.org/10.1007/978-3-031-27527-2_1

Bilek, G., & Ulas, E. (2019). Predicting match outcome according to the quality of opponent in the English Premier League using situational variables and team performance indicators. *International Journal of Performance Analysis in Sport, 19*(6), 930–941. https://doi.org/10.1080/24748668.2019.1684773

Brooks, J., Kerr, M., & Guttag, J. (2016). Using machine learning to draw inferences from pass location data in soccer. *Statistical Analysis and Data Mining: The ASA Data Science Journal, 9*(5), 338–349. https://doi.org/10.1002/sam.11318

Decroos, T., Bransen, L., Van Haaren, J., & Davis, J. (2019). Actions speak louder than goals: Valuing player actions in soccer. In *Proceedings of the 25th ACM SIGKDD International Conference on Knowledge Discovery & Data Mining* (pp. 1851–1861). Association for Computing Machinery. https://doi.org/10.1145/3292500.3330758

Decroos, T., Bransen, L., Van Haaren, J., & Davis, J. (2020a). Vaep: An objective approach to valuing on-the-ball actions in soccer. In *Proceedings of the Twenty-Ninth International Joint Conference on Artificial Intelligence*, IJCAI-20 (pp. 4696–4700). https://doi.org/10.24963/ijcai.2020/648

Decroos, T., Roy, M. V., & Davis, J. (2020b). Soccermix: Representing soccer actions with mixture models. In *Joint European Conference on Machine Learning and Knowledge Discovery in Databases* (Vol. 12461, pp. 459–474). https://doi.org/10.1007/978-3-030-67670-4_28

Dick, U., & Brefeld, U. (2019). Learning to rate player positioning in soccer. *Big Data, 7*(1), 71–82. https://doi.org/10.1089/big.2018.0054

Fayyad, U., Piatetsky-Shapiro, G., & Smyth, P. (1996). From data mining to knowledge discovery in databases. *AI Magazine, 17*(3), 37–37. https://doi.org/10.1609/aimag.v17i3.1230

Goes, F., Kempe, M., Meerhoff, L., & Lemmink, K. (2019). Not every pass can be an assist: A data-driven model to measure pass effectiveness in professional soccer matches. *Big Data, 1*(7), 57–70. https://doi.org/10.1089/big.2018.0067

Goes, F. R., Brink, M. S., Elferink-Gemser, M. T., Kempe, M., & Lemmink, K. A. (2021). The tactics of successful attacks in professional association football: Large-scale spatiotemporal analysis of dynamic subgroups using position tracking data. *Journal of Sports Sciences, 39*(5), 523–532. https://doi.org/10.1080/02640414.2020.1834689

Hao, J., & Ho, T. K. (2019). Machine learning made easy: A review of SCIKIT-learn package in python programming language. *Journal of Educational and Behavioral Statistics, 44*(3), 348–361. https://doi.org/10.3102/1076998619832248

Knauf, K., Memmert, D., & Brefeld, U. (2016). Spatio-temporal convolution kernels. *Machine Learning, 102*(2), 247–273. https://doi.org/10.13140/RG.2.1.2572.9129

Lee, G., & Jung, J. (2022). DNN-based multi-output model for predicting soccer team tactics. *PeerJ Computer Science, 8*, e853. https://doi.org/10.7717/peerj-cs.853

Leser, R., Hoch, T., Tan, X., Moser, B., Kellermayr, G., & Baca, A. (2019). Finding efficient strategies in 3-versus-2 small-sided games of youth soccer players. *Kinesiology, 51*(1), 110–118. https://doi.org/10.26582/k.51.1.7

Link, D., & Hoernig, M. (2017). Individual ball possession in soccer. *PLoS One, 12*(7), e0179953. https://doi.org/10.1371/journal.pone.0179953

Memmert, D. (2021). *Match analysis*. Routledge.

Memmert, D. (2023). Reale Datensätze–Positionsdaten. In *Sportinformatik: Modellbildung, simulation, Datenanalyse und Visualisierung von sportbezogenen Daten* (pp. 47–53). Springer.

Memmert, D., Raabe, D., Schwab, S., & Rein, R. (2019). A tactical comparison of the 4-2-3-1 and 3-5-2 formation in soccer: A theory-oriented, experimental approach based on positional data in an 11 vs. 11 game set-up. *PLoS One, 14*, e0210191. https://doi.org/10.1371/journal.pone.0210191

Montoliu, R., Martín-Félez, R., Torres-Sospedra, J., & Martínez-Usó, A. (2015). Team activity recognition in association football using a bag-of-words-based method. *Human Movement Science, 41*, 165–178. https://doi.org/10.1016/j.humov.2015.03.007

Pappalardo, L., Cintia, P., Rossi, A., Massucco, E., Ferragina, P., Pedreschi, D., & Giannotti, F. (2019). A public data set of spatio-temporal match events in soccer competitions. *Scientific Data, 6*(1), 236. https://doi.org/10.1038/s41597-019-0247-7

Phatak, A. (2023). *Data-driven performance analysis in soccer: A compilation of data science and machine learning techniques for pre-processing and knowledge discovery*. Deutsche Sporthochschule.

Phatak, A., Wieland, G., Vempala, K., & Memmert, D. (2021). Artificial intelligence based body sensor network framework—Narrative review: Proposing an end-to-end framework using wearable sensors, real-time location systems and artificial intelligence/machine learning algorithms for data collection, data mining and knowledge discovery in sports and healthcare. *Sports Medicine-Open, 7*, 79. https://doi.org/10.1186/s40798-021-00372-0

Power, P., Ruiz, H., Wei, X., & Lucey, P. (2017). Not all passes are created equal: Objectively measuring the risk and reward of passes in soccer from tracking data. In *Proceedings of the 23rd ACM SIGKDD International Conference on Knowledge Discovery and Data Mining* (pp. 1605–1613). Association for Computing Machinery. https://doi.org/10.1145/3097983.3098051

Raabe, D., Biermann, H., Bassek, M., Wohlan, M., Komitova, R., Rein, R., et al. (2022). floodlight—A high-level, data-driven sports analytics framework. *arXiv preprint*. https://doi.org/10.48550/arXiv.2206.02562

Rein, R., & Memmert, D. (2016). Big data and tactical analysis in elite soccer: Future challenges and opportunities for sports science. *Springerplus, 5*(1), 1410. https://doi.org/10.1186/s40064-016-3108-2

Rein, R., Raabe, D., & Memmert, D. (2017). "Which pass is better?" Novel approaches to assess passing effectiveness in elite soccer. *Human Movement Science, 55*, 172–181. https://doi.org/10.1016/j.humov.2017.07.010

Rico-González, M., Pino-Ortega, J., Méndez, A., Clemente, F., & Baca, A. (2023). Machine learning application in soccer: A systematic review. *Biology of Sport, 40*(1), 249–263. https://doi.org/10.5114/biolsport.2023.112970

Roy, R., Paul, A., Bhimjyani, P., Dey, N., Ganguly, D., Das, A. K., & Saha, S. (2020). A short review on applications of big data analytics. In *Emerging Technology in Modelling and Graphics: Proceedings of IEM Graph 2018* (pp. 265–278). https://doi.org/10.1007/978-981-13-7403-6_25

Swartz, T. B. (2020). Where should I publish my sports paper? *The American Statistician, 74*(2), 103–108. https://doi.org/10.1080/00031305.2018.1459842

Wagenaar, M., Okafor, E., Frencken, W., & Wiering, M. A. (2017). Using deep convolutional neural networks to predict goal-scoring opportunities in soccer. In *International Conference on Pattern Recognition Applications and Methods* (Vol. 2, pp. 448–455). https://doi.org/10.5220/0006194804480455

Zago, M., Sforza, C., Dolci, C., Tarabini, M., & Galli, M. (2019). Use of machine learning and wearable sensors to predict energetics and kinematics of cutting maneuvers. *Sensors, 19*(14), 3094. https://doi.org/10.3390/s19143094

Zhang, J., Liang, D., & Sun, Z. (2022). Study on the sustainable development strategy of school soccer based on the background of big data era. *Mobile Information Systems, 2022*, 1–12. https://doi.org/10.1155/2022/7932684

Machine Learning in Basketball

9

Kehang Chen and Marc Garnica Caparrós

Abstract

Basketball ranks among the world's most popular team invasion sports, captivating millions with its dynamic play and intense competition. In recent decades, the advent of Artificial Intelligence has opened new pathways for enhancing the performance of basketball players, indicating a transformative era in basketball analytics and athlete development. Artificial Intelligence encompasses a variety of cutting-edge technologies, each contributing uniquely to the data-driven evolution of basketball. This chapter delves into two of the pivotal Artificial Intelligence technologies, including Machine Learning and Deep Learning, which have found extensive and successful applications within the realm of basketball. Consequently, this chapter aims to equip stakeholders in basketball, from players to coaches, with a deeper comprehension of how Artificial Intelligence technologies can enhance athletic performance and strategic decision-making, thereby translating complex data into actionable insights.

9.1 Introduction

Nowadays, marked by rapid evolution in Artificial Intelligence (AI), the domain of sports has undergone a significant transformation, stepping into the digital age (Stein et al., 2017). Undoubtedly, participants in basketball are also reaping significant benefits from this progress. The motivation behind this innovative development can largely be attributed to the increasing adoption of data-driven analytical paradigms brought about by AI. In the realm of basketball, cutting-edge Machine

K. Chen (✉) · M. G. Caparrós
German Sport University Cologne, Köln, Germany
e-mail: kehang.chen@stud.dshs-koeln.de; m.garnica@dshs-koeln.de

D. Memmert (ed.), *Artificial Intelligence and Machine Learning in Sports Science*,
https://doi.org/10.1007/978-3-662-70155-3_9

Learning (ML) and Deep Learning (DL) technologies such as computer vision, smart wearable devices, big data, etc., have been widely utilized, boosting advancements in critical areas of basketball like human motion recognition and tracking, injury prevention and rehabilitation, and data-driven performance analysis (Yan et al., 2023). The utility of these advanced AI techniques extends beyond the mere extraction of information from historical basketball matches. They serve an important role in enabling basketball practitioners to derive deeper insights and conduct predictive analyses about future trends and outcomes (Richter et al., 2021). Concurrently, ML technologies have been instrumental in transforming raw data into a reservoir of actionable knowledge. This transformation facilitates practical applications of ML and DL in basketball, allowing for the identification of intricate patterns and trends underlying the data. Their implementation ranges from enhancing tactical and strategic decision-making processes to optimizing training regimens based on sophisticated analyses of athlete performance data. Given the increasingly integral role of ML and DL in basketball, it becomes essential to outline the key applications and impacts of these technologies within basketball.

▶ Despite the broad and sometimes overlapping definitions of AI in the academic realm, for the purposes of this chapter, AI is specifically regarded as the ensemble of data-driven techniques through ML and DL. These techniques span the entire spectrum from data collection and processing to sophisticated analysis, with the overarching aim of augmenting player performance and refining tactical strategies.

▶ ML is an umbrella term for algorithms that learn from data, whereas DL is a subset of ML that uses neural networks with many layers. DL is designed to handle and interpret vast amounts of data, making it more suited for complex tasks like image and speech recognition.

9.2 Tracking System Based on ML in Basketball

Computer Vision (CV) technology, a pivotal branch within the realm of AI, has seen widespread application in basketball, enhancing the analysis and understanding of basketball games. By utilizing high-definition cameras, this technology captures intricate tracking data on players and the ball, which is processed using DL algorithms, like Openpose (Cao et al., 2021), Yolo (Redmon et al., 2016), etc. for comprehensive analysis of player movements and team dynamics. This synergy between high-definition cameras and ML algorithms significantly enhances the analytical capabilities of game footage, enabling more accurate and effective identification of movement trajectories. The advent of the SportsVU system, now widely adopted in the National Basketball Association (NBA), exemplifies the technological advancement in this area, offering real-time tracking of players and the ball at a rate of 25 frames per second. This system outperforms traditional data collection methods by capturing precise spatial-temporal data, including both 2D and 3D spatial coordinates of players and the ball. Such advancements have led to the creation of a rich type of statistical data, deeply impacting tactical analysis and performance evaluation by incorporating critical metrics like speed, distance, player spacing, and ball possession (Li & Xu, 2021). This novel approach to data collection and analysis furnishes a more comprehensive and detailed understanding of basketball games, allowing for a deeper insight into player and team performance. Moreover, the application of DL techniques based on graph technology further enables analysts to extract spatial-temporal data from video broadcasts of basketball games through the combination of techniques like camera calibration, object detection, pose estimation, and player tracking (Patton et al., 2021). This is particularly beneficial in 3 × 3 basketball contexts, where fewer players and smaller playing fields provide optimal conditions for data capture. The use of these advanced methodologies breaks free from the constraints of proprietary hardware, enabling a broader range of teams to access valuable tracking data in basketball matches. Undoubtedly, this DL-based tracking system leverages tracking data's potential in basketball, accurately reflecting players and the ball positions in Cartesian coordinates. It enables basketball analysts to unpack player interactions in both spatial and temporal dimensions, offering a quantitative understanding of previously abstract concepts such as game style, team dynamics, and player synergy (Johnson, 2020). This comprehensive analysis contributes significantly to the refined appreciation of basketball's strategic and tactical intricacies, marking the significance of introducing advanced DL techniques into basketball.

► In professional basketball, the ban on wearable sensors during games challenges the collection of tracking data. Computer Vision technology emerges as a crucial solution, allowing non-intrusive, accurate tracking of player and ball movements directly from game footage. This technology utilizes advanced DL algorithms to analyze videos, identifying positions, velocities, and trajectories with an accuracy that competes with wearable sensors. Its application marks a significant leap in basketball analytics, combining respect for the game's integrity with cutting-edge technological advances.

9.3 Injury Prediction Based on ML in Basketball

Injury prevention, performance enhancement, and extending the careers of athletes are fundamental areas of focus within basketball research. The inherently physically demanding nature of basketball, characterized by frequent, high-impact jumping manoeuvres, markedly elevates the incidence of lower extremity injuries, notably anterior cruciate ligament (ACL) ruptures and stress fractures among participants (Taborri et al., 2021). In response to these risks, studies have progressively embraced ML technologies, such as neural networks, decision trees, Markov processes, and support vector machines (Claudino et al., 2019). The adoption of such ML models constitutes a substantial advancement in the field of sports medicine, providing detailed insights into the underlying mechanisms and risk factors associated with common basketball injuries (Lu et al., 2022). Neural networks, in particular, demonstrated considerable potential by leveraging their capacity to learn complex patterns from extensive datasets, thereby identifying subtle indicators of ligament and bone injuries common in the sport (Claudino et al., 2019). Parallel to the advancements in ML, the development and integration of wearable biosensors have also emerged as a powerful tool in basketball analytics. These data-powered body sensor networks offer real-time monitoring of physiological and biomechanical data, providing a continuous stream of insights into an athlete's health and performance status (Phatak et al., 2021). By analyzing data such as heart rate variability, movement speed, jump height, and landing mechanics, these technologies can identify signs of fatigue, overtraining, or improper form that may lead to injuries.

A notable example of these technological innovations in practice is the Connexion platform, which has gained attraction within the NBA (Joe Lemire, 2017). This platform, which represents a synergistic integration with the renowned performance and training platform Fusionetics, incorporates an array of sensors to facilitate non-invasive biometric monitoring. Parameters such as heart rate, respiratory rate, oxygen saturation, hydration levels, and electrolyte concentrations are meticulously tracked. Subsequently, this data is processed through an adaptive ML system designed to swiftly evaluate players' health statuses, thereby enabling teams to be promptly informed of potential injuries and setbacks. This emerging platform exemplifies the transformative potential of data-driven tools in enhancing athlete health management and injury prevention strategies. By leveraging real-time data and sophisticated ML algorithms, basketball teams can achieve a proactive stance towards injury prevention, ensuring athletes remain at the pinnacle of health and performance. This paradigm shift towards data-driven athlete management underscores the evolving landscape of sports medicine, where technological innovation plays a pivotal role in safeguarding the well-being and career longevity of professional athletes.

The integration of ML and wearable sensors stands at the forefront of transforming injury prediction and prevention in basketball. This combination harnesses the analytical power of ML algorithms with the precise, real-time data collection capabilities of wearable technology. By analyzing the extensive physiological and biomechanical datasets gathered from sensors, ML models can identify subtle patterns and risk factors associated with common basketball injuries. This paradigm shift toward a data-driven methodology in sports medicine exemplifies how technology can redefine the limits of athlete care and performance optimization.

9.4 Performance Analysis Based on ML in Basketball

The early basketball analytics relied heavily on manually recorded statistical data (Oliver, 2004), offering a quantifiable measure of key events within games but falling short in depth and breadth. The integration of ML and DL technologies into the realm of basketball has revolutionized the evaluation method of players' performance. As mentioned previously, the rapid advancement of DL technologies has expanded the scope of data collection in basketball games significantly, now encompassing real-time player and ball position information, thereby capturing the dynamic changes and complex interactions on the court (Kovalchik, 2023). Importantly, the significant breakthrough brought about by advanced ML algorithms is their unparalleled capacity to delve into these complex datasets. They are adept at uncovering hidden patterns and trends that elude traditional statistical methods, thereby offering insights into aspects of the game that were previously inaccessible (Claudino et al., 2019). This enhanced analytical capability stems from a robust data-driven framework that encompasses a series of critical steps: data gathering, data preprocessing, feature engineering, model selection, training, parameter optimization, and thorough evaluation of the model performance (Richter et al., 2021). The process begins with the meticulous collection and preprocessing of data in which raw data is cleaned and structured, making it suitable for analysis. Feature engineering then comes into the next step where data scientists and analysts identify and select specific attributes that could significantly influence the outcomes of the analysis. The model selection follows, involving the choice of an appropriate ML model that aligns with the data's characteristics and the analysis objectives. The training phase involves feeding the model with a large volume of data, allowing it to learn and adapt to the patterns within the dataset. Parameter optimization is then conducted to fine-tune the model, enhancing its accuracy and efficiency. Finally, the model undergoes rigorous evaluation, where its performance is assessed through various metrics to ensure its reliability and validity in real-world applications (Ofoghi et al., 2013).

Nowadays, techniques such as logistic regression, decision trees, and k-means clustering, have been extensively applied in basketball, and have enabled basketball analysts to conduct more refined clustering analyses, classifications, and relationship modelling. For example, the discrimination of match outcomes and key performance indicators was the main research theme with the application of supervised learning algorithms (Giovanini et al., 2021; Sampaio et al., 2013). This has not only enhanced the understanding of game strategies but also provided a scientific basis for player performance evaluation, opponent analysis, and training methodology optimization. Furthermore, the interpretability of ML models plays a crucial role in basketball analytics. Interpretability technologies, including but not limited to feature importance assessment, visualization of the model decision-making process, and locally interpretable models (LIME) and SHAP value analysis based on model predictions (Garnica-Caparrós & Memmert, 2021; Wang et al., 2022), provide a method for coaches and athletes with non-professional backgrounds. A way to understand the underlying patterns behind complex data. This in-depth understanding can help them grasp the development process of the game and guide the optimization of training and game strategies.

Example

The Milwaukee Bucks took advantage of the power of data analytics to uncover that Giannis Antetokounmpo exhibits his peak efficiency when positioned as a center on the court. This insight was gained through a detailed analysis of his performance metrics across various game situations. The analysis revealed that Antetokounmpo's blend of size, agility, and skill set becomes particularly advantageous in the center position, where he can maximize his impact on both ends of the floor. Recognizing his enhanced effectiveness and productivity in this role, the Bucks strategically capitalized on this positioning to bolster their overall offensive strategy. This adjustment not only leveraged Antetokounmpo's unique talents more effectively but also played a pivotal role in optimizing the team's offensive dynamics, contributing significantly to their tactical versatility and competitive edge. ◄

9.5 Conclusion

In conclusion, the integration of cutting-edge ML and DL technologies, encompassing computer vision, intelligent sensors, and big data analytics, has seen widespread application across various contexts of basketball, particularly within the top-level leagues, such as the NBA. These technological developments played a pivotal role in elevating the performance of players and teams. By harnessing the power of DL to capture tracking data, monitor athletes in real time, and visually extract valuable knowledge behind the related-match data, basketball has experienced a significant transformation in how games are played, analyzed, and strategized. This fusion of basketball with ML not only underscores the sport's evolution towards a more data-driven and scientifically informed practice but also highlights the potential for further innovations to shape the future of basketball.

References

Cao, Z., Hidalgo, G., Simon, T., Wei, S.-E., & Sheikh, Y. (2021). OpenPose: Realtime multi-person 2D pose estimation using part affinity fields. *IEEE Transactions on Pattern Analysis and Machine Intelligence, 43*(1), 172–186. https://doi.org/10.1109/TPAMI.2019.2929257

Claudino, J. G., Capanema, D. O., de Souza, T. V., Serrão, J. C., Machado Pereira, A. C., & Nassis, G. P. (2019). Current approaches to the use of artificial intelligence for injury risk assessment and performance prediction in team sports: A systematic review. *Sports Medicine—Open, 5*(1), 28. https://doi.org/10.1186/s40798-019-0202-3

Garnica-Caparrós, M., & Memmert, D. (2021). Understanding gender differences in professional European football through machine learning interpretability and match actions data. *Scientific Reports, 11*(1), 10805. https://doi.org/10.1038/s41598-021-90264-w

Giovanini, B., Conte, D., Ferreira-Junior, A., & Nascimento, V. B. (2021). Assessing the key game-related statistics in Brazilian professional basketball according to season phase and final score difference. *International Journal of Performance Analysis in Sport, 21*(2), 295–305. https://doi.org/10.1080/24748668.2021.1881358

Joe Lemire. (2017, December 7). *'iPhone Of Health Care' debuts at NBA summer league*. https://www.sportsbusinessjournal.com/Daily/Issues/2017/07/12/Technology/iphone-health-care-debuts-nba-summer-league.aspx

Johnson, N. (2020, March). Extracting player tracking data from video using non-stationary cameras and a combination of computer vision techniques. In *Proceedings of the 14th MIT Sloan Sports Analytics Conference, Boston, MA, USA* (Vol. 218).

Kovalchik, S. A. (2023). Player tracking data in sports. *Annual Review of Statistics and Its Application, 10*(1), 677–697. https://doi.org/10.1146/annurev-statistics-033021-110117

Li, B., & Xu, X. (2021). Application of artificial intelligence in basketball sport. *Journal of Education, Health and Sport, 11*(7), 54. https://doi.org/10.12775/JEHS.2021.11.07.005

Lu, Y., Pareek, A., Lavoie-Gagne, O. Z., Forlenza, E. M., Patel, B. H., Reinholz, A. K., Forsythe, B., & Camp, C. L. (2022). Machine learning for predicting lower extremity muscle strain in National Basketball Association athletes. *Orthopaedic Journal of Sports Medicine, 10*(7), 23259671221111742. https://doi.org/10.1177/23259671221111742

Ofoghi, B., Zeleznikow, J., MacMahon, C., & Raab, M. (2013). Data mining in elite sports: A review and a framework. *Measurement in Physical Education and Exercise Science, 17*(3), 171–186. https://doi.org/10.1080/1091367X.2013.805137

Oliver, D. (2004). Basketball on paper: Rules and tools for performance analysis. Potomac Books Incorporated.

Patton, A., Scott, M., Walker, N., Ottenwess, A., Power, P., Cherukumudi, A., & Lucey, P. (2021). Predicting NBA talent from enormous amounts of college basketball tracking data. In *MIT Sloan Sports Analytics Conference, Boston*.

Phatak, A. A., Wieland, F.-G., Vempala, K., Volkmar, F., & Memmert, D. (2021). Artificial intelligence based body sensor network framework—Narrative review: Proposing an end-to-end framework using wearable sensors, real-time location systems and artificial intelligence/machine learning algorithms for data collection, data mining and knowledge discovery in sports and healthcare. *Sports Medicine—Open, 7*(1), 79. https://doi.org/10.1186/s40798-021-00372-0

Redmon, J., Divvala, S., Girshick, R., & Farhadi, A. (2016). You only look once: Unified, real-time object detection. In *2016 IEEE Conference on Computer Vision and Pattern Recognition (CVPR), Las Vegas, NV, USA* (pp. 779–788). https://doi.org/10.1109/CVPR.2016.91

Richter, C., O'Reilly, M., & Delahunt, E. (2021). Machine learning in sports science: Challenges and opportunities. *Sports Biomechanics, 23*(8), 961–967. https://doi.org/10.1080/14763141.2021.1910334

Sampaio, J., Lago-Peñas, C., & Gómez, M. A. (2013). Brief exploration of short and mid-term timeout effects on basketball scoring according to situational variables. *European Journal of Sport Science, 13*(1), 25–30.

Stein, M., Janetzko, H., Seebacher, D., Jäger, A., Nagel, M., Hölsch, J., Kosub, S., Schreck, T., Keim, D. A., & Grossniklaus, M. (2017). How to make sense of team sport data: From acquisition to data modeling and research aspects. *Data, 2*(1), 2. https://doi.org/10.3390/data2010002

Taborri, J., Molinaro, L., Santospagnuolo, A., Vetrano, M., Vulpiani, M. C., & Rossi, S. (2021). A machine-learning approach to measure the anterior cruciate ligament injury risk in female basketball players. *Sensors, 21*(9), 3141. https://doi.org/10.3390/s21093141

Wang, Y., Liu, W., & Liu, X. (2022). Explainable AI techniques with application to NBA gameplay prediction. *Neurocomputing, 483*, 59–71. https://doi.org/10.1016/j.neucom.2022.01.098

Yan, W., Jiang, X., & Liu, P. (2023). A review of basketball shooting analysis based on artificial intelligence. *IEEE Access, 11*, 87344–87365. https://doi.org/10.1109/ACCESS.2023.3304631

Machine Learning in Handball

10

Manuel Bassek

Abstract

This book chapter explores the application of machine learning techniques in handball. Research in handball is a historically reliant on manual data annotation. However, recent advancements in ML, particularly in action recognition, performance prediction, tactical pattern recognition, expected goals modeling, and sequence detection, offer new possibilities for in-depth game analysis. By utilizing video, event, and sensor data, researchers and practitioners can now automate the extraction of complex game phases, predict future performances, and identify tactical patterns. The introduction of frameworks like FAUPA-ML demonstrates the potential to scale domain-specific knowledge to large datasets, enabling the analysis of large quantities of data without the limiting costs of manual annotation. This chapter underscores the importance of ML in enhancing our understanding of handball dynamics, providing tools for coaches and analysts to refine training processes and prepare strategically for opponents, and marking a shift towards data-driven approaches in sports analytics.

10.1 Introduction

10.1.1 Data in Handball

In recent years, the growing interest in team sports has led to the introduction of technology in sports research and practice (Gudmundsson & Horton, 2017; Memmert, 2021, 2024; Rein & Memmert, 2016). Soccer, due to its popularity, is currently under investigation by many researchers. A PubMed search in 2024 for

M. Bassek (✉)
German Sport University Cologne, Köln, Germany
e-mail: m.bassek@dshs-koeln.de

D. Memmert (ed.), *Artificial Intelligence and Machine Learning in Sports Science*,
https://doi.org/10.1007/978-3-662-70155-3_10

"(machine learning OR artificial intelligence) AND (soccer OR football)", results in 251 articles with an increasing trend over time. A similar query for handball "(machine learning OR artificial intelligence) AND handball" returns only 15 results since 2011. However, five articles were published in 2023 alone, indicating a rising interest in applying machine learning in handball.

The application of machine learning in handball, as in any domain, requires data. For match analysis in sport, three different data sources can be defined: (1) video data, (2) event data, (3) sensor data. Video data is the most prominent data source and can be available for professional matches in the form of a broadcast stream. Although simple to collect, video data can be costly to process (Bassek et al., 2023a; Manafifard et al., 2017). Usually, domain experts watch the videos and annotate specific actions with semantic information, resulting in event data.

Event data typically contain a timestamp referencing the source material and information about the event, e.g., a shot. These events can be further contextualized by adding attributes about players (e.g., shot taker and goalkeeper), result (e.g., missed, saved, successful), or even contain coordinates of players during that event (Hassan et al., 2017b; Pappalardo et al., 2019). Due to the practicality of this approach, event data in handball has been analyzed with a focus on the frequency of actions and their relationship to success (Ferrari et al., 2019; Foreti et al., 2013; Rogulj et al., 2004).

The improvement of wearable technology has finally introduced sensor data to team sports (Lutz et al., 2019). By wearing a small device, usually in a pouch between the shoulders, players can be tracked with high spatio-temporal resolution. The device contains different sensors such as Global Positioning Systems (GPS) or Inertial Measurement Units (IMU). Therefore, sensor data can further be divided into position data, i.e., x/y-coordinates of the players (and ball) on the pitch, and IMU data, i.e., accelerometer and gyroscope measurements. Initially, sensor data has been used to characterize physiological demands of handball matches and playing positions in small datasets (Büchel et al., 2019). However, since 2019, sensor data is systematically collected in professional handball competitions like the German Handball Bundesliga (HBL), increasing the amount of data available for analysis (Bassek et al., 2023a, 2023b; Manchado et al., 2021; Saal et al., 2023).

▶ Despite more data being collected in recent years, only limited data is available to the public, mainly because of the proprietary nature of the data and competitive interests of the stakeholders. The "Events in Invasion Games Dataset" (EIGD) includes a handball dataset (EIGD-H) that was published by Biermann et al. (2021) with the permission of the HBL. The dataset contains 125 min of video, event, and position data. It is publicly available and licensed under Creative Commons. Easy access and tools for data manipulation and initial analyses can be found in the software package *floodlight* for python (Raabe et al., 2022a). This dataset can be an introduction for new analysts in research and practice trying to enter data analysis in handball.

Although the main body of handball research currently uses frequentist or inferential analysis methods (Ferrari et al., 2019), the increasing amount of available data has led to different applications of machine learning in the handball domain. This chapter focuses on various tasks in handball like scaling of performance analysis, action recognition, and pattern recognition, and provides an overview of the existing literature.

10.2 Action Recognition

10.2.1 Video Data

Handball is recognized for its multidirectional and dynamic nature, characterized by repeated runs, changes of direction, jumps, tacklings, and throws (Karcher & Buchheit, 2014). Although position data may provide accurate information regarding players' locations on the pitch, it offers limited insights into these actions. However, video data can facilitate the automatic detection of such actions, as part of a broader range of methods used in the computer vision domain (Ivasic-Kos & Pobar, 2018). A significant body of research in computer vision has focused on the localization of players on the pitch from video data, utilizing either multiple or single camera setups (Manafifard et al., 2017; Theiner et al., 2022). Deep learning models are often employed to identify shot types, register the pitch, and identify players (Theiner et al., 2022).

For precise action recognition, further pose estimation is necessary (Šajina & Ivašić-Kos, 2022). Pose estimation aims at identifying not only the players' positions but also the locations of their skeletons and joints in 3D space. Šajina and Ivašić-Kos (2022) tested 12 different pipelines for pose estimation in handball players. The top-performing models were able to detect body parts with over 90% accuracy, with an error in joint positions of less than 0.08%. These results demonstrate that pose estimation, as a component of action recognition, is feasible and accurate, even for the dynamic actions inherent in handball.

10.2.2 Sensor Data

Unlike video data, the extraction of signals from sensors, such as IMUs, does not rely on complex methods like computer vision. Thus, few studies have attempted to predict events from sensor data. Lentz-Nielsen et al. (2023) used the magnitude of acceleration, angular velocity, and velocity extracted from IMUs to predict different types of locomotion, including standing and walking, linear and curved running, accelerations and decelerations, as well as overhead throws. The actions were classified with an xgboost model with moderate to high accuracy; F1-scores for linear and curved running were 0.66, however, the classification of throws was achieved with an F1-score of 0.95 (Lentz-Nielsen et al., 2023).

Van Den Tillaar et al. (2021) used IMUs attached to players' wrists to classify different throw types and ball velocity after throwing. They employed four different supervised machine learning models to classify the throw type (circular or whip), throw approach (from running, standing, or jumping), and the resulting ball velocity in an experimental setting. The gradient boosting machine performed best in classifying throw type and approach. The most important features were the acceleration in the y-axis (up) for detecting the approach, the angular rate in the x-axis (horizontal plane) for detecting throw type, and acceleration in the x-axis (towards the goal) for predicting ball velocity.

10.3 Performance Prediction

10.3.1 Tactical Performance

The ability to predict future performance based on recent observations is an interesting task in sports research (Klemp et al., 2021). For instance, early anticipation of an opponent's tactical behavior could lead to a strategic advantage when countering it (Hassan et al., 2017b). Hassan et al. (2017b) explored this by analyzing the emergence of attackers' final shot positions. They curated a dataset of 723 action sequences leading to a shot on target. For each sequence, the locations for the final five passes and receptions leading to the shot were annotated. Using a radial basis function neural network approach, they predicted the final shot location, first based on the five previous passes and second based on the four passes leading to the final pass. The first approach predicted the shot position with an average distance of 1.20 m from the actual shot position, while the second approach resulted in an average distance of 1.42 m. Given the assumption that each defender covers about 3.6 m on the 6 m line, this could be useful for anticipating defensive actions and could be implemented when preparing for an opponent's offensive behavior (Hassan et al., 2017b).

10.3.2 Physical Performance

A different angle to performance prediction is forecasting an athlete's physical performance based on known anthropometric data and diagnostics, which may be valuable for training evaluation and player selection (Oytun et al., 2020). In an experimental approach, Oytun et al. (2020) investigated the potential to predict performance in six different performance tests for female handball players. Anthropometric data such as age, body mass, height, skinfold thickness, as well as results from jumping, sprinting, and power diagnostics, were used to train six machine learning models. The best-performing radial-basis neural network was capable of predicting performance, achieving R^2 values between 0.86 and 0.97. These findings underscore the potential of machine learning in exercise science, diagnostics, and coaching, illustrating its capability in predicting athletic performance and developing new performance indicators (Oytun et al., 2020).

10.4 Tactical Pattern Recognition

The ability to identify and analyze patterns in tactical behavior, such as positioning for shots, extends beyond simple prediction, offering valuable insights for game preparation against opponents (Hassan et al., 2017b). Schrapf et al. (2017) approached this problem by clustering a dataset of pass locations from 723 action sequences using neural networks. They identified six distinct clusters and analyzed them regarding their scoring rate, distance to goal, and distance between the shooting position and the nearest defensive player. Although no statistically significant differences in scoring rate were found between the clusters, this study underscores the potential of machine learning to model tactical interactions between teams based on player positions (Schrapf et al., 2017).

This potential was further explored by Hassan et al. (2017a), who aimed to evaluate the effectiveness of tactical training based on patterns extracted through machine learning. Using a self-organizing feature map, they clustered sequences from the previously generated action sequences. A total of 27 clusters were identified, and 12 were selected as target patterns for training. A junior handball team was trained for 6 weeks with an emphasis on implementing these target patterns. The effectiveness of this tactical training was then evaluated by clustering patterns of play in a follow-up match. 58% of the executed patterns could be directly related to the trained patterns, indicating a successful transfer of training to game performance. This study shows how machine learning can support coaches by refining their training processes and preparing for opponents' offensive behavior more effectively.

10.5 Expected Goals

The concept of expected goals (xG) is a subtask of assigning value to players actions. Generally, xG is the probability that a shot will result in a goal. This metric is derived by analyzing historical data to group shots into similar categories based on various features such as distance and angle to the goal, positioning of the goalkeeper and defenders, and the execution of the shot (Bauer & Anzer, 2021). When a new shot is taken, its xG value is estimated based on the success rate of historically similar shots.

In handball, Adams et al. (2023) developed an xG model utilizing both event and position data. They implemented a pipeline for synchronizing event data with position data to identify relevant features for more than 29,000 throwing events during the 2022/23 season of the German Handball Bundesliga (HBL). A variety of machine learning models were trained, with the CatBoost model demonstrating the highest accuracy at 70% in predicting shot outcome. The importance of different features was assessed using the Shapley Additive exPlanations (SHAP) method. The most important features were the thrower's distance to goal, angle to goal, distance from the goalkeeper to the goal, and distance from the goalkeeper to the thrower.

Applying the xG model to the entire 2022/23 HBL season, the researchers showed that expected performance closely correlates with actual performance, both at the team and player levels, indicating the model's robust generalization capability. As the first model to assess xG in handball, this work provides the basis for future investigations to build upon its performance and explore the value of other in-game actions, such as assists.

10.6 Sequence Detection

The analysis of handball matches traditionally relies on manual labeling of video data by domain experts. Notational analysis can provide valuable insights into the effectiveness of shots or attack types but is limited by the available resources for acquiring reliable expert labels (Bassek et al., 2023a; O'Donoghue, 2007). An automated approach to extracting relevant phases from matches would not only reduce the workload of analysts but also enable large-scale analysis of data.

An early approach to that problem was proposed by Hervieu et al. (2009), who used hierarchical parallel semi-Markov models to extract activity phases from players' trajectories. These trajectories were manually extracted from 10 min of video data and enriched with the audio signal of the referee's whistle. The model was trained to identify eight different activity phases, including offensive and defensive phases against organized opponents, transitions to offense and defense, counter attacks, and match interruptions. The model achieved a high classification accuracy of 92% on the test dataset, indicating that position data and machine learning could be effectively used to automate subtasks of match analysis in handball, such as the annotation of activity phases.

The systematic collection of position data in handball has since resulted in hundreds of matches' worth of data and allow novel possibilities for analysis relevant to both sports science and practice. For that purpose, Bassek et al. (2023a) introduced the Framework for Automatic Upscaled Performance Analysis with Machine Learning (FAUPA-ML) in handball. This framework aims to scale domain-specific knowledge to large datasets by curating a small dataset of desired manual annotations, training a machine learning model, and applying it to a larger dataset. Throughout this process, the quality of each individual step is controlled by reliability checks and face validation. The workflow of the general framework is shown in Fig. 10.1.

In their application of FAUPA-ML, counter and position attacks in ten matches from the HBL were labeled, and the resulting graph-based deep learning model, TGNet (Raabe et al., 2022b), achieved a balanced classification accuracy of 86%. The model was then applied to 539 matches of the HBL, extracting over 80,000 ball possession phases. They further compared the intensity of counter attacks and position attacks while attacking and defending. The FAUPA-ML pipeline offers novel possibilities for investigating the tactical, technical, and physical dynamics of handball matches with big data sources, without relying on extensive expert labeling (Bassek et al., 2023a).

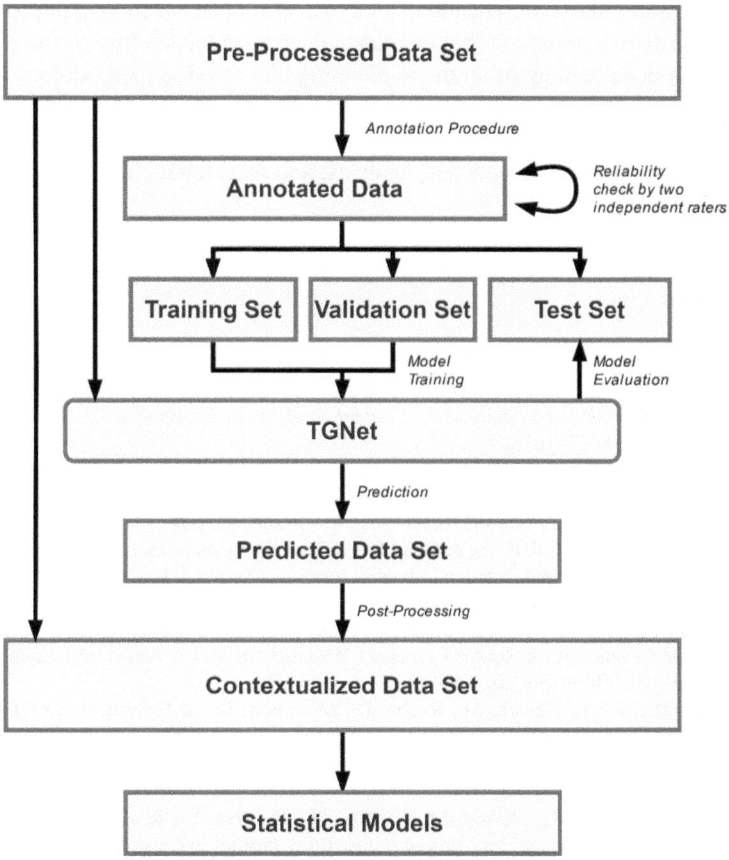

Fig. 10.1 The FAUPA-ML workflow (Bassek et al., 2023a)

10.7 Conclusion

The application of machine learning techniques in handball marks a significant advancement in the sport's analytical methods. It allows moving beyond the limitations of manual data annotation to automated analyses. This chapter has highlighted the key areas where machine learning contributes to handball analytics, including action recognition, performance prediction, tactical pattern recognition, expected goals (xG) modeling, and sequence detection. These advancements demonstrate the ability of machine learning to process complex game data efficiently, providing actionable insights for practitioners.

One notable innovation, the FAUPA-ML framework, exemplifies how ML can be used to extend domain-specific knowledge to larger datasets, significantly reducing the resource constraints associated with traditional analysis methods. This shift not only makes advanced analytics more accessible to teams at different levels but also opens up new possibilities for research and strategy development in handball.

The integration of ML into handball analytics is not just a technical improvement but a transformative approach that enables a deeper understanding of the game. It facilitates the development of strategic planning and training methodologies based on empirical data. As machine learning models continue to evolve, their application in handball is expected to uncover even more insights into player performance and game dynamics, supporting coaches, analysts, and researchers in making informed decisions.

References

Adams, M., David, A., Hesse, M., & Rückert, U. (2023). Expected goals prediction in professional handball using synchronized event and positional data. In *Proceedings of the 6th International Workshop on Multimedia Content Analysis in Sports* (pp. 83–91). https://doi.org/10.1145/3606038.3616152

Bassek, M., Raabe, D., Banning, A., Memmert, D., & Rein, R. (2023a). Analysis of contextualized intensity in Men's elite handball using graph-based deep learning. *Journal of Sports Sciences, 41*(13), 1299–1308. https://doi.org/10.1080/02640414.2023.2268366

Bassek, M., Raabe, D., Memmert, D., & Rein, R. (2023b). Analysis of motion characteristics and metabolic power in elite male handball players. *Journal of Sports Science and Medicine, 22*(2), 310–316. https://doi.org/10.52082/jssm.2023.310

Bauer, P., & Anzer, G. (2021). A goal scoring probability model for shots based on synchronized positional and event data in football (soccer). *Frontiers in Sports and Active Living, 3*, 53. https://doi.org/10.3389/fspor.2021.624475

Biermann, H., Theiner, J., Bassek, M., Raabe, D., Memmert, D., & Ewerth, R. (2021). A unified taxonomy and multimodal dataset for events in invasion games. In *Proceedings of the 4th International Workshop on Multimedia Content Analysis in Sports* (pp. 1–10). https://doi.org/10.1145/3475722.3482792

Büchel, D., Jakobsmeyer, R., Döring, M., Adams, M., Rückert, U., & Baumeister, J. (2019). Effect of playing position and time on-court on activity profiles in German elite team handball. *International Journal of Performance Analysis in Sport, 19*(5), 832–844. https://doi.org/10.1080/24748668.2019.1663071

Ferrari, W. R., Sarmento, H., & Vaz, V. (2019). Match analysis in handball: A systematic review. *Montenegrin Journal of Sports Science and Medicine, 8*(2), 63–76. https://doi.org/10.26773/mjssm.190909

Foreti, N., Rogulj, N., & Papi, V. (2013). Empirical model for evaluating situational efficiency in top level handball. *International Journal of Performance Analysis in Sport, 13*(2), 275–293. https://doi.org/10.1080/24748668.2013.11868648

Gudmundsson, J., & Horton, M. (2017). Spatio-temporal analysis of team sports. *ACM Computing Surveys, 50*(2), 1–34. https://doi.org/10.1145/3054132

Hassan, A., Schrapf, N., Ramadan, W., & Tilp, M. (2017a). Evaluation of tactical training in team handball by means of artificial neural networks. *Journal of Sports Sciences, 35*(7), 642–647. https://doi.org/10.1080/02640414.2016.1183804

Hassan, A., Schrapf, N., & Tilp, M. (2017b). The prediction of action positions in team handball by non-linear hybrid neural networks. *International Journal of Performance Analysis in Sport, 17*(3), 293–302. https://doi.org/10.1080/24748668.2017.1336688

Hervieu, A., Bouthemy, P., & Cadre, J.-P. L. (2009). Trajectory-based handball video understanding. In *Proceedings of the ACM International Conference on Image and Video Retrieval* (pp. 1–8). https://doi.org/10.1145/1646396.1646447

Ivasic-Kos, M., & Pobar, M. (2018). Building a labeled dataset for recognition of handball actions using mask R-CNN and STIPS. In *2018 7th European Workshop on Visual Information Processing (EUVIP)* (pp. 1–6). https://doi.org/10.1109/EUVIP.2018.8611642

Karcher, C., & Buchheit, M. (2014). On-court demands of elite handball, with special reference to playing positions. *Sports Medicine, 44*(6), 797–814. https://doi.org/10.1007/s40279-014-0164-z

Klemp, M., Wunderlich, F., & Memmert, D. (2021). In-play forecasting in football using event and positional data. *Scientific Reports, 11*(1), 24139. https://doi.org/10.1038/s41598-021-03157-3

Lentz-Nielsen, N., Hart, B., & Samani, A. (2023). Prediction of movement in handball with the use of inertial measurement units and machine learning. *Sports Biomechanics*, 1–14. https://doi.org/10.1080/14763141.2023.2224279

Lutz, J., Memmert, D., Raabe, D., Dornberger, R., & Donath, L. (2019). Wearables for integrative performance and tactic analyses: Opportunities, challenges, and future directions. *International Journal of Environmental Research and Public Health, 17*(1), 1–26. https://doi.org/10.3390/ijerph17010059

Manafifard, M., Ebadi, H., & Abrishami Moghaddam, H. (2017). A survey on player tracking in soccer videos. *Computer Vision and Image Understanding, 159*, 19–46. https://doi.org/10.1016/j.cviu.2017.02.002

Manchado, C., Pueo, B., Chirosa-Rios, L. J., & Tortosa-Martínez, J. (2021). Time–motion analysis by playing positions of male handball players during the European Championship 2020. *International Journal of Environmental Research and Public Health, 18*(6), 2787. https://doi.org/10.3390/ijerph18062787

Memmert, D. (Ed.). (2021). *Match analysis*. Routledge.

Memmert, D. (2024). *Computer science in sport: Modeling, simulation, data analysis and visualization of sports-related data*. Springer.

O'Donoghue, P. (2007). Reliability issues in performance analysis. *International Journal of Performance Analysis in Sport, 7*(1), 35–48. https://doi.org/10.1080/24748668.2007.11868386

Oytun, M., Tinazci, C., Sekeroglu, B., Acikada, C., & Yavuz, H. U. (2020). Performance prediction and evaluation in female handball players using machine learning models. *IEEE Access, 8*, 116321–116335. https://doi.org/10.1109/ACCESS.2020.3004182

Pappalardo, L., Cintia, P., Rossi, A., Massucco, E., Ferragina, P., Pedreschi, D., & Giannotti, F. (2019). A public data set of spatio-temporal match events in soccer competitions. *Scientific Data, 6*(1), 236. https://doi.org/10.1038/s41597-019-0247-7

Raabe, D., Biermann, H., Bassek, M., Wohlan, M., Komitova, R., Rein, R., Groot, T. K., & Memmert, D. (2022a). *floodlight—A high-level, data-driven sports analytics framework*. https://doi.org/10.48550/ARXIV.2206.02562.

Raabe, D., Nabben, R., & Memmert, D. (2022b). Graph representations for the analysis of multi-agent spatiotemporal sports data. *Applied Intelligence, 53*(4), 3783–3803. https://doi.org/10.1007/s10489-022-03631-z

Rein, R., & Memmert, D. (2016). Big data and tactical analysis in elite soccer: Future challenges and opportunities for sports science. *Springerplus, 5*(1), 1410. https://doi.org/10.1186/s40064-016-3108-2

Rogulj, N., Srhoj, V., & Srhoj, L. (2004). The contribution of collective attack tactics in differentiating handball score efficiency. *Collegium Antropologicum, 28*(2), 739–746.

Saal, C., Baumgart, C., Wegener, F., Ackermann, N., Sölter, F., & Hoppe, M. W. (2023). Physical match demands of four LIQUI-MOLY Handball-Bundesliga teams from 2019–2022: Effects of season, team, match outcome, playing position, and halftime. *Frontiers in Sports and Active Living, 5*, 1183881. https://doi.org/10.3389/fspor.2023.1183881

Šajina, R., & Ivašić-Kos, M. (2022). 3D pose estimation and tracking in handball actions using a monocular camera. *Journal of Imaging, 8*(11), 308. https://doi.org/10.3390/jimaging8110308

Schrapf, N., Alsaied, S., & Tilp, M. (2017). Tactical interaction of offensive and defensive teams in team handball analysed by artificial neural networks. *Mathematical and Computer Modelling of Dynamical Systems, 23*(4), 363–371. https://doi.org/10.1080/13873954.2017.1336733

Theiner, J., Gritz, W., Muller-Budack, E., Rein, R., Memmert, D., & Ewerth, R. (2022). Extraction of positional player data from broadcast soccer videos. In *2022 IEEE/CVF Winter Conference on Applications of Computer Vision (WACV)* (pp. 1463–1473). https://doi.org/10.1109/WACV51458.2022.00153

Van Den Tillaar, R., Bhandurge, S., & Stewart, T. (2021). Can machine learning with IMUs be used to detect different throws and estimate ball velocity in team handball? *Sensors, 21*(7), 2288. https://doi.org/10.3390/s21072288

Risk, Reward, and Reinforcement Learning in Ice Hockey Analytics

11

Sheng Xu, Oliver Schulte, Yudong Luo, Pascal Poupart, and Guiliang Liu

Abstract

What makes many decisions in sports difficult is that they involve a trade-off between risk and reward. Actions such as taking a three-point shot, carrying a puck, or dribbling with a ball carry a higher risk of failure and require exceptional skill to pull off, but also bring a higher potential reward. This paper describes computational tools for *risk analytics* to model the risk inherent in the choices faced by teams and athletes. We leverage *distributional reinforcement learning (RL)* as a source of concepts and techniques for computational risk analytics. Distributional RL techniques allow us to model a dynamic distribution of outcomes for 1000+ games in the National Hockey League. We find strong evidence that strong teams take many risks (0.90 correlation between team season standing and team season standard/Gini deviation). For players, we also find strong evidence that stronger players take more risks (e.g., 0.86 correlation between a player's season goals and their value-at-risk metric).

S. Xu · G. Liu
School of Data Science, The Chinese University of Hong Kong, Shenzhen, Guangdong, China
e-mail: shengxu1@link.cuhk.edu.cn; liuguiliang@cuhk.edu.cn

O. Schulte (✉)
School of Computing Science, Simon Fraser University, Burnaby, Vancouver, BC, Canada
e-mail: oschulte@cs.sfu.ca

Y. Luo · P. Poupart
Cheriton School of Computer Science, University of Waterloo, Waterloo, ON, Canada
e-mail: yudong.luo@uwaterloo.ca; ppoupart@uwaterloo.ca

D. Memmert (ed.), *Artificial Intelligence and Machine Learning in Sports Science*,
https://doi.org/10.1007/978-3-662-70155-3_11

11.1 Introduction: Taking Chances in Sports

Many decisions by athletes and coaches involve accepting higher risk for potentially higher rewards. A well-known example from basketball is whether to take a long-distance shot, which potentially nets three points, versus a shot from a shorter distance for two points. A more complex example from ice hockey is pulling the goalie—substituting an attacker for the goalie—when a team is trailing. Pulling the goalie earlier increases the chances of equalizing, but also increases the chances of the leading team scoring, which in practice decides the game immediately. While the decision-theoretically optimal choice is to *maximize the average success*, that is the average number of points and wins, several sports analysts have observed that players and coaches are often influenced by a secondary goal, which is to *minimize the probability of failure*, or generally the probability of bad outcomes. To illustrate the point in the basketball scenario, consider a player in a situation where the chance of scoring a three-pointer is 20% and the chance of scoring a two-pointer is 30%. Then the expected number of points is the same for each choice (namely 0.6). However, the probability of failure is 80% for the three-pointer and only 70% for the two-pointer; bad outcomes are less likely for the two-point shot. Now consider a different situation where the long-distance shot has a 25% chance of success. In this case, the expected number of long-distance points is 0.75 versus 0.6, and the optimal decision is to take the long-distance shot. However, the chance of failure is still 75% versus only 70% for the two-pointer, so a *risk-averse* player may still prefer the safer two-point shot.

Several sports analysts have argued that players and teams tend to take risk-averse decisions at the expense of their total success averaged over many games and match situations. Pelechrinis (2016) provides evidence that American football coaches aim to minimize the variance of expected points, rather than the expected points directly, perhaps to avoid public criticism for failure. Beaudoin and Swartz (2010) argued that trailing hockey teams should pull their goalies earlier to maximize winning chances. Indeed NHL teams have recently started pulling their goalies earlier.[1] Another piece of evidence for a trend towards more risk-taking is the rise of 3-point attempts in basketball, rising from 22.2% in the 2010–2011 season to 39.2% in 2020–2021.[2] The trend towards riskier actions over time is evidence that risk-taking affords an advantage over more cautious tactics.

In this paper, we study risk-taking in the National Hockey League. We examine different ways to quantify how risky a decision is, including traditional notions such as the variance/standard deviation of outcomes, as well as the Gini deviation, an alternative variability concept well-suited to multi-modal distributions (Luo et al., 2023). From a technical viewpoint, *modeling risk requires modeling higher-order moments of the distribution of possible action outcomes*. To model the outcome distribution beyond its mean, we leverage recent work in *distributional reinforcement learning* (RL). Reinforcement learning is a branch of machine learning that

[1] https://www.nhl.com/news/coaches-room-paul-maclean-nhl-pulling-goalies-trend-301099580

[2] https://www.nba.com/news/3-point-era-nba-75

studies how to act in sequential decision-making scenarios, such as we encounter in sports analytics. Reinforcement learning has developed an extensive set of methods for estimating expected outcomes from actions, known as prediction or policy evaluation methods (Sutton & Barto, 1998). Distributional reinforcement learning is a more recent development that provides methods for modelling the distribution of action outcomes. A recent paper by Liu et al. (2022) developed a distributional RL method for estimating the distribution of action outcomes for play-by-play (event) data. We utilize the computational tools from their work to estimate action outcome distributions from large play-by-play event data (1000+ games, 1M+ events).

We apply their framework to quantify and study the riskiness of actions by professional players in the NHL. The main question is how performance relates to risk-taking, for both teams and players. We examine three different variability concepts for quantifying the risk associated with a distribution of outcomes: standard deviation, Gini deviation, and value at risk. The risk impact of an action is the extent to which it increases/decreases the variability of the game outcomes for the acting player's team.

Our main findings are as follows: For *team performance*, the total risk of the actions taken by a team in a season displays a very high correlation with the team performance, measured by the number of total season points. Using standard deviation or Gini deviation as our risk measure, the correlation reaches 0.90. For the value-at-risk metric studied previously by Liu et al. (2022) the correlation is only 0.51, given a confidence level 0.2 that represents risk-seeking (see Sect. 11.6 for further details).

For *player performance*, we use the total risk of the actions taken by a player in a season as a measure of the player's risk-taking. All player risk metrics show a high degree of temporal consistency, with their round-by-round totals essentially converging less than halfway through the season. The Gini deviation and standard deviation player metrics achieve substantial correlations of 0.56 and 0.51, respectively, with the player's total goals in a season. The value-at-risk metric achieves an even higher goal correlation of 0.86 (with the risk-seeking confidence level of 0.2). These results provide evidence that variability risk metrics are very good at predicting team success, but less suitable for predicting player success. However, because of the lack of a ground truth ranking for players, we do not consider this finding conclusive, and investigating risk-taking by players is a valuable direction for future research. While our study focuses on ice hockey data from the National Hockey League (NHL), our methods apply to any play-by-play dataset; see Liu et al. (2022) for an application to soccer data.

Paper Outline Our paper is organized as follows. We begin with an overview of the rules of ice hockey and our play-by-play dataset. Then we review the background of reinforcement learning (RL), especially the distributional RL techniques for learning the distribution of action outcomes from play-by-play event data. Our discussion focuses on the main principles and intuitions; details on our learning methods may be found in the appendix and in the references. Given a dynamic distribution over future action outcomes at each point in a match, we define the risk

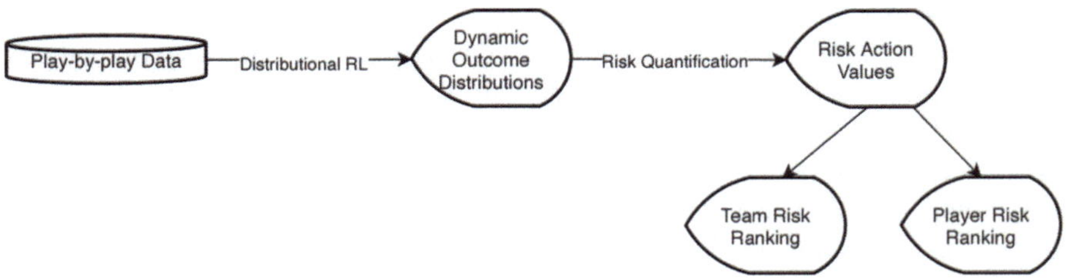

Fig. 11.1 System components in our risk analytics framework

impact of an action as the increase/decrease in the risk associated with the outcome distribution, after the action. The total risk impact of all actions is used to quantify risk-taking by teams and by players. Figure 11.1 summarizes our system components.

11.2 Hockey Rules and Hockey Data

NHL Rules We give a brief overview of rules of play in the NHL (National Hockey League, 2014). NHL games consist of three periods, each 20 min in duration. A team has to score more goals than their opponent within three periods in order to win the game. Teams have five skaters and one goalie on the ice during even strength situations. Penalties result in a player sitting in the penalty box for 2, 4, 5 or 10 min and the penalized team will be shorthanded, creating a manpower differential between the two teams. The period where one team is penalized is called a powerplay for the opposing team with a manpower advantage. A shorthanded goal is a goal scored by the penalized team, and a powerplay goal is a goal scored by the team on the powerplay.

Dataset In this paper, we use a play-by-play proprietary dataset constructed by Sportlogiq.[3] The data are constructed through a combination of computer vision and manual annotation. The dataset contains a total of 1196 games played between October 3rd, 2018 and April 6th, 2019. The training dataset for constructing our model contains 956 games (from October 3rd, 2018 to February 24th, 2019). Table 11.1 lists the features used in our analysis (Liu et al., 2022) and Fig. 11.2 illustrates the adjusted coordinates in the Sportlogiq dataset. Figure 11.2 shows a schematic layout of the ice hockey rink. The units are feet. Adjusted Y-coordinates run from −42.5 at the bottom to 42.5. The goal line is at X = 89.

The dataset records event data known as **play-by-play** data. Play-by-play data specifies the timing and location of actions, identifies the player responsible for each action, and includes the **contextual event features** outlined in Table 11.1. Table 11.2 lists the most frequent action types in our dataset. To help visualize play-by-play data, Table 11.3 provides a partial sample.

[3] https://sportlogiq.com

Table 11.1 The complete list of game features for the ice hockey dataset

	Type	Name	Range
	Spatial features	X Coordinate of Puck	[−100, 100]
		Y Coordinate of Puck	[−42.5, 42.5]
		Velocity of Puck	$(-\infty, +\infty)$
		Angle between the puck and the goal	[−3.14, 3.14]
Ice Hockey	Temporal features	Game Time Left	[0, 3600]
		Event/Action Duration	$(0, +\infty)$
	In-Game features	Score Differential	$(-\infty, +\infty)$
		Manpower Situation	{Even Strength, Short-Handed, Power Play}
		Home or Away Team Action	{Home, Away}
		Outcome	{successful, failure}

The table utilizes adjusted spatial coordinates where negative numbers denote the defensive zone of the acting player and positive numbers denote the offensive zone

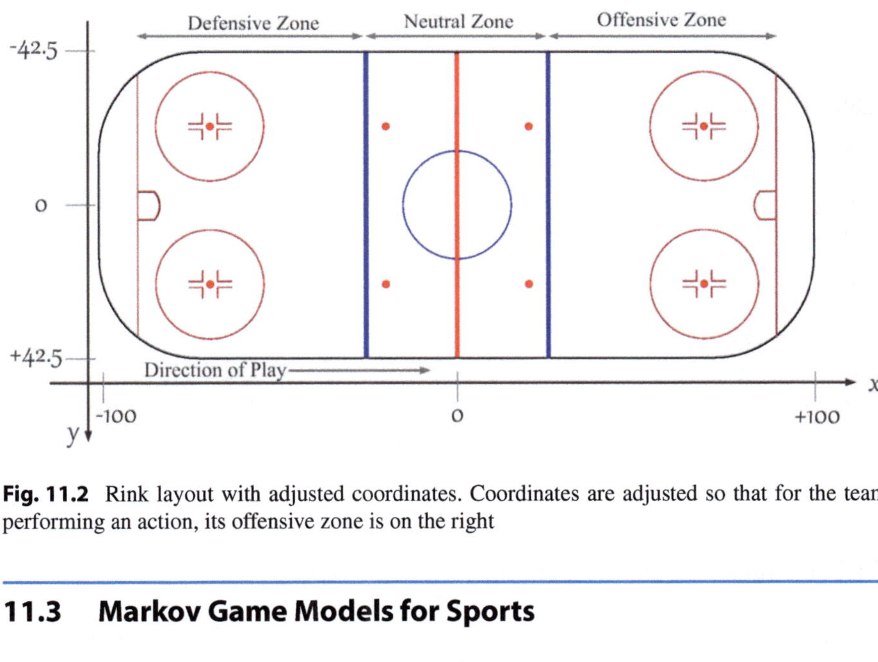

Fig. 11.2 Rink layout with adjusted coordinates. Coordinates are adjusted so that for the team performing an action, its offensive zone is on the right

11.3 Markov Game Models for Sports

Reinforcement learning provides a rich toolkit for estimating the chances of future success for strategic agents. Schulte (2022) gives a short accessible introduction to applying RL in sports analytics. For single-agent problems, RL is based on the fundamental Markov decision process model. Generalizing Markov decision process to multiple decision makers leads to the Markov game model. Markov game models have been developed for several sports, such as ice hockey, soccer, and American football (Chan et al., 2021; Liu et al., 2020; Liu & Schulte, 2018). We utilize the ice hockey model of Liu and Schulte (2018).

Table 11.2 Definition of the most frequent action types in the dataset

Action	Description
Block	A block attempt on the puck's trajectory
Carry	Controlled carry over a blue line or the red center line
Check	When a player attempts to use his body to remove possession from an opponent
Dump in	When a player sends the puck into the offensive zone
Dump out	When a defending player dumps the puck up the boards without targeting a teammate for a pass
lpr	Loose puck recovery. The player recovered the puck as it was out of possession of any player
Offside	When a player is caught over the offensive blue line before their teammate brings the puck in
Pass	The player attempts a pass to a teammate
Puck protection	When a player uses their body to protect the puck along the boards
Reception	When a player receives a pass from a teammate
Shot	A player shoots on goal
Shot against	A shot was taken by the opposing team

Table 11.3 Sample excerpt of play-by-play data

Game ID	Player ID	Period	Team ID	x Coord	y Coord	Manpower	Action type
849	402	1	15	−9.5	1.5	Even	lpr
849	402	1	15	−24.5	−17	Even	Carry
849	417	1	16	−75.5	−21.5	Even	Check
849	402	1	15	−79	−19.5	Even	Puckprot
849	413	1	16	−92	−32.5	Even	lpr
849	413	1	16	−92	−32.5	Even	Pass
849	389	1	15	−70	42	Even	Block
849	389	1	15	−70	42	Even	lpr
849	389	1	15	−70	42	Even	Pass
849	425	1	16	−91	34	Even	Block
849	395	1	15	−97	23.5	Even	Reception

11.3.1 Markov Game Model for NHL Ice Hockey

Similar to Liu and Schulte (2018), we apply the Markov Game Framework to model the play dynamics for sports games. A Markov Game (Littman, 1994), sometimes called a stochastic game, is defined by a set of states $\mathcal{S}$, and a collection of action sets $\mathcal{A}$, one for each agent in the environment. State transitions are controlled by the current state and a list of actions, one action from each agent. For each agent, there is an associated reward function mapping a state transition to a reward. An overview of how a hockey Markov Game model fills in this schema is as follows.

- There are two agents, *Home* and *Away*, representing their respective teams.
- The **action** a_t denotes the movements of players who control the puck. Our model applies a discrete action vector using a one-hot representation.

Fig. 11.3 A hockey match is segmented into goal-scoring episodes

- An **observation** is a feature vector x_t specifying a value of the features listed in Table 11.1 at a discrete time step t. We use the complete sequence $s_t \equiv (x_t, a_t, x_{t-1}, \ldots, x_0)$ to represent the **state** (Mnih et al., 2015) (Fig. 11.3).
- Since we have two agents, we have two **reward functions**, one for the home team and one for the away team. Conceptually, the reward at time t is 1 for a team that scores a goal at time t, 0 if there is no goal; we write $\text{goal}_{t, Home}$, $\text{goal}_{t, Away}$. For example, $\text{goal}_{t, Home} = 1$ indicates that the home team scores at time t. It is technically useful to introduce a virtual "none" agent for the eventuality that neither team scores until the end of a game. If neither team scores at the end of the match, we write $goal_{T, Neither} = 1$ where T is the last time step.

The **expected goal model** $R_k(s_t, a_t) = P(\text{goal}_{t, k} = 1 | s_t, a_t)$ specifies the probability that a team scores a goal after an action in a given match state. This model makes the **Markov assumption**, which implies that the state information available at time t is informative enough that scoring chances can be estimated based on the current state only, independent of the current time t and previous states. Technically, the Markov game model is stationary and the goal scoring probability is independent of the current time index t. Similarly, transition probabilities $P(s_{t+1}, a_{t+1} | s_t, a_t)$ are assumed to be stationary and depend on the current match state only. This means that for a given match state s_t and action a_t at time t, the *dynamics of NHL play define a distribution over future game trajectories that depends only on the current state and action.*

11.3.2 The Expected Value Function

We divide a sports game into **goal-scoring episodes**, so that each episode: (1) starts at the beginning of the game, or immediately after a goal, and (2) terminates with a goal or at the end of the game (s_H). Episodes extend through period breaks.

A key quantity in reinforcement learning is the *expected reward* with respect to future trajectories. Given our binary reward (score goal or not), the expected reward for a team k is the chance of scoring the next goal, denoted as $Q_k(s_t, a_t)$. To explain the basic RL approach to learning a Q-function, consider first the expected reward for a bounded *horizon*, that is, a fixed look-ahead length H . The chance of scoring within the next H steps is then defined by the expression:

$$Q_k^H(s_t, a_t) = \sum_{h=0}^{H} \gamma^h P(\text{goal}_{t+h,k} = 1 | s_t | a_t) \quad (11.1)$$

Following previous studies (Liu & Schulte, 2018; Liu et al., 2020), we set $\gamma = 1$. In this case the Q_k^H value simply denotes the probability that team k scores a goal within H steps.

The Q-value satisfies an important recurrence relation known as the dynamic programming update:

$$Q_k^{H+1}\left(s_t, a_t\right) = R_k\left(s_t, a_t\right) + \gamma E_{s_{t+1}, a_{t+1} \sim P\left(s', |, a', |, s_0 = s_t, |, a_0 = a_t\right)}\left[Q_k^H\left(s_{t+1}, a_{t+1}\right)\right] \quad (11.2)$$

The principle behind dynamic programming is that a team scores a goal in $H + 1$ steps if and only if they (1) score immediately or (2) take another step and then score within H steps. The value iteration algorithm uses dynamic programming to estimate Q-values from a dataset $\mathcal{D}$ of observed trajectories as follows.

1. Initialize the Q-values for $H = 0$ with the expected goal model R.
2. Iteratively apply Eq. (11.2) through $H = 1$, $H = 2$, …, until convergence where $Q_k^{H+1}\left(s_t, a_t\right) = Q_k^H\left(s_t, a_t\right)$ for each team and state-action pair. We denote the **convergent Q-value** as $Q_k(s_t, a_t)$.

For continuous state spaces, such as we have in ice hockey, the expectation $E_{s_{t+1}, a_{t+1} \sim P\left(s', |, a', |, s_0 = s_t, |, a_0 = a_t\right)}$ can be estimated by averaging over all transitions $(s_t, a_t; s_{t+1}, a_{t+1})$ observed in the data set. In the NHL model of Schulte et al. (2017b), convergence occurred with a lookahead of $H = 13$. The fact that expected values in RL incorporate lookahead means that they can capture the medium-term effects of actions on goals scoring.

With an unbounded lookahead $H \rightarrow \infty$, under mild conditions value iteration converges to a fixed point that satisfies:

$$Q_k\left(s_t, a_t\right) = R_k\left(s_t, a_t\right) + \gamma E_{s_{t+1}, a_{t+1} \sim P\left(s_{t+1}, |, a_{t+1}, |, s_t, |, a_t\right)}\left[Q_k\left(s_{t+1}, a_{t+1}\right)\right] \quad (11.3)$$

which is known as the **Bellman equation** for policy evaluation. In the Appendix we discuss how the Bellman equation can be applied to learn a neural net model of the Q-function.

Remark For readers familiar with reinforcement learning models, we briefly situate our NHL model with respect to other RL models. Other readers can skip this paragraph without loss of continuity. Our learning setting is *off-line* learning where we learn a value function from a dataset without executing actions; that is, our problem is prediction not control. In the off-line perspective, the observed actions can be treated as another feature similar to states. Formally, what we have defined is a *Markov reward process* in an expanded state space $\mathcal{S} \times \mathcal{A}$ where an expanded state is a pair (s, a) (cf. Sutton & Barto, 1998, Ch.6.4). The Q-function as we have defined it is the value function of this Markov reward process. We have used the Q-notation, rather than V for value function, because its meaning is the same as in policy evaluation: the expected cumulative reward for an agent given a current action and a current state. An equivalent model would be to first estimate a policy π_{Home} for the home team and another policy π_{Away} for the away team. For example, if in state s the home team passes the puck 30% of the time, we might estimate $\pi_{Home}(pass| s) = 30\%$. The Q-function as we have defined it represents $Q^{\pi_{Home}, \pi_{Away}}$, which is the Q-function of the NHL Markov game where the home and the away team follow the behavioral policies with action frequencies shown in the data. While the Markov reward model for off-line data is perhaps less familiar than the policy evaluation formulation, we use it because it is conceptually simpler and in fact fits naturally the position of the sports analyst who is passively watching the matches: decisions by the players are events for the sports analyst to analyze, not choices to control. For more discussion of reducing off-line Markov game analysis to other RL models please see (Luo et al., 2020).

11.3.3 Learning Reward Distributions

Distributional RL learns the distribution of the random variable $Z^k(s_t, a_t)$ that returns the sum of (discounted) rewards for future episode trajectories starting with state s_t and a_t (Bellemare et al., 2017). Therefore the Q-value is the expectation of the Z variable: $Q_k(s_t, a_t) = E(Z_k(s_t, a_t))$. Similar to the Q-value, the distribution of the total rewards Z follows the **distributional Bellman equation**:

$$Z_k(s_t, a_t) :\triangleq R_k(s_t, a_t) + \gamma E_{s_{t+1}, a_{t+1} \sim P(s_{t+1}, |, a_{t+1}, |, s_t, |, a_t)} \left[Z_k(s_{t+1}, a_{t+1}) \right] \tag{11.4}$$

where $X :\overset{\Delta}{=} Y$ indicates that random variables X and Y follow the same distribution. Given a computationally tractable representation of the Z_k distributions, we can iteratively apply the Bellman update Eq. (11.4) to update the distribution for different look-ahead lengths $H = 1$, $H = 2$, …, until we arrive at a **convergent distribution** Z_k.

Computational Representation To make the distributional Bellman equation operational, the question is how to choose a computationally tractable representation that supports learning. One option is to choose a parametric family, such as a Gaussian distribution. The issue with such parametric families is that they are typically unimodal. Unimodal distributions are not appropriate for the complex dynamics of sports, where events typically have a high branching factor, with different branches corresponding to different modes. For example, if a player attempts a pass, three different possible outcomes are that the pass is (1) intercepted, (2) reaches the intended recipient, or (3) turns into a loose puck. Each of these outcomes has alternative subsequent events, etc. Different possible event sequences determine different scoring probabilities, leading to a highly multi-modal distribution.

Bellemare et al. (2017) proposed modelling reward distributions using **quantile regression**; Liu et al. (2022) applied quantile regression to ice hockey and soccer data.

The quantile-regression (QR)-DQN method represents the conditional distribution of Z by a uniform mixture of N supporting quantiles as $\widehat{Z_k}(s_t, a_t) = \frac{1}{N}\sum_{i=1}^{N} \delta_{\theta_{k,i}(s_t, a_t)}$, where $\theta_{k,i}$ estimates the quantile at the quantile level (or quantile index) $\tau_i = i/N$ for $1 \leq i \leq N$ and $\delta_{\theta_{k,i}}$ denotes a Dirac distribution at $\theta_{k,i}$. For example, suppose we take $N = 4$ and estimate quantiles at 25%, 50%, 75%, 100% as 0.1, 0.4, 0.7, 0.9. Then the cumulative density function (cdf) of $\widehat{Z_k}(s_t, a_t)$ has 25% of values at 0.1 or less, 50% (the median) of values at 0.4 or less, 75% of values at 0.7 or less, and 100% of values at 0.9 or less (so none above 0.9). Within each pair of quantiles, the cdf is approximated as uniform over the quantiles (e.g., the cdf has value 0.1 between 0 and 25%).[4]

Given a fixed number N of target quantiles ($N = 4$ in our example), we can train a neural network to take as input a state-action pair (s, a), and output 4 numbers corresponding to the quantiles. By increasing the number N, this procedure provides a non-parametric approximation to the cumulative density function of $\hat{Z}$ and therefore to the distribution of $\hat{Z}$. For further details please see (Liu et al., 2022).

Choice of Outcome Variable: Expected Goals Vs. Actual Goals We obtained good empirical results with actual goals as rewards. As a further refinement, we follow (Decroos et al., 2017) and decompose goal scoring probabilities into the probability of managing a shot and the probability that a goal leads to a shot:

$$P\left(\text{goal}_{t,k} | s_{t_0}\right) = P\left(\text{goal}_{t,k} | s_t, \text{shot}\right) \times P\left(s_t, \text{shot} | s_{t_0}\right)$$

which can be read as saying that probability of scoring a goal from an initial state s_{t_0} is the probability of managing a shot times the probability of the shot leading to a goal. This equation is true in hockey because the only way to score a goal is to first take a shot. In our application of distributional RL, we take the goal scoring probabilities $P(\text{goal}_{t,k} | s_t, \text{shot})$ as the outcomes (virtual rewards) whose distribution is to be modelled.

The motivation for using the shot-goal decomposition is as follows:

1. A team can largely control whether they achieve a shot, whereas the success of the shot depends on factors such as the skill of the opposing goalie that are less under the control. So a model of team/player strength should reward teams for managing shots.
2. Shots are sparse but not as sparse as goal.
3. With actual goals as rewards, our outcome variable Z_k is binary and the outcome distribution is basically a Bernoulli distribution. With expected goals as rewards, the outcome variable Z_k ranges over the interval [0,1], and the outcome distribution is an informative distribution over goal scoring probabilities.

[4]We have slightly simplified notation compared to Bellemare et al. (2017) where a quantile is associated with the midpoint of a bins, rather than the higher endpoint if the bin.

So far, we have described how to build a machine learning model that estimates the distribution of possession outcomes for a given team, given a specific time and context in a match. We now show how to apply the model to gain analytical insights for a sport. Specifically, we discuss how to evaluate risk-taking by teams, the riskiness of actions, and ranking players by how risky their actions are.

11.4 Risk Measures for an Outcome Distribution

Luo et al. (2023) consider in depth the properties of different measures of risk for a distribution of outcomes. In this study we employ three of these measures: standard deviation, Gini deviation, and Value-at-Risk (VaR). Variance, the square of standard deviation, and VaR are commonly used risk measures in portfolio analysis. Gini deviation is recommended by Luo et al. (2023) for multi-modal distributions. The formal definitions are as follows.

Definition 1 (Risk Measures) *For a random variable* Z, *let* Z_1 *and* Z_2 *be two i.i.d. copies of* Z, *that is,* Z_1 *and* Z_2 *are independent and follow the same distribution as* Z.

- *The variance is defined as* $\mathbb{V}[Z] = \frac{1}{2}\mathbb{E}\left[(Z_1 - Z_2)^2\right]$
- *The standard deviation is the square root of the variance* STD $[Z] = \sqrt{\mathbb{V}[Z]}$
- *The Gini deviation is defined as* $\mathbb{D}[Z] = \frac{1}{2}\mathbb{E}\left[|Z_1 - Z_2|\right]$.

Thus the Gini deviation replaces the L2 norm of variance by the L1 norm. The definition **value-at-risk** (VaR) depends on the choice of a **confidence level** $c \in (0, 1]$. The VaR for level c is defined as the $(1 - c)^{th}$ quantile in the distribution. Thus in the hypothetical example from above, the value at risk for $c = 25\%$ is 0.7, and for $c = 50\%$ it is 0.4. Intuitively, VaR provides a kind of worst-case analysis with respect to a user-controlled risk-level. For example, choosing $c = 0.8$ corresponds to *risk-aversion* since it focuses on bad outcomes. In contrast, choosing $c = 0.2$ corresponds to *risk-seeking* with better sensitivity to positive outcomes. Figure 11.4 illustrates how different risk concepts apply to different outcome distributions.

Computing Risk Measures from Quantile Regression VaR is defined in terms of quantiles and thus can be naturally computed from a quantile regression model. Given a quantile representation, we can estimate the variance and Gini deviation as follows (Luo et al., 2023).

$$\mathbb{V}\left[\hat{Z}_k\right] \approx \frac{1}{2N^2}\sum_{i=1}^{N}\sum_{j=1}^{N}\left(\theta_{k,i} - \theta_{k,j}\right)^2 \tag{11.5}$$

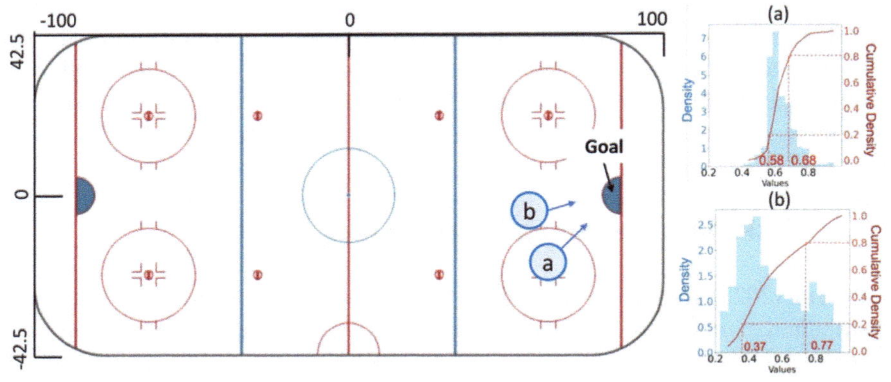

Fig. 11.4 The predicted distribution of future goals in an ice hockey game between Blues and Coyotes, 2018–2019 NHL season. The shots are made in the positions (**a**) and (**b**). Next-goal scoring distributions (**a**) and (**b**) have *the same expectation* (around 0.6), but the first shot has a much lower variance and Gini deviation of outcomes. Shot (**a**) also displays a larger risk-averse estimate (at the confidence 0.8, we find a larger next-goal chance with 0.58 > 0.37) and a smaller risk-seeking estimate (at the confidence 0.2, we find a smaller next-goal chance of 0.68 < 0.77)

$$\mathbb{D}\left[\hat{Z}_k\right] \approx \frac{1}{2N^2}\sum_{i=1}^{N}\sum_{j=1}^{N}\left|\theta_{k,i} - \theta_{k,j}\right| \tag{11.6}$$

Figure 11.5 illustrates the Q-values and standard deviations that the trained model assigns during a game between the Flyers and Maple Leafs.

11.5 Measuring the Riskiness of an Action

Using the techniques described in the previous section, we can compute from an estimated outcome distribution $\hat{Z}_k(s_t, a_t)$ a risk measure $\rho_k(s_t, a_t)$, where ρ is one of the risk measures described above (standard/Gini deviation, VaR(c)). A simple approach would be to measure the riskiness of an action in a match state simply by the risk measure $\rho_k(s_t, a_t)$. The problem with this approach is that it measures the general match context of the action, rather than the specific *impact* of the action. For example if a player makes a pass when his team has an empty net, the risk measure will be high regardless of how risky his pass is. Intuitively, the pass is taking place in a risky place at a risky time, but may itself not contribute to a team's risk.

The same issue arises with respect to expected action values (i.e., Q-values): A team playing against an empty net has a high chance of scoring the next goal, but a particular action by a player may not be increasing his team's scoring chances beyond playing an empty net. Routley (2015) proposed to address this issue by computing the **action impact**, which is measured by how much an action *changes* the scoring chances of the team in possession. The action **goal impact** is defined as follows for an action a_{t+1} and state s_{t+1} comprising the $t + 1$-th event:

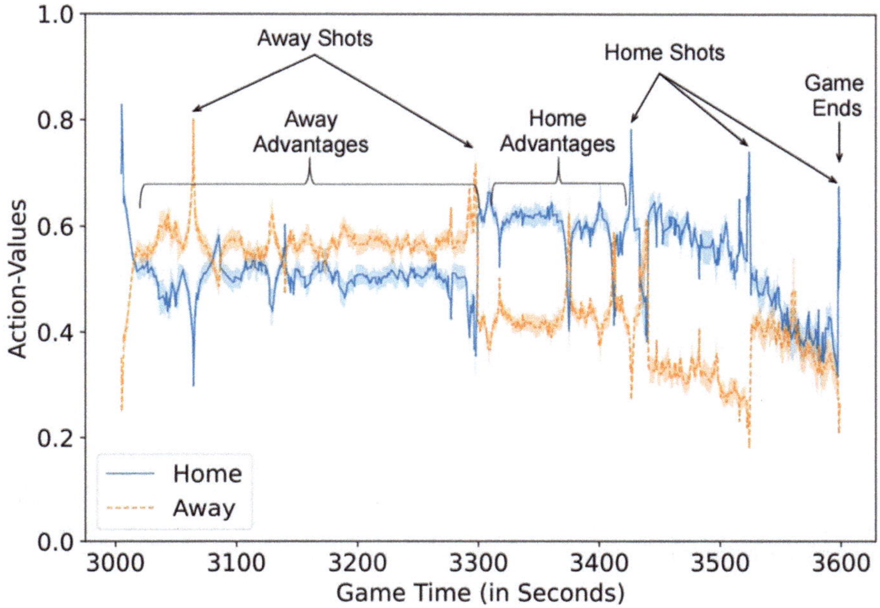

Fig. 11.5 Illustrating the dynamic distribution of next-goal scoring chances by showing the corresponding mean ± standard deviation of the action values at each time step in a match between the Flyers (Home team) and the Maple Leafs (Away team) on March 15, 2019

$$impact_k(s_{t+1}, a_{t+1}) = \left[Q_k(s_{t+1}, a_{t+1}) - Q_k(s_t, a_t)\right]\mathbb{I}_{p(s_{t+1}, a_{t+1}) \geq \epsilon} \tag{11.7}$$

Here p is a density estimator for state-action pairs, and $\mathbb{I}_{p(\cdot)\geq\epsilon}$ is an indicator value that returns 0 if the probability of the event falls below a threshold ϵ. The idea is to filter out rare events because the Q-value estimates for rare events are often biased due to small sample sizes. Using a discrete Markov game model, Routley (2015) removed all state-action pairs from consideration that occur less than 10 times in the data. For our continuous state space, we follow Liu et al. (2022) and eliminate anomalous events with $p < 20\%$. Anomaly filtering is not an essential component of our risk analysis framework.

We can adapt the goal impact approach by measuring how much an action changes the risk of an action, which we call the **risk impact**:

$$Rimpact_k(s_{t+1}, a_{t+1}) = \left[\rho_k(s_{t+1}, a_{t+1}) - \rho_k(s_t, a_t)\right]\mathbb{I}_{p(s_{t+1}, a_{t+1}) \geq \epsilon} \tag{11.8}$$

where ρ is one of our risk measures (standard/Gini deviation, value at risk).

Figures 11.6 and 11.7 show the box plots for the standard/Gini deviation impacts for different actions. Game-changing events such as shots tend to have a high impact on the variability of team outcomes. Defensive faceoffs tend to decrease a team's risk, likely because they follow a successful defense. It is interesting to note that exerting pressure tends to increase a team's risk. Box plots for value-at-risk are in the appendix.

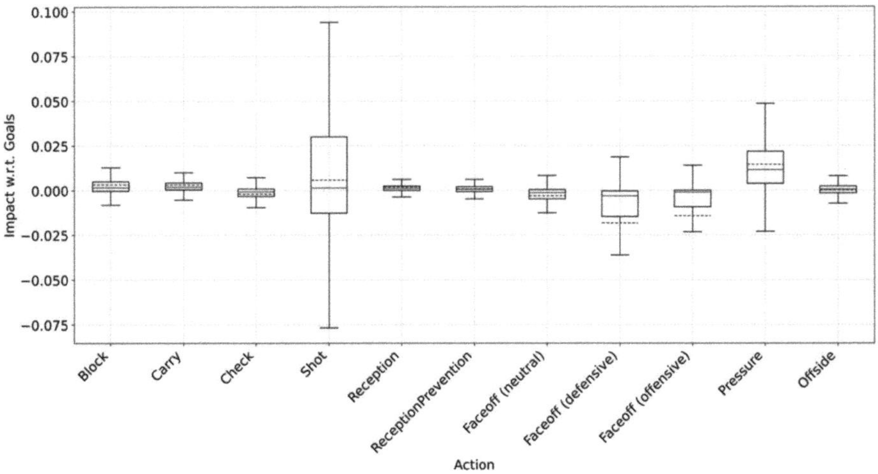

Fig. 11.6 Box Plot for the risk impact of an action on the standard deviation of next-goal scoring chances. The red line indicates the median value and the blue line indicates the mean value. Each point represents the impact of the given action in a state (outliers removed)

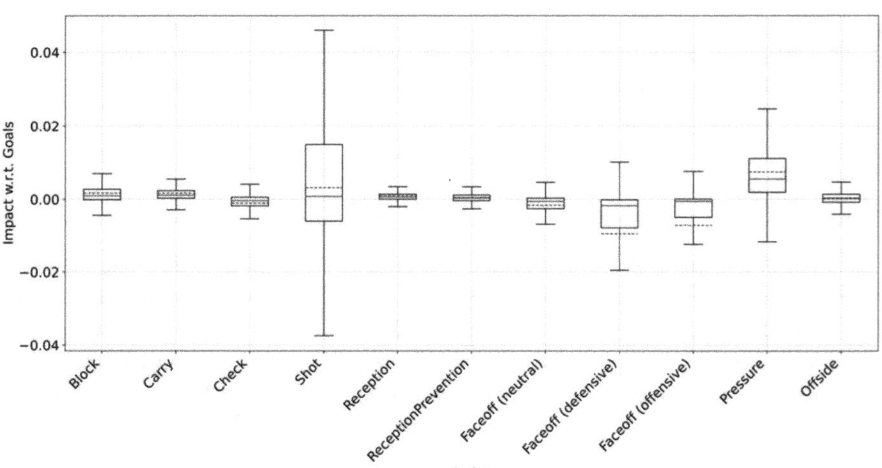

Fig. 11.7 Box Plot for the risk impact of an action on the Gini deviation of next-goal scoring chances. The red line indicates the median value and the blue line indicates the mean value. Each point represents the impact of the given action in a state (outliers removed)

11.6 Team Performance and Team Risk-Tasking

We apply our outcome distribution model to teams, by evaluating how much risk a team takes on aggregate. We start with teams because we can use the total team performance over a season as a ground-truth metric of team success. *The main question in this section is whether risk-taking by a team correlates with team success.* To quantify risk-taking by teams, we add up the risk impact of the team's total actions.

For a dataset $\mathcal{D}$, let g, t be a generic instance for event number t in game number g. Also let $team_{gt}$ be the team in possession at event t in game g, and similarly for the state s_{gt} and the action a_{gt}. Then the total team risk impact for team T in the dataset is given by

$$RIM_k = \sum_{g,t:team_{gt}=T} Rimpact_{k_t}\left(s_{gt}, a_{gt}\right) \tag{11.9}$$

where k_t denotes the appropriate agent at time t (Home or Away). Equation (11.9) shows how to define a team risk impact metric for each risk measure, which we abbreviate as follows: StdRIM = risk impact for standard deviation, GdRIM = risk impact for Gini deviation, RIM(0.2) = risk impact for VaR with risk-seeking confidence level 0.2, RIM(0.8) = risk impact for VaR with risk-averse confidence level 0.8. RIM(c) is denoted as RiGIM(c) by Liu et al. (2022) using the same notation.

A team's season league standing is determined by the number of points the team earns in each match. We therefore measure the correlation between a team's season risk impact and season total points as a measure of team performance. As Figs. 11.8 and 11.9 show, both the standard and Gini deviations provide measures of risk-taking that are excellent predictors of team performance. This shows that *stronger teams take more risks.* Table 11.4 provides the Pearson correlations between team total points and team risk metrics. Value-at-risk metrics are substantially less informative about team performance, especially at the high confidence level of 0.8. Table 11.5 shows the top 10 teams in the league and their risk metrics.

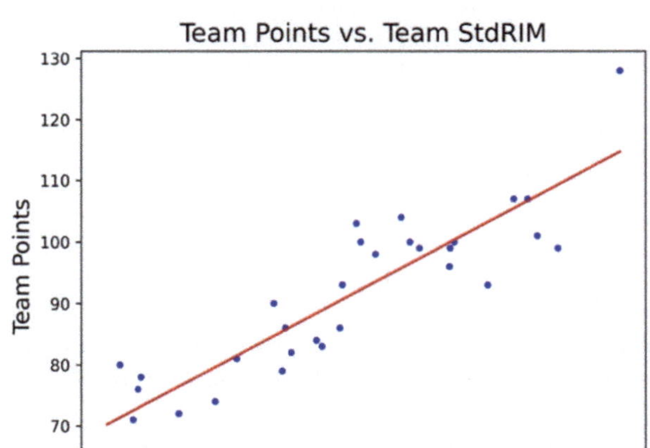

Fig. 11.8 Team points vs. Team StdRIM

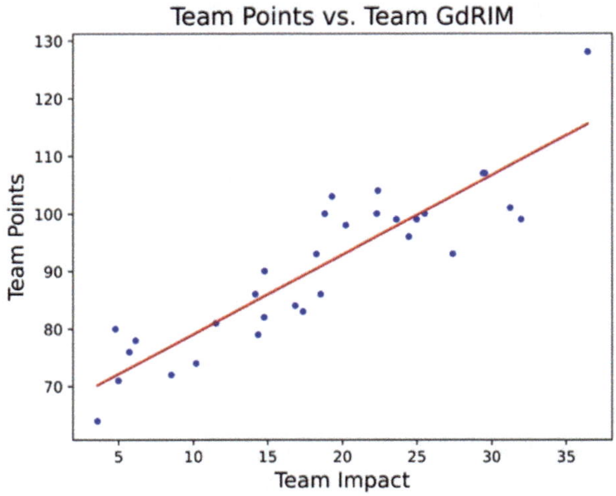

Fig. 11.9 Team points vs. Team GdRIM

Table 11.4 Correlations between a team's risk-impact metric and their season totals

StdRIM	GdRIM	RIM(0.2)	RIM(0.8)
0.90	0.90	0.51	−0.24

The standard/Gini deviations show a very high predictive ability for team season performance, which shows that stronger teams take more risks

Table 11.5 Top 10 teams with StdRIM and GdRIM based on the entire season

Team name	Predicted rank	Real rank	Total points	StdRIM	GdRIM
Lightning	1	1	128	64.43	36.42
Blues	2	12	99	57.01	31.98
Sharks	3	6	101	54.56	31.25
Flames	4	2	107	53.43	29.53
Bruins	5	3	107	51.76	29.41
Golden Knights	6	16	93	48.66	27.39
Maple Leafs	7	7	100	44.67	25.52
Hurricanes	8	11	99	44.17	24.98
Canadiens	9	14	96	44.09	24.45
Jets	10	10	99	40.47	23.62

The real rank is based on Total Points. The predicted rank is based on the risk impact metrics (both agree on the ranking), which measures a team's total risk taking over the season

11.7 Ranking Hockey Players by Risk-Taking

As with teams, we use the total risk impact of a player's actions to evaluate their risk-taking. Let pl_{gt} be the player in possession at event t in game g . Then the **total player risk impact** for player l in the dataset is given by

$$RIM_l = \sum_{g,t:pl_{gt}=l} Rimpact_{k_t}\left(s_{gt}, a_{gt}\right) \tag{11.10}$$

where we use the same notation as in Sect. 11.6. Tables 11.6 and 11.7 show the top 20 players according to their risk ranking. Applying the "eye test", risk-seeking VaR(0.2) identifies many stars, such as Connor McDavid, Leon Draisaitl, Sidney Crosby. The standard deviation metric also identifies several stars, such as Alex Ovechkin and Johnny Gaudreau. But it also highlights several less-heralded players,

Table 11.6 Top 20 players according to the risk metric value-at-risk with a risk-seeking confidence level 0.2 (i.e., RIM(0.2)) based on the entire season

Player name	Position	Team	P	A	G	RIM(0.2)
Nikita Kucherov	RW	TBL	128	87	41	61.12
Mitchell Marner	RW	TOR	94	68	26	60.81
Johnny Gaudreau	LW	CGY	99	63	36	59.71
Patrick Kane	RW	CHI	110	66	44	56.55
Brad Marchand	LW	BOS	100	64	36	53.34
Mark Stone	RW	VGK	73	40	33	51.29
Connor McDavid	C	EDM	116	75	41	51.00
Leon Draisaitl	C	EDM	105	55	50	50.16
Timo Meier	RW	SJS	66	36	30	49.67
Blake Wheeler	RW	WPG	91	71	20	48.96
Sidney Crosby	C	PIT	100	65	35	48.56
Jonathan Huberdeau	LW	FLA	92	62	30	48.19
Kyle Connor	LW	WPG	66	32	34	47.85
Artemi Panarin	LW	CBJ	87	59	28	47.34
Evgenii Dadonov	RW	FLA	70	42	28	45.83
Cam Atkinson	RW	CBJ	69	28	41	45.70
Matthew Tkachuk	LW	CGY	77	43	34	45.32
Brendan Gallagher	RW	MTL	52	19	33	44.95
Jake Guentzel	LW	PIT	76	36	40	44.58
Brandon Saad	LW	CHI	47	24	23	43.69

Table 11.7 Top 20 players with StdRIM based on the entire season

Player name	Position	Team	P	A	G	StdRIM
Johnny Gaudreau	LW	CGY	99	63	36	13.07
Patrick Kane	RW	CHI	110	66	44	12.04
Nikita Kucherov	RW	TBL	128	87	41	11.68
Alex Ovechkin	LW	WSH	89	38	51	11.30
Mitchell Marner	RW	TOR	94	68	26	11.14
Cam Atkinson	RW	CBJ	69	28	41	11.05
Timo Meier	RW	SJS	66	36	30	10.77
Vladimir Tarasenko	RW	STL	68	35	33	9.31
Matthew Tkachuk	LW	CGY	77	43	34	9.30
Brad Marchand	LW	BOS	100	64	36	9.12
Jaden Schwartz	LW	STL	36	25	11	9.10
Brendan Gallagher	RW	MTL	52	19	33	9.04
David Pastrnak	RW	BOS	81	43	38	9.01
Kyle Connor	LW	WPG	66	32	34	8.95
Filip Forsberg	LW	NSH	50	22	28	8.72
Josh Anderson	RW	CBJ	47	20	27	8.30
Mikko Rantanen	RW	COL	87	56	31	8.28
Evgenii Dadonov	RW	FLA	70	42	28	7.91
Jason Zucker	LW	MIN	42	21	21	7.85
Jonathan Huberdeau	LW	FLA	92	62	30	7.73

such as Jason Zucker and Jaden Schwartz. Another difference is that the standard deviation metric shows a bias towards forwards, whereas the Var(0.2) risk impact metric includes some centres. The top 20 tables for the other risk metrics are in the appendix.

A difficulty in evaluating player rankings is that unlike with teams, there are few suitable ground-truth metrics for performance (Franks et al., 2016). We follow previous work (Liu & Schulte, 2018; Decroos et al., 2019) and consider the correlations between our risk metrics and other meaningful player statistics such as goals, assists, and points (= goals + assists). Figure 11.10 plots round-by-round correlations between these metrics and the deviation risk impact metrics. For each round in the season, for each player, we compute their total risk impact so far (e.g., standard deviation impact over all games up to round 30), and correlate it with their statistics (e.g., total goals scored up to round 30). We observe a substantive correlation for goal-based statistics, reaching 0.51 for StdRIM and 0.56 for GdRIM at season's end. Note that the correlation is already relatively high after about 30 rounds, less than half-way through the season. This means that *the risk-impact metrics have high predictive power for future player performance.* The auto-correlation plot in the bottom right directly confirms the temporal consistency of the risk metrics. This plot correlates the value of the risk metric after n rounds with the final season value.

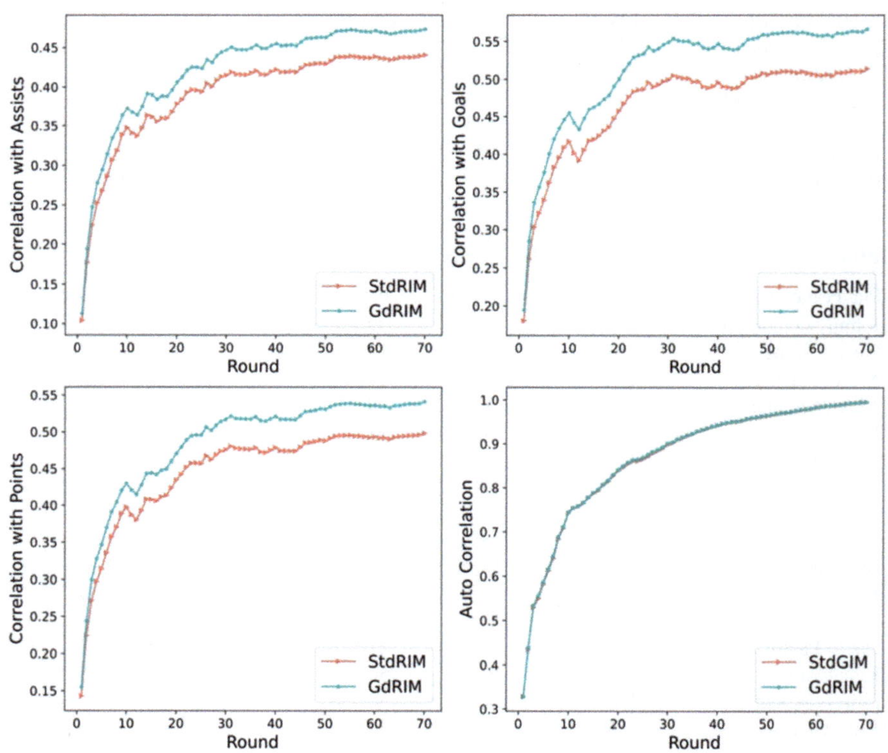

Fig. 11.10 Round by round correlations between different player metrics and risk metrics. The figure plots correlations for the standard deviation and Gini deviation risk metrics (StdRIM and GdRIM)

We see that already after 25 rounds, a player's risk impact observed so far predicts their final risk impact with correlation above 0.8. The strong temporal auto-correlation is evidence that risk-taking measures capture a stable player's characteristics (Pettigrew, 2015).

Figure 11.11 shows the correlations between value-at-risk metrics and goal-related statistics. The correlations are even higher than with the deviation metrics (0.86 vs. 0.56 with goals). As the auto-correlation figure shows, a player's risk-taking as measured by value-at-risk is stable throughout a season. Our observations for team and player rankings therefore point in different directions: for teams, we have strong evidence that standard/Gini deviation measures risk-taking that indicates team strength, whereas for players, value-at-risk with a low confidence level seems to correlate better. Correlations with goal-based statistics are only a superficial signal of player strength, as goals occur rarely and cover only a small part of relevant actions. Further research into risk-taking by players seems warranted.

A possible explanation for why the deviation-based metrics correlate less with goals, and do worse by the eye test, is to distinguish two ways in which a player's actions can increase outcome variability: (1) *Deliberate Risks*, and (2) *Unforced*

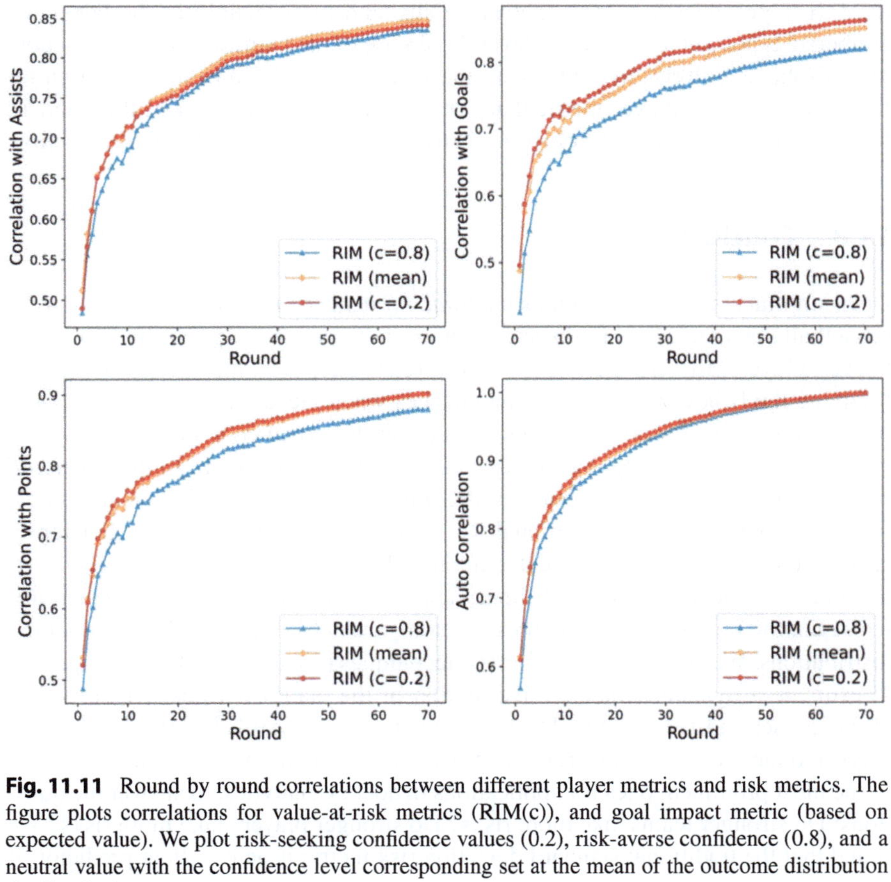

Fig. 11.11 Round by round correlations between different player metrics and risk metrics. The figure plots correlations for value-at-risk metrics (RIM(c)), and goal impact metric (based on expected value). We plot risk-seeking confidence values (0.2), risk-averse confidence (0.8), and a neutral value with the confidence level corresponding set at the mean of the outcome distribution (its Q-value)

Errors. Strong players take controlled risks, based on confidence in their skills. For example, in ice hockey, a strong player will often carry the puck into their offensive zone, drawing defenders to themselves, which increases the risk of losing the puck but also increases the chance of a successful attack. In contrast, dumping the puck behind the defending team's goal is a safer move, but tends to lead to fewer goals (Schulte et al., 2017a). An analogue in soccer would be dribbling the ball towards the defenders' goal rather than passing it to a teammate. An example of an unforced error would be losing the ball during a promising attack, which causes the range of likely outcomes to spread from a concentration on a successful attack to both teams being likely to score. For a simple statistical model of this distinction, consider a Bernoulli model for a binary event with probability p. For example, an expected goals model might assign a probability p to a shot succeeding at time t. The variance of success is given by $p(1-p)$ and is maximal at $p = 50\%$. A deliberate risk may increase the scoring chance from $p < 50\%$ towards $p' > p$ where $p' < 50\%$; for example, a deliberate risk may increase the scoring chance from 30% to 40%. *Such a move increases both scoring chance and variance.* On the other hand, an error may move the scoring chance from p towards $p' < p$ with $p' < 50\%$, for example from 70% to 60%. Such a move *decreases the scoring chance and increases variance*. It is possible that strong players' risks tend to be mainly of the beneficial deliberate type, whereas weaker players' risks are of the harmful error type, so both may display a high risk measure. This analysis suggests that a fruitful direction for future research is a player metric that combines risk and reward, e.g., the standard deviation and the mean of the outcome distribution associated with a player's action.

11.8 Conclusion

Decision-makers in sports often face a *trade-off between risk and reward.* Should a basketball player take a long-distance 3-point shot or a safer 2-point shot? Should a hockey player carry the puck, pass it to a teammate, or dump it behind the net? Studying the behavior of athletes and coaches when faced with risk-reward trade-offs requires tools for *risk analytics*. This paper described computational tools for risk analytics in sports by leveraging concepts and techniques from *distributional reinforcement learning*. Distributional RL aims to model the distribution of possible future outcomes, whereas traditional RL estimates the expected value. For representing the complex multi-modal outcome distributions that stem from sports dynamics, we adapted a state-of-the-art approach to distributional RL, which utilizes quantile regression as an expressive non-parametric framework for modelling distributions. We applied distribution RL techniques based on the Bellman equation to estimate a dynamic outcome distribution in the National Hockey League, for 1000+ games and 1M+ events. The literature on risk analysis has proposed different ways to quantify the risk inherent in an outcome distribution. We evaluated several of the most prominent ones for applications in hockey analytics, to answer the questions: Do stronger teams take more risks? Do stronger players take more risks? We found that the traditional standard deviation risk metric is an excellent predictor of team success: Both the standard deviation and Gini deviation of a team's outcome distribution, aggregated over a season, show a 0.90 correlation with the team's league standing at the end of a season (determined by their total points). Value-at-risk with confidence level 0.2 shows a lesser but still strong correlation of 0.51.

For player ranking, we found that value-at-risk with risk-seeking confidence level 0.2, aggregated over all actions by a player in a season, shows a very high correlation of 0.86 with the player's total season goals. Standard deviation and Gini deviation, in contrast, show slightly lower but still strong correlations of 0.51 and 0.56, respectively. We suggest that modelling the risk-taking behavior of players may require a more fine-grained metric that distinguishes between deliberate risks, which are incurred by actions requiring high skill, and risks stemming from errors (e.g., losing possession of the puck). A promising source of such fine-grained metrics are measures from portfolio theory that combine risk and reward (expected outcome). For example, we might investigate in sport analytics a version of the famous Sharpe ratio that divides expected value by standard deviation.

In sum, risk analytics is a promising new approach to sports analytics that focuses on the difficult trade-offs between taking risks and maximizing the chance of success that decision-makers in sports face. Distributional reinforcement learning provides the computational tools for estimating both risks and expected rewards in large sports data sets. Our NHL study shows that strong teams take big risks, and confirms to a lesser degree, that strong players are risk-takers as well.

Acknowledgements This research was supported by discovery grants from Canada's Natural Sciences and Engineering Research Council, awarded to Schulte and Poupart. Yudong Luo is also supported by a David R. Cheriton Graduate Scholarship, and a Waterloo AI Institute Graduate Scholarship. Resources used in this work were provided by the Province of Ontario and the Government of Canada through CIFAR and companies sponsoring the Vector Institute (https://vectorinstitute.ai/partners/). The dataset analyzed in this paper was generously provided by Sportlogiq.

Appendix

Learning Value Functions and Value Distributions

Figure 11.12 illustrates our recurrent neural network architecture for learning Q-values and distributional quantiles. Given a demonstration dataset of observed trajectories (i.e., $\mathcal{D} = \{\tau\}$), a Q-function satisfying the Bellman equation can be learned by minimizing the 2-Norm of the Temporal difference (TD) error, which is defined by:

$$\mathcal{L}(\theta) = \mathbb{E}_{\tau\sim\mathcal{D}}\left(\hat{Q}_k\left(s_{t+1}, a_{t+1}\right) + r_{k,t} - \hat{Q}_k\left(s_t, a_t\right)\right)^2 \tag{11.11}$$

where the expectation represents the average over all transitions $(s_t, a_t; s_{t+1}, a_{t+1})$ observed in the dataset. The term $\widehat{Q}_k\left(s_t, a_t\right)$ represents the current value estimate,

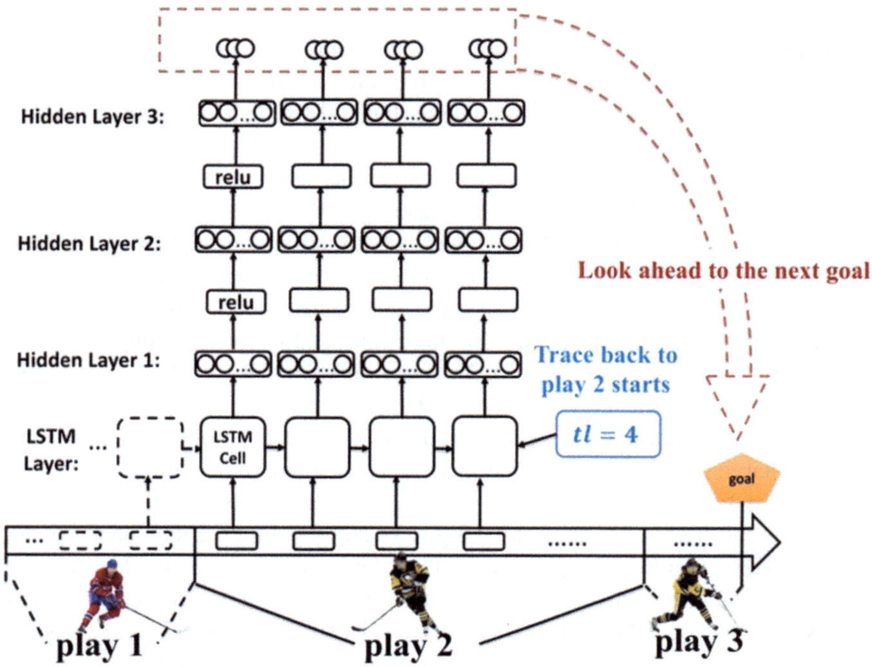

Fig. 11.12 Our recurrent neural network architecture for learning Q-values and a distribution of action outcomes. At each time step, the RNN receives as input a pair s_t, a_t where the state s_t is a vector of features shown in Table 11.1. It outputs an estimate of the Q-value (chance of scoring the next goal) or a set of quantiles representing the distribution of action outcomes (goals scored). We utilize an LSTM architecture. For more details please see Liu et al. (2022)

and $\widehat{Q_k}\left(s_{t+1}, a_{t+1}\right) + r_{k,t}$ a look-ahead step. Minimizing their squared difference drives the neural network to satisfy the Bellman equation (Eq. 11.3). For more details on implementing the temporal Bellman equation with quantile regression, please see Liu et al. (2022).

Action Impact on Value-at-Risk

Figure 11.13 shows the action impacts for Var(0.2) with confidence level 0.2 Game-changing events have a high impact on risk, as we observed in Sect. 11.5. In general the changes in risk are less than with standard/Gini deviations as this metric focuses on low probability outcomes.

Figure 11.14 shows the impact of actions on risk as measured by Var(0.8) with confidence level 0.8. At this confidence level, risk impact is similar to goal impact, that is, tends to measure increase in goal scoring chances rather than the increase in the variability of goal-scoring chances.

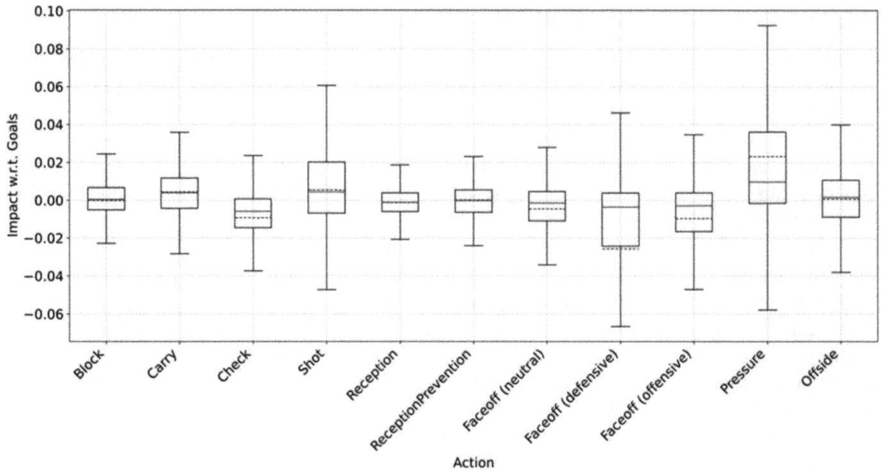

Fig. 11.13 Box Plot with confidence 0.2 (risk-seeking) The red line indicates the median value and the blue line indicates the mean value. Each point represents the risk impact of the given action in a state

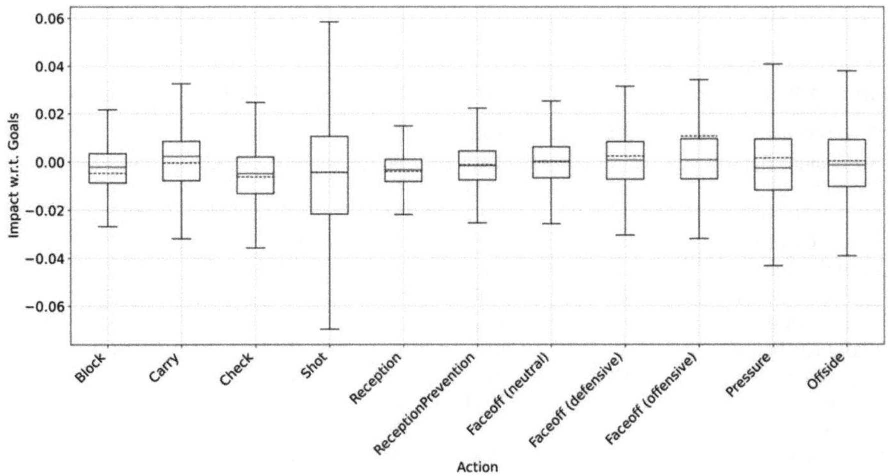

Fig. 11.14 Box Plot with for RIM with confidence 0.8 (risk-averse) The red line indicates the median value and the blue line indicates the mean value. Each point represents the risk impact of the given action in a state

Top 20 Player Tables for Other Risk Metrics

Table 11.8 shows the top 20 players for the Gini impact risk metric, which is very similar to the top 20 for the standard deviation risk metric.

As shown in Table 11.9, for the risk-averse confidence level 0.8, value-at-risk identifies many stars, such as Scheifele and Crosby. This metric shows a bias towards centres.

Table 11.8 Top 20 players with GdRIM based on the entire season

Player name	Position	Team	P	A	G	GdRIM
Johnny Gaudreau	LW	CGY	99	63	36	7.15
Patrick Kane	RW	CHI	110	66	44	6.61
Nikita Kucherov	RW	TBL	128	87	41	6.53
Cam Atkinson	RW	CBJ	69	28	41	6.06
Alex Ovechkin	LW	WSH	89	38	51	6.00
Mitchell Marner	RW	TOR	94	68	26	5.98
Timo Meier	RW	SJS	66	36	30	5.97
Matthew Tkachuk	LW	CGY	77	43	34	5.18
Brad Marchand	LW	BOS	100	64	36	5.06
Kyle Connor	LW	WPG	66	32	34	5.03
Vladimir Tarasenko	RW	STL	68	35	33	4.96
Brendan Gallagher	RW	MTL	52	19	33	4.91
David Pastrnak	RW	BOS	81	43	38	4.87
Jaden Schwartz	LW	STL	36	25	11	4.87
Filip Forsberg	LW	NSH	50	22	28	4.64
Mikko Rantanen	RW	COL	87	56	31	4.62
Josh Anderson	RW	CBJ	47	20	27	4.52
Evgenii Dadonov	RW	FLA	70	42	28	4.44
Jason Zucker	LW	MIN	42	21	21	4.27
Jonathan Huberdeau	LW	FLA	92	62	30	4.22

Table 11.9 Top 20 players with RIM at risk-averse confidence 0.8 based on the entire season

Player name	Position	Team	P	A	G	RIM(0.8)
Aleksander Barkov	C	FLA	96	61	35	50.50
Leon Draisaitl	C	EDM	105	55	50	49.67
Mark Scheifele	C	WPG	84	46	38	48.73
Sidney Crosby	C	PIT	100	65	35	47.24
Jonathan Toews	C	CHI	81	46	35	45.22
Mitchell Marner	RW	TOR	94	68	26	42.71
Dylan Larkin	C	DET	73	41	32	41.82
Nikita Kucherov	RW	TBL	128	87	41	41.00
Max Domi	LW	MTL	72	44	28	40.47
Connor McDavid	C	EDM	116	75	41	40.45
Bo Horvat	C	VAN	61	34	27	39.92
Mika Zibanejad	C	NYR	74	44	30	39.29
Artemi Panarin	LW	CBJ	87	59	28	38.93
Sebastian Aho	C	CAR	83	53	30	38.55
Mark Stone	RW	VGK	73	40	33	38.38
Claude Giroux	C	PHI	85	63	22	37.98
Johnny Gaudreau	LW	CGY	99	63	36	37.86
Mathew Barzal	C	NYI	62	44	18	37.83
Nicklas Backstrom	C	WSH	74	52	22	37.78
Brad Marchand	LW	BOS	100	64	36	37.55

References

Beaudoin, D., & Swartz, T. B. (2010). Strategies for pulling the goalie in hockey. *The American Statistician, 64*(3), 197–204.

Bellemare, M. G., Dabney, W., & Munos, R. (2017, July). A distributional perspective on reinforcement learning. In *International Conference on Machine Learning* (pp. 449–458). PMLR.

Chan, T. C., Fernandes, C., & Puterman, M. L. (2021). Points gained in football: Using Markov process-based value functions to assess team performance. *Operations Research, 69*(3), 877–894.

Decroos, T., Dzyuba, V., Haaren, J. V., & Davis, J. (2017). Predicting soccer highlights from spatio-temporal match event streams. In *AAAI 2017* (pp. 1302–1308).

Decroos, T., Bransen, L., Van Haaren, J., & Davis, J. (2019, July). Actions speak louder than goals: Valuing player actions in soccer. In *Proceedings of the 25th ACM SIGKDD International Conference on Knowledge Discovery & Data Mining* (pp. 1851–1861).

Franks, A. M., D'Amour, A., Cervone, D., & Bornn, L. (2016). Meta-analytics: Tools for understanding the statistical properties of sports metrics. *Journal of Quantitative Analysis in Sports, 12*(4), 151–165.

Littman, M. L. (1994). Markov games as a framework for multi-agent reinforcement learning. In *Machine learning proceedings* (pp. 157–163). Morgan Kaufmann.

Liu, G., & Schulte, O. (2018). Deep reinforcement learning in ice hockey for context-aware player evaluation. *arXiv preprint arXiv:1805.11088.*

Liu, G., Luo, Y., Schulte, O., & Kharrat, T. (2020). Deep soccer analytics: Learning an action-value function for evaluating soccer players. *Data Mining and Knowledge Discovery, 34*, 1531–1559.

Liu, G., Luo, Y., Schulte, O., & Poupart, P. (2022). Uncertainty-aware reinforcement learning for risk-sensitive player evaluation in sports game. *Advances in Neural Information Processing Systems, 35*, 20218–20231.

Luo, Y., Schulte, O., & Poupart, P. (2020, July). Inverse reinforcement learning for team sports: Valuing actions and players. In *IJCAI* (pp. 3356–3363).

Luo, Y., Liu, G., Poupart, P., & Pan, Y. (2023). An alternative to variance: Gini deviation for risk-averse policy gradient. *Advances in Neural Information Processing Systems, 36*, 60922–60946.

Mnih, V., Kavukcuoglu, K., Silver, D., Rusu, A. A., Veness, J., Bellemare, M. G., et al. (2015). Human-level control through deep reinforcement learning. *Nature, 518*(7540), 529–533.

National Hockey League. (2014). *National Hockey League official rules 2014–2015.* http://www.nhl.com/nhl/en/v3/ext/rules/2014-2015-rulebook.pdf

Pelechrinis, K. (2016). Decision making in American football: Evidence from 7 years of NFL data. In *MLSA@ PKDD/ECML.*

Pettigrew, S. (2015). Assessing the offensive productivity of NHL players using in-game win probabilities. In *9th Annual MIT Sloan Sports Analytics Conference* (Vol. 2(3), p. 8).

Routley, K. D. (2015). *A Markov game model for valuing player actions in ice hockey.*

Schulte, O. (2022, September). Valuing actions and ranking hockey players with machine learning. In *Linköping Hockey Analytics Conference* (pp. 2–9).

Schulte, O., Khademi, M., Gholami, S., Zhao, Z., Javan, M., & Desaulniers, P. (2017a). A Markov game model for valuing actions, locations, and team performance in ice hockey. *Data Mining and Knowledge Discovery, 31*, 1735–1757.

Schulte, O., Zhao, Z., Javan, M., & Desaulniers, P. (2017b, March). Apples-to-apples: Clustering and ranking NHL players using location information and scoring impact. In *Proceedings of the MIT Sloan Sports Analytics Conference.*

Sutton, R. S., & Barto, A. G. (1998). The reinforcement learning problem. In *Reinforcement learning: An introduction* (pp. 51–85).

Part III

Metrics in Individual Sports

Machine Learning for Contemporary Dance

12

Daniel Bisig

Abstract

In contemporary dance, experimentation with novel and idiosyncratic forms of creating and staging bodily movements lies at the center of artistic attention. Many of these experiments also involve the use of generative techniques. For this reason, contemporary dance provides a fertile environment for exploring creative applications of generative machine learning. This article presents the author's own work on four machine-learning-based systems, each of which addresses a specific application domain in contemporary dance. The system *Granular Dance* generates synthetic motions that support choreographic ideation. The system *Puppeteering AI* creates an artificial dancer that human dancers can interact with. The system *Expressive Aliens* provides insights into the relationship between body morphology and movement expressivity. The system *RAMFEM* demonstrates how a dancer's creative movements made while improvising to music can be adopted for movement sonification. For each of these systems, the article describes the technical implementation and first applications. This information is complemented with a description of the context that is most relevant for this research. This includes how contemporary dance distinguishes itself from other dance practices, the types of generative techniques that are popular in contemporary dance, and the main application domains of machine-learning in contemporary dance.

D. Bisig (✉)
Institute for Computer Music and Sound Technology, Zürcher Hochschule der Künste, Zürich, Switzerland
e-mail: daniel.bisig@zhdk.ch

D. Memmert (ed.), *Artificial Intelligence and Machine Learning in Sports Science*,
https://doi.org/10.1007/978-3-662-70155-3_12

12.1 Introduction

Contemporary dance is a unique art form that sets itself apart from other types of dance through its focus on experimentation, collaborations across disciplines, artistic individualism, and a constant desire to innovate. Among all forms of dance, contemporary dance shows the greatest interest in using technology and generative methods. Because of this, contemporary dance provides a fertile environment for experimenting with the creative use of machine learning. However, several aspects of contemporary dance make the use of machine learning and generative techniques challenging. These include a focus on experiential aspects of movement, reliance on tacit knowledge, resistance towards formalization and generalization, and the importance of ephemerality.

The author of this article collaborates among others as a software developer and media artist with various dance companies and helps them create new stage performances. For these performances, the author implements interactive generative systems that augment the stage with media-rich and responsive scenographies. Through the Marie Curie Fellowship entitled *E2-Create* which ran from April 2020 to October 2022, the author had the opportunity to complement these practical activities with an extension of his disciplinary background. During the fellowship, the author gained a deeper understanding of embodied creativity, the generative approaches that are most popular in dance, and the potential of machine-learning-based creative tools for dance. Also during the fellowship, the author explored how principles of embodied creativity can inform the development of generative systems and how generative systems can be combined with data-driven approaches. Among the outcomes of the fellowship were four generative systems, each designed to address one of four specific interests in contemporary dance: gaining new insights into embodied creativity, supporting creative ideation in choreography, creating artificial dancers, and translating movement into other modalities. Originally, these systems were developed for proof of concept or for application by artists who possess programming skills. Now, in the context of a new Horizon Europe project entitled *Premiere*, the practical applicability of these generative systems and the range of technical skills and artistic interests of the target users are being expanded. This article hopefully contributes to the ongoing discussion about how to extend the artistic appeal of generative systems for a larger dance community and how to develop generative systems that have a more sustainable impact on dance creation than a one-time application for a specific production.

The main focus of this article is on the implementation and application of the four generative systems that have been developed during the fellowship. The article also provides background information about contemporary dance, the role and types of generative techniques that are popular in dance, and an overview of the main applications of machine learning in dance. Finally, the article concludes with a discussion and outlook concerning additional applications of the four generative systems beyond those they were originally devised for, and the acquisition of user feedback to improve the functionality and usability of generative systems for a wider range of dance practitioners.

12.2 Background

The following section provides some context for understanding the potential and challenges of using machine learning for creative applications in contemporary dance. It begins with a brief explanation of the history and central principles of contemporary dance. It then provides an overview of the applications of generative systems in dance and some insights into how generative methods relate to the ambivalent relationship between dance and technology. Finally, it discusses how machine learning differs from more traditional generative approaches and identifies five application domains that are of particular interest to dance.

12.2.1 Contemporary Dance

Contemporary Dance emerged in the 1980s as a reaction to the rigid and restrictive nature of classical dance forms, especially Ballet (Coton et al., 1975). To distance itself from these conventions, Contemporary Dance embraces creative experimentation, improvisation, and self-expression (Sagiv et al., 2020).

Contemporary Dance originated by borrowing and merging elements from different theatrical dances such as Ballet, Classical, Modern Dance, and Jazz Dance. Later, dance practitioners became more liberal in drawing inspiration from additional sources and creating their own distinctive styles. They took inspiration not only from the realm of dance but also from other areas. These sources included non-Western dance forms such as those from India, Africa, and Japan (Hardt, 2011), community dances like folk (Fishwick, 2010) and street dances, and non-dance movements like martial arts or everyday actions.

Contemporary Dance has a distinct attitude towards the human body and its movements, which are considered an artistic art form in their own right (Kuzian, 2021). This perspective has led to the emancipation of dance from its subservient role and instrumental use for storytelling, exhibiting athleticism, or supporting social conventions. By situating bodily movements at the center of artistic expression, dancers were attributed greater creative roles and authorship when collaborating with choreographers on new performance works (Gilfillan, 2022).

The focus on body movement has led to the development of various movement vocabularies and methods for ideating movements by dancers and choreographers. Some of these vocabularies and methods have become codified as techniques. Among the most influential techniques are the Release technique (Lepkoff, 1999), Graham technique (Bragato et al., 2023), Cunningham technique (Miller & Wood, 2017), Limon Technique (Sandvik, 2023), and Gaga Technique (Galili, 2015). Several techniques employ unconventional movements that are for example rapid and contrasting (Cunningham technique), unstable or temporally suspended (Limon Technique), or alternate between contraction and release (Graham technique). Often, techniques explore and push the limits of the human body to break with habitual movement patterns (Gaga technique) or exploit the body's full movement capabilities (Release Technique). The use of and refinement of the dancers' kinesthetic awareness also plays a central role in many techniques. These techniques use guided improvisation to direct bodily attention (Gaga technique), focus on the perception of the body's center of gravity or employ systematic breathing (Limon technique), or work with mental imagery of kinesthetic experiences (Release technique).

Contemporary Dance is an art form that constantly evolves and values experimentation over establishing a stable and common stylistic repertoire. Many dance practitioners feel the need to continuously revise and reinvent their approach to dance creation. As a result, dance performances are highly individualistic and vary not only between choreographers but also between different works of the same choreographer, and sometimes even between re-stagings of the same performance. This focus on individual artistic expression and bodily expressivity makes Contemporary Dance a challenging field of application for generative and machine learning-based approaches.

12.2.2 Generative Dance

Many practitioners in Contemporary Dance employ generative techniques to experiment with new creative workflows and/or present performances in unconventional settings. In this context, the term generative refers to artistic approaches that delegate the creation of an artwork fully or partially to an automated process that is frequently but not necessarily executed by a computational device (Boden & Edmonds, 2009; Dorin et al., 2012; Galanter, 2003).

The range of applications and types of generative systems developed in the context of dance is very large. For the sake of brevity, only a few examples are presented here. A more thorough survey has been published by the author (Bisig, 2022b). Some dance works experiment with performance environments other than the stage such as artificial ecosystems (e.g. Antunes & Leymarie, 2012), virtual reality (e.g. Gromala & Sharir, 1994), or across telematically connected locations (e.g. Zannos & Dimitrakopoulou, 2023). Other dance works change the relationships between performers and audiences by creating intimate relationships (e.g. Pitcher, 2015) or by blurring their respective roles (e.g. Glynn, 2008). Some dance works manifest dance through other media than the human body to reveal hidden aspects of dance (e.g. Palazzi et al., 2009) or to elicit new forms of kinesthetic empathy (e.g. Leach & Delahunta, 2017). Other dance works employ co-creative tools to foster movement ideation by providing ambiguous and incomplete choreographic instructions (e.g. Sicchio et al., 2020) or by proposing impossible movements that force dancers to break habits (e.g. Schiphorst, 2013).

Generative approaches can also point towards less contested applications of technology in dance. An unreflected application of technology is often perceived as being in opposition to core characteristics of Contemporary Dance, such as its reliance on tacit knowledge, its interdisciplinary perspective, the importance of bodily experience, and its ephemerality.

Due to the importance of implicit and embodied knowledge, the usefulness of formalisation and abstraction is questioned (Blades, 2012). Generative Art as an artistic discipline is uniquely predisposed to highlight the artistic benefits of working with formalisations to inspire new forms of choreographic thinking (DeLahunta, 2002) or to render implicit knowledge accessible for external perspectives and thereby facilitate the collaboration across disciplinary boundaries (Shaw, 2011). Generative Art frequently adopts algorithmic methods from other scientific fields, particularly the synthetic natural or social sciences such as artificial intelligence, artificial live, and computational creativity. These methods can further bolster the interdisciplinary characteristics of Contemporary Dance. Contemporary Dance places a specific focus on the sensorial, behavioral, and physical properties and capabilities of the human body. Digital technology is often considered dismissive of the human body (Sutil, 2015). Generative Art's focus on the processual underpinnings of natural phenomena rather than their surface appearance offers the possibility to explore and exploit the causal connections between digital representations of a body's physicality and its capabilities for movement making (Leach & Delahunta, 2017). The ephemerality of dance is considered one of its central features. The use of digital technology is feared to undermine this essential characteristic (Blades, 2012). Generative systems are often non-deterministic in their behaviour and sensitive to initial conditions. These systems need to be executed to create a perceivable result and multiple executions lead to different results. Accordingly, Generative Art shares some of the ephemeral characteristics of dance.

12.2.3 Machine Learning and Dance

Generative Art has long been inspired by artificial intelligence, which offers a variety of techniques for modeling cognitive and behavioural capabilities. However, it is only with the advent of powerful deep learning methods that generative approaches have gained significant attention within the arts. Most artistic applications of deep learning have been within the visual arts and music, but Contemporary Dance is also seeing a growing community of artists experimenting with deep learning.

Traditional approaches to Generative Art focus on the design of explicit algorithms as a central element of the creative process. In contrast, machine-learning models are data-driven, shifting the focus of artistic attention towards the selection and preparation of the data used for training the models. This shift brings unique benefits and challenges to Contemporary Dance.

The benefits of using machine learning in the context of Contemporary Dance are at least twofold. Firstly, it allows for a shift in artistic focus from algorithmic implementation to data curation. This approach is similar to established dance practices that involve the mixing and juxtaposition of different movement vocabularies and techniques. Therefore, machine learning can appeal to dance practitioners who lack the skills and/or interests for developing generative algorithms on their own. Secondly, machine learning models can capture and reproduce the unique characteristics of the data they have been trained on. This is particularly prominent when the model is trained on a small amount of data and overfits. As a result, artists can adapt generative systems to the specifics of their creative language with relative ease.

Machine learning presents its own unique challenges to any artistic domain, including Contemporary Dance. Several challenges need to be addressed when using machine learning creatively. One challenge is the originality of the output generated by machine learning models. This is because many machine learning systems excel at imitating the data on which they have been trained, which limits their ability to create novel and original output. Another challenge is the capability of artists to master the use of machine learning models beyond simply training them on their own data. This requires specialized expertise to make informed decisions when modifying existing machine-learning models or designing new ones. Since many artists lack this expertise, they resort to using off-the-shelf models as black box mechanisms. A further challenge is the resources required to train and use state-of-the-art machine learning models. These models often possess a very high number of parameters and therefore require a large dataset and significant computational power to be trained. These resources are normally out of reach for artists. This challenge also brings with it a problem of idiosyncrasy. While machine learning can preserve the idiosyncratic characteristics of an artistic corpus, this only applies to models that don't require larger amounts of training data than an artist can provide from their own work. State-of-the-art models typically exceed this amount and training data can only be obtained from standardized databases or by scraping the internet. As a result, these models' output does not represent the personal style of a single artist.

When surveying applications of machine learning in contemporary dance, five main motivations can be identified. These are the use of machine learning for gaining insights into principles of embodied creativity, enhance the creativity of choreographers, provide intuitive and gestural interfaces for interaction, translate media across different modalities, and create artificial dancers that can either perform solo or alongside human dancers. Again, for the sake of brevity, only a few examples are presented here.

Machine learning can support the acquisition of insights into principles of embodied creativity, or as Marc Downie mentioned in his PhD thesis, serve as foundations for 'developing a theoretical, technical, and aesthetic framework for the innovative art form of digitally augmented human movement' (Downie, 2005). In 2000, choreographer Wayne McGregor and scholar Scott deLahunta initiated an interdisciplinary research project with the goal of studying the potential of AI to 'broaden understanding of the unique blend of physical and mental processes that constitute dance and dance making' (DeLahunta, 2009). Based on her own practical work with machine learning, Pettee and colleagues argue that machine learning can be used as tool to 'spark introspection and exploration of our embodied knowledge'. Among others, they suggest that machine learning can shift our description of movement away from culturally centered opinions and encourage 'normative discussion about what it means to choreograph' (Pettee et al., 2019).

Machine learning is also used to enhance the creativity of choreographers. For their system Chor-rnn, the authors propose a form of creativity facilitation that involves the system and the choreographer taking turns in creating movement material (Crnkovic-Friis & Crnkovic-Friis, 2016). Similarly, Pettee et al. (2019) present a suite of deep-learning-based tools whose output is meant to be used more or less directly for creating a new choreography (Pettee et al., 2019).

A popular use of machine learning in the context of interactive media performance is to design interactivity through demonstration rather than by specifying rules and algorithms (Gillies et al., 2016). The authors argue that this approach is particularly suitable for creative practices in that it emphasizes the exploratory, playful, embodied, and expressive aspects of the design process (Fiebrink & Caramiaux, 2016).

The application of machine learning for translating media content across different modalities has gained significant attention with the advent of models that convert textual prompts into images, music, or video. Of interest for dance are models that use body movements and music as translation modalities. There exist several engineering-focused research projects, that explore the application of machine learning for translating music into movement. Lee et al. (2019) developed a machine-learning-based synthesis-by-analysis system. This system learns to compose dance by discretising movements into multiple basic units and align them with the beat of music. Li et al. (2020) employ two models as part of a sequence-to-sequence architecture to learn the contexts of movement and music. This architecture is trained on a wide range of movement objectives such as beat synchronisation, physical plausibility, and diversity.

The most popular application of machine learning is for the realisation of artificial dancers. This application has been extensively studied by Berman and James (2018) and McCormick and colleagues (2014). In both cases, an artificial dancer that acts as an improvisation partner is trained on the movement vocabulary of a single human dancer. The improvisation is based on the recognition of movement patterns executed by the human dancer which the artificial dancer then employs as a starting point for navigating a topological representation of pose or motion encodings. The training of the artificial dancer takes place either before (McCormick et al., 2014) or during the performance (McCormick et al., 2014).

12.3 Systems

The following section introduces four machine-learning-based generative systems that have been developed by the author for creative applications in Contemporary Dance. Each of these systems combines a different machine learning architecture with a different application domain. The first system is entitled *Granular Dance* (Bisig, 2021). This system employs an autoencoder architecture for generating synthetic motions. Its application is mainly for choreographic ideation. The second system is entitled *Puppeteering AI* (Bisig & Wegner, 2021, 2022). This system employs the same model architecture as *Granular Dance* but can be controlled in real-time through gestural interaction. Its application is mainly for the creation of an artificial dancer that can be improvised with. The third system is entitled *Expressive*

Aliens (Bisig, 2022a). This system combines a reinforcement learning algorithm with a physics simulation to train agents that possess non-human morphologies to exhibit expressive movements. Its application is mainly for gaining insights into principles of embodied creativity. The fourth system is entitled *RAMFEM* (Bisig & Tatar, 2021). This system employs a sequence-to-sequence architecture to translate movement into music. Its application is mainly for interactive movement sonification. Each of these systems and their applications will be described in more detail.

12.3.1 Granular Dance

Granular Dance can be employed by choreographers as co-creative tool that generates synthetic movements for a single dancer. The system can be trained on motion capture recordings of solo performances and then subsequently used in an exploratory manner to discover dance movements that balance stylistic similarity and novelty. Two core ideas form the basis for *Granular Dance*: the combination of a machine-learning-based model with a more conventional generative algorithm, and a creative strategy for movement ideation that is based on navigating the model's internal representational space of dance movements.

12.3.1.1 Model

The machine-learning part of *Granular Dance* employs an adversarial autoencoder (AAE) architecture (see Fig. 12.1, left side) that consists of three models, an encoder,

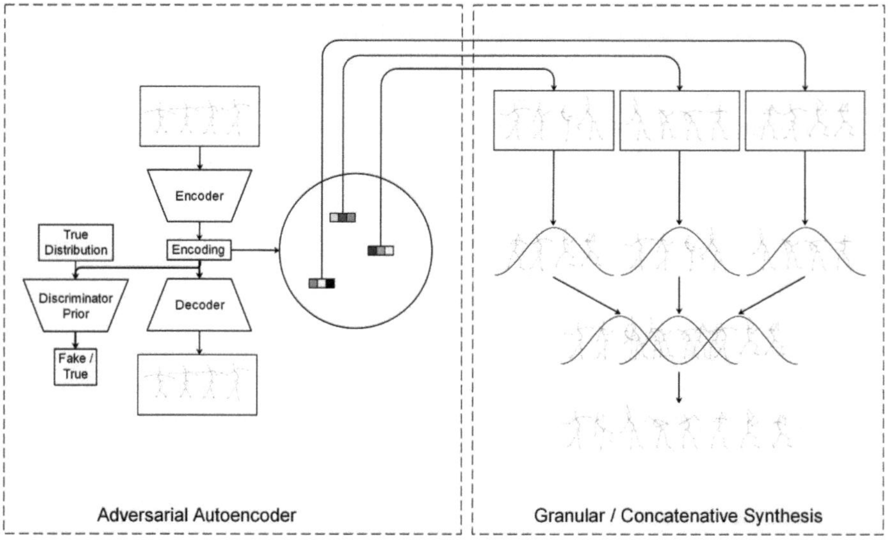

Fig. 12.1 Architecture of an adversarial autoencoder for pose sequences. The inputs and outputs of the autoencoder correspond to pose sequences. The trapezoid shapes represent the dimensionality reduction and expansion that are conducted by the encoder and decoder, respectively. After training, only the decoder is required for pose sequence generation

a decoder, and a discriminator. This architecture follows one of the designs proposed by Wang et al. (2020).

The autoencoder compresses (encodes) dance movements into lower dimensional representations (latent vectors) and then decompresses (decodes) the representations back into dance movements. After training, the latent vectors represent the statistically most salient aspects of the movements necessary for a faithful reconstruction. The adversarial part forces latent vectors to follow a specific probability distribution (in this case a Gaussian Distribution). This ensures that the distribution of latent vectors is free of gaps and that distances between them represent a measure of similarity. This in in turn ensures that arbitrarily chosen latent vectors can be decompressed into meaningful dance movements. The model used for the applications described here employs movement sequences of 128 poses and a latent encoding of 64 dimensions.

The autoencoder operates on sequences of dance poses in which each pose is represented by joint orientations in the form of unit quaternions. Both the encoder and decoder parts of the autoencoder are autoregressive and employ a combination of a Long Short-Term Memory (Hochreiter & Schmidhuber, 1997) and fully connected artificial neural network. The encoder part of the autoencoder takes as input a sequence of poses and produces as output a latent vector. The decoder part of the autoencoder operates in reverse. It takes as input a latent vector and produces as output a sequence of poses. The discriminator consists of a fully connected artificial network only. It takes a latent vector as input and produces a scalar value as output that represents the classification of the vector as following a Gaussian Distribution or not.

Granular Dance is trained on multiple loss functions. Three loss functions quantify the autoencoder's error in reconstructing dance movements through the encoding and decoding process. The loss functions associated with the reconstruction error are based on the deviation of quaternions from unit length, and the difference in joint orientations and joint positions between original and reconstructed movements. Two loss functions are involved in the adversarial game between encoder and discriminator. One loss function rewards the discriminator for successfully distinguishing between latent vectors coming from a true Gaussian distribution and latent vectors generated by the encoder. Another loss function rewards the encoder for fooling the discriminator.

Granular Dance employs an explicit generative algorithm for concatenating the short synthetic dance movements generated by the autoencoder into long movement sequences (see right side of Fig. 12.1). This algorithm is inspired by two methods from computer music that combine short audio waveforms to generate longer sounds: Granular Synthesis and Concatenative Synthesis. Granular Synthesis typically employs very short (microseconds to milliseconds) waveforms which it concatenates by cross-fading their amplitude envelopes (Roads, 2004). Concatenative Synthesis employs longer waveforms (milliseconds to seconds) which it concatenates based on finding best acoustic matches between the beginning and ending parts of the waveforms (Schwarz et al., 2004; Zils & Pachet, 2001). Granular Dance adopts these two algorithms to obtain smooth transitions between concatenated pose sequences. Similar to Granular Synthesis, successive pose sequences are cross-faded by using an envelope. This envelope controls the interpolation factor for a spherical linear interpolation (Shoemake, 1985) between joint rotations. Similar to Concatenative Synthesis, similar pose sequences are obtained for concatenation by choosing latent vectors that possess short Euclidean distances among each other.

12.3.1.2 Application

As a co-creative tool for choreographers, *Granular Dance* supports creative workflows that favour experimentation and discovery over goal-oriented production. *Granular Dance* offers three main approaches for experimentation. Experimenting with the relationship between movement and its encoding, experimenting with methods for navigating latent space, and experimenting with concatenating generated movements.

The relationship between movement and its encoding can be explored by manipulating movement in the domain of its representation by the model rather than in the domain of movement itself. The encoding of movement captures the statistically most significant attributes of the movement material that the choreographer provided for training the model. These attributes reflect both bio-mechanical constraints of the human body and idiosyncratic stylistic principles inherent in the choreographer's work. By directly working with encodings, a choreographer can generate new movements that bring to the forefront the model's representation of these attributes. An example of directly manipulating movement encodings is shown in Fig. 12.2 and available as animation online.[1] Here, an encoding of a given dance movement is varied by incrementing or decrementing each value at each encoding dimension one at a time.

Navigating latent space offers another approach to generating dance movements. Navigation involves following a trajectory within latent space, sampling encodings at locations along the trajectory, decoding these encodings into short movement segments, and concatenating the movement segments into longer movements. This approach is interesting for two reasons. First, the principle of latent space navigation offers a choreographer a high level of control for balancing the level of familiarity and novelty of the generated movements. The navigation can be interpreted as traveling across and between regions on a map, some of which are densely populated with encodings of existing movements whereas others are devoid of such encodings. Depending on whether the trajectory stays close to the former regions or ventures into the latter ones, the probability of encountering encodings that represent familiar movements changes. Second, contrary to the direct manipulation of encodings, following a trajectory doesn't require the choreographer to understand the relationship between encodings and movements. Instead, a trajectory can be created from a given dance movement which is split into multiple short movements and subsequently encoded. Depending on how faithfully this trajectory is navigated, the choreographer can control the level of similarity and dissimilarity of the generated movements to the original movement. Figure 12.3 depicts all the trajectories in

[1] Granular Dance—Encoding Variations: https://vimeo.com/940858550

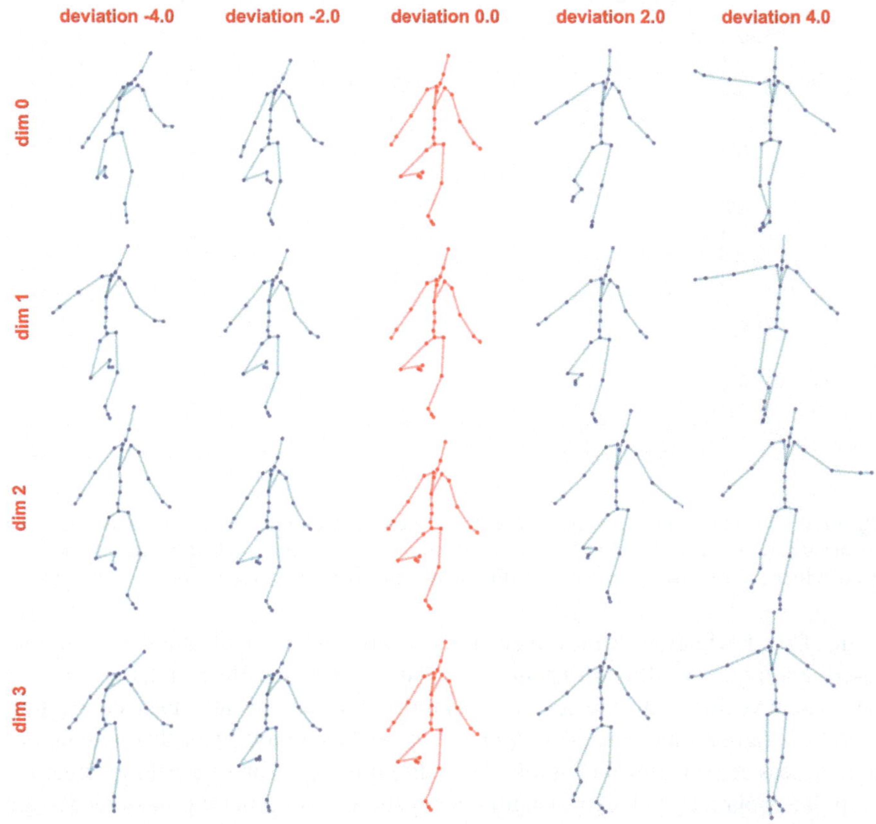

Fig. 12.2 Direct variation of encoding values. In the figure, value changes run along the horizontal axis with no change in the center (red poses), and increasingly negative and positive changes (blue poses) to the left and right, respectively. The dimension increases from top to bottom. For space reasons, only changes for the first four dimensions are shown

latent space that correspond to the movements used for training. Figure 12.4 depicts movement sequences that have been created from different approaches for latent space navigation. Example animations are available online for each approach to latent space navigation.[2,3,4,5]

In random walk navigation, the beginning of a movement is encoded and chosen as starting point for navigation. The navigation then consists of repeatedly adding a random offset to this encoding. This approach generates a movement that becomes increasingly dissimilar from a given initial movement.

[2] Granular Dance—Random Walk: https://vimeo.com/940862571

[3] Granular Dance—Trajectory Offset: https://vimeo.com/940865558

[4] Granular Dance—Trajectory Interpolation: https://vimeo.com/940873910

[5] Granular Dance—Trajectory Extrapolation: https://vimeo.com/940875625

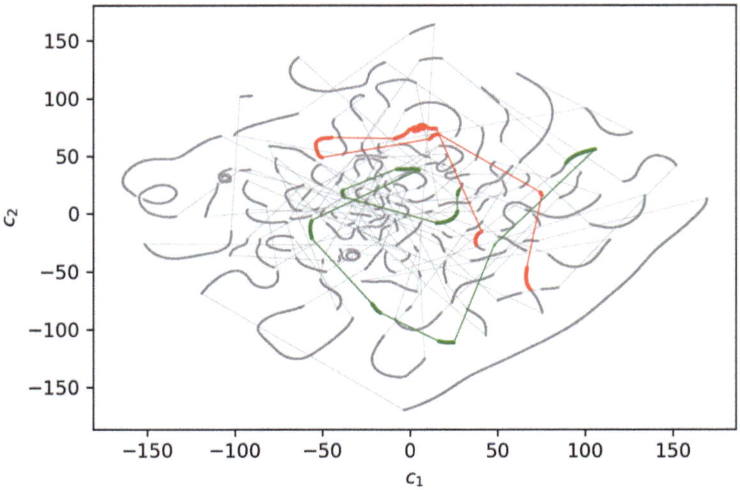

Fig. 12.3 Trajectories in latent space. The figure shows a two-dimensional representation of all the encoded movements that have been used for training. The trajectories highlighted in color correspond to the movement sequences used for the navigation experiments explained in the text

In offset navigation, a movement is encoded into a list of latent vectors that describe a trajectory through latent space. Then a user specified offset is added to these latent vectors. This creates a second trajectory that runs at a distance in parallel to the original trajectory. This approach generates a movement that maintains a more or less constant level of similarity or dissimilarity to an original movement.

In interpolation and extrapolation navigation, two different movements are encoded into two lists of latent vectors. Each list represents an individual trajectory through latent space. A new trajectory can then be created through interpolation by following the given trajectories while gradually approaching one trajectory and withdrawing from the other. Alternatively, a new trajectory can be created through extrapolation, which involves following the given trajectories while moving away from one trajectory in the opposite direction of the other trajectory. Interpolation of trajectories generates movements whose similarity gradually changes from one movement to another. Extrapolation of trajectories generates movements that become increasingly dissimilar from one movement by amplifying its differences to another movement.

12.3.2 Puppeteering AI

Puppeteering AI realises an artificial dancer with which a human dancer can interact and improvise. The artificial dancer acquires its movement vocabulary through training on motion capture recordings of solo performances. Two core ideas form the basis of *Puppeteering AI*: a direct gestural control of a subset of the artificial

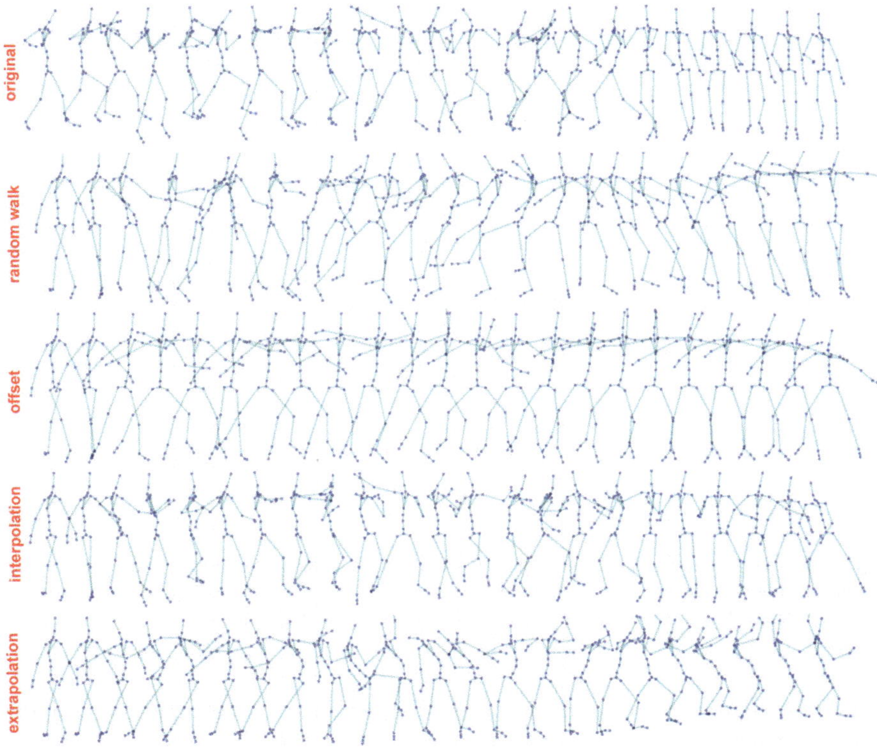

Fig. 12.4 Experiments in latent space navigation. The figure depicts different approaches of navigating latent space in combination with sequence concatenation. From top to bottom, the sequences represent: an original sequence, a sequence created from a random walk, a sequence created by adding an offset of 2.0 to all the encodings, a sequence obtained by interpolating between the encodings of two original sequences, a sequence obtained by extrapolating between the encodings of the same two original sequences

dancer's limbs that is inspired by puppeteering and the creation of spatial movement trajectories according to Rudolf Laban's concept of the Kinesphere (Christian, 2021).

12.3.2.1 Model

Puppeteering AI employs the same machine-learning architecture and movement concatenation mechanism as *Granular Dance*. But it extends the functionality of *Granular Dance* by operating in real-time and by providing interactive control of spatial target positions that guide the generated movements.

The mechanism of relating spatial positions to movements is based on a nearest neighbors search in the latent space of movement encodings in combination with simple vector arithmetic. This mechanism is depicted in Fig. 12.5. The mechanism operates as follows. At first, a dictionary is created for all the movement excerpts used in training that maps the position of a selected joint to the encoding of the corresponding movement. A nearest neighbor search (KD-Tree in this case) identifies

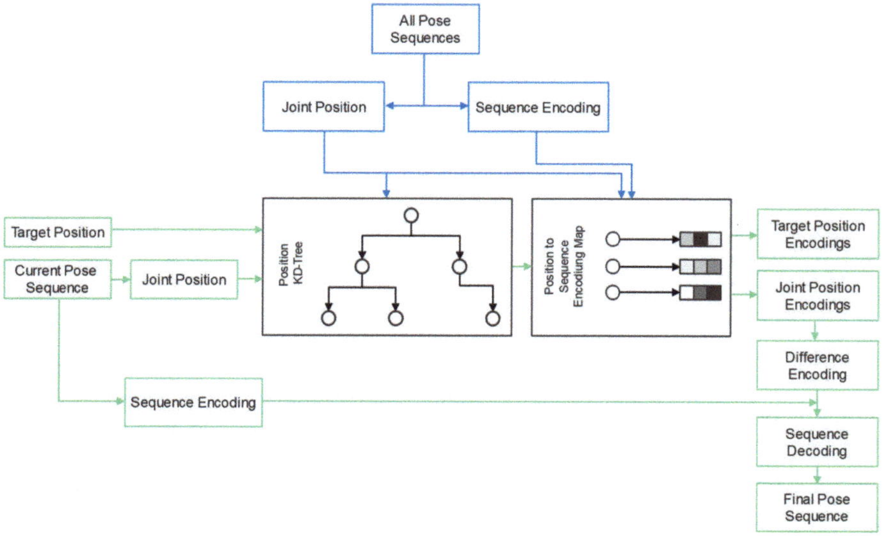

Fig. 12.5 *Puppeteering AI* nearest neighbour search for movement encodings. The figure depicts the processing steps involved in combining nearest neighbour search and vector arithmetic to generate encodings that correspond to movements that bring a specific joint closer to a target location in space

those movement excerpts whose joint positions are closest to a target position, the latter of which is directly controlled through interaction. The encodings of these movement excerpts are retrieved from the dictionary and their difference to the encoding of the artificial dancer's current movement is calculated. The mean of these differences is added to the artificial dancer's current movement encoding and the resulting encoding is decoded. This final decoded movement represents that trajectory that the selected joint follows to get closer to the target position.

This mechanism possesses several control parameters that affect the resulting movements: the selection of the joint that should move towards a target position, the location of the target position, the neighbour count used in the nearest neighbor search, and the scaling factor that is applied to the mean of the encoding differences prior to being added to the current movement encoding. Figure 12.6 illustrates the effect of the neighbour count and scaling factor. In this illustration, the selected joint (the right hand) and the location of the target position remain unchanged. A corresponding animation is available online.[6] Figure 12.7 illustrates the effect of changing the location of the target position while keeping the other parameters fixed. A corresponding animation is available online.[7]

[6] Puppeteering AI—Target Position Parameter Control: https://vimeo.com/941180614

[7] Puppeteering AI—Target Position Location Control: https://vimeo.com/941183120

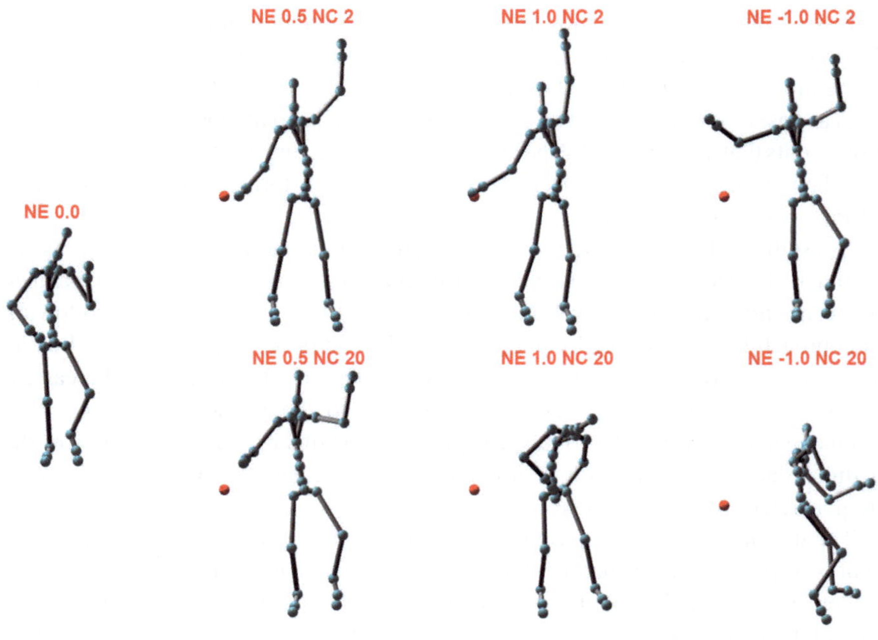

Fig. 12.6 Variation of joint target position control parameters. A graphical representation of the artificial dancer is shown alongside a red ball that indicates the target position of the selected joint (the right hand). The following abbreviations are used: NE stands for neighbour effect scaling, NC stands for neighbour count

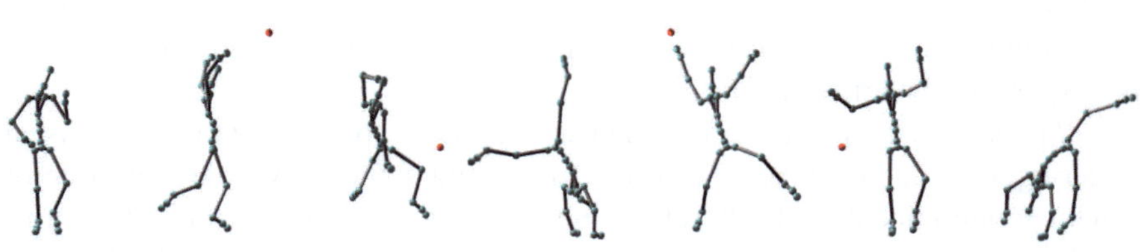

Fig. 12.7 Variation of target joint position. A graphical representation of the artificial dancer is shown alongside a red ball that indicates the target position of the selected joint. The fixed parameters are set as follows: the right hand as joint, a nearest neighbor count of 20, and a scaling factor of 1.0

12.3.2.2 Application

The main application of *Puppeteering AI* is the creation of an artificial dancer that can serve as an improvisation partner for a human dancer. For this reason, *Puppeteering AI* provides a form of interaction that lends itself better to live performance situations than choreographic workflows. It does so by focusing on the following aspects of interaction: intuitive affordances based on gestural interaction, balance between control and autonomy of the artificial dancer, connection with a well-known movement principle, provision of multimodal feedback.

Instead of manipulating movement via its representation in latent space, as was the case with *Granular Dance*, *Puppeteering AI* situates interactivity in the domain of the movement itself. This allows a dancer to employ his or her own bodily gestures as basis for interaction. The artificial dancer reacts to interaction by generating movements that are readable by the human dancer as gestural responses. Accordingly, both human and artificial dancer can employ their gestural vocabulary as shared affordances for interaction.

To establish the artificial dancer as an improvisation partner that is rewarding for a human dancer to engage with, the artificial dancer needs to strike a balance between autonomy and control. *Puppeteering AI* strikes this balance by following an approach that is inspired by puppeteering. In traditional puppeteering, a human performer controls only a subset of the joints of an articulated puppet whereas the remaining joints move on their own according to the physical principles of a mechanical system. Similar to this, *Puppeteering AI* makes only a few joints of the artificial dancer accessible for interaction. The movements of the remaining joints are generated by the movement synthesis model.

The use of a location in space that the artificial dancer can reach with one of its joint is inspired by the principle of the Kinesphere that has been coined by choreographer and scholar Rudolf Laban. Laban defines the Kinesphere as "the sphere around the body whose periphery can be reached by easily extended limbs without stepping away from that place which is the point of support when standing on one foot" (Laban & Ullmann, 1966). He employs this principle as part of his Choreutics system to analyze the spatial trajectories inscribed by the movements of the human body. This connection between interaction principle employed in *Puppeteering AI* and Rudolf Laban's Kinesphere establishes an interaction scenario that dancer's are familiar with and provides them with clear guidelines for the movements that are at the disposal of the artificial dancer.

In *Puppeteering AI*, the artificial dancer is rendered perceivable through visual rendering and sound synthesis. The use image and sound serves the dual role of providing multimodal feedback for the human dancer and of integrating the artificial dancer as a mediated object into a stage scenography. Acoustic feedback in particular has been found to play a crucial role in guiding a dancer's engagement with an interactive system (Akerly, 2015). Visual feedback complements acoustic feedback in that it establishes the artificial dancer as a distinct entity with its own visual appearance and location on stage.

The generation of sound is based on a digital instrument that is played by the artificial dancer. The digital instrument simulates a vibrating surface via a bank of resonating filters following a modal synthesis approach (Morrison & Adrien, 1993). These filters possess locations in space that are distributed on the same surface of a sphere that the target point for interaction is traveling across (see Fig. 12.8). Playing this instrument is based on the artificial dancer approaching the filters with some of its joints. Those filters that are sufficiently close to the dancer's joints are excited through an external audio signal. External audio signals employed here include impulses, impulse chains, and pink noise.

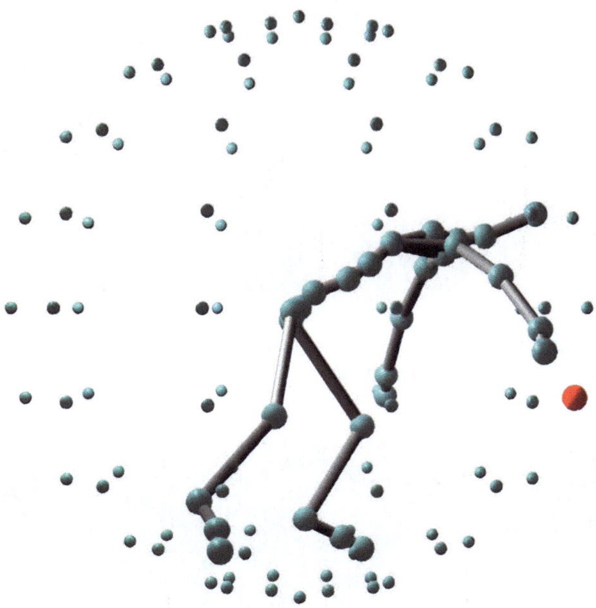

Fig. 12.8 Distribution of filters for modal synthesis. The figure depicts an artificial dancer surrounded by filters for modal synthesis. These filters lie on the surface of a sphere and are depicted by small blue balls. The current target location for the dancer's right hand is indicated by a red ball

The visual appearance of the artificial dancer is generated based on the Ray Marching rendering technique (Tomczak, 2012). Ray Marching generates an image by passing a ray through each image pixel into a space. Positions along the ray are tested for collisions with geometrical objects that are described by mathematical functions representing the distance between a point in space and the surface of the object. While computationally demanding, this rendering method excels at generating smooth surfaces that seamlessly blend and morph into each other. By representing the joints and their connections as mathematical functions, a wide diversity of visual appearances can be generated for the artificial dancer. These appearances vary from depicting a humanoid body to displaying abstract amorphous shapes (see Fig. 12.9).

Puppeteering AI has been used in two dance performances. A first performance entitled *Skin Diving* was choreographed by Emi Myoshi and involved two professional dancers, one controlling *Puppeteering AI.* A second performance entitled *Parzival Game* On was choreographed by Graham Smith and involved lay persons as dancers (see Fig. 12.10). In this piece, two dancers simultaneously interacted with two instances of *Puppeteering AI.* In both performances, a human dancer was controlling the target position for the artificial dancer by wearing an orientation sensor. The sensed orientation was mapped to a location on the surface of the Kinesphere. Some video excerpts of the performance are available online.[8]

[8] Parzival Game On—Performance Excerpts: https://vimeo.com/941185650/7ec6b167ce

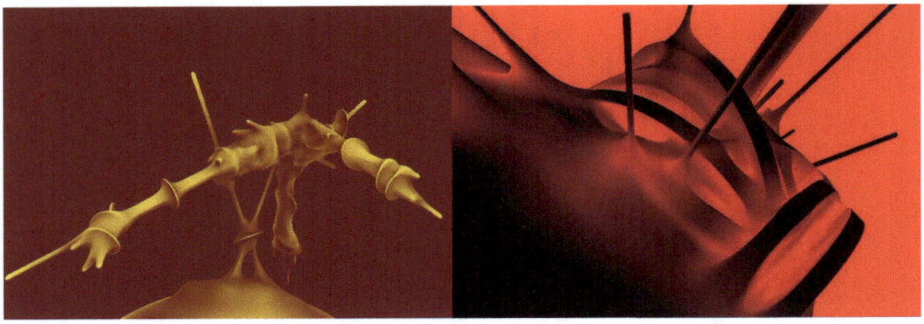

Fig. 12.9 Raymarching visualisations. The image depicts two different visualisations of an artificial dancer using the Raymarching Technique

Fig. 12.10 *Parzival Game On*. The image shows a performance situation in the dance piece Parzival Game On. (Image Credits: Marc Doradzillo, reprint with permission of the author)

12.3.3 Expressive Aliens

Expressive Aliens is an experimental tool to explore the connection between body morphology and expressive movement capabilities. It combines a physics simulation and reinforcement learning to train agents with arbitrary articulated body morphologies to exhibit movement behaviors that match the expressive qualities described by Laban Effort Factors. The purpose of *Expressive Aliens* is twofold: firstly, to provide insights into the principles of embodied creativity by studying the relationship between bodily properties and expressivity, and secondly, to experiment with extended forms of choreography that include the conception of non-anthropomorphic agents as dancers.

12.3.3.1 Model

The articulated morphologies of the agents that are trained to exhibit expressive movements are designed using the software onShape.[9] These morphologies consist of rigid body parts connected via joints that possess a single rotational degree of freedom. The morphologies that have been experimented with so far are depicted in Fig. 12.12. The agent's behavior is simulated using the rigid body dynamics functionality of the PyBullet physics engine.[10]

The training of the agent is based on reinforcement learning, a type of learning often applied for agent-based simulations in which an agent needs to develop a strategy for successfully interacting with an environment. The level of success is measured as a reward that an agent receives from the environment after performing an action. The goal of reinforcement learning is for the agent to learn a strategy (policy) that maximizes its long-term reward. The main components of reinforcement learning are the environment, the agent, the (observable) state of the environment, the actions available to the agent, a reward, and a mechanism for mapping the state of the environment into a choice for action (Sutton & Barto, 2018). The actions and states can either be discrete or continuous, the mapping mechanism can be as simple as a lookup table or as involved as a deep neural network, and the policy that an agent uses to select actions can either be directly improved through learning (on-policy learning) or the optimal policy is learned independently of the policy the agent uses to select actions (off-policy learning).

In *Expressive Aliens*, training is based on the Soft Actor Critic (SAC) reinforcement learning algorithm (Haarnoja et al., 2018) (see Fig. 12.11, left side). Actor-Critic methods employ a policy function and a value function, both of which are typically implemented as neural networks. The policy function is called the actor, and the value function is called the critic. The actor decides which action to take, and the critic informs the actor how good the action taken was and how it should be adjusted. Among the benefits that SAC offers over other Actor-Critic methods are its operation on continuous representations of actions and states and its exploitation of the efficiency of off-policy methods while combating their tendency for premature convergence.

In *Expressive Aliens*, the observation state comprises the rotations and rotational velocities of the agent's body joints, the contacts between body parts and the ground plane, and the overall position, velocity, and rotation of the entire body. The action state includes the rotational torques for all the actuated joints of the agent's body. The reward is calculated by combining several individual rewards. What sets *Expressive Aliens* apart from other simulation-based reinforcement learning systems are the rewards based on Laban Effort Factors. These factors are part of the Laban Movement Analysis (LMA) notation system introduced by Rudolf Laban (1975), which defines four categories to formalize different aspects of the human body (Maletic, 1987): *Body*, *Space*, *Shape*, and *Effort*. The Effort category describes aspects that contribute to the expressivity of movement (Bartenieff & Lewis, 2013)

[9] OnShape: https://www.onshape.com/

[10] PyBullet: https://pybullet.org/

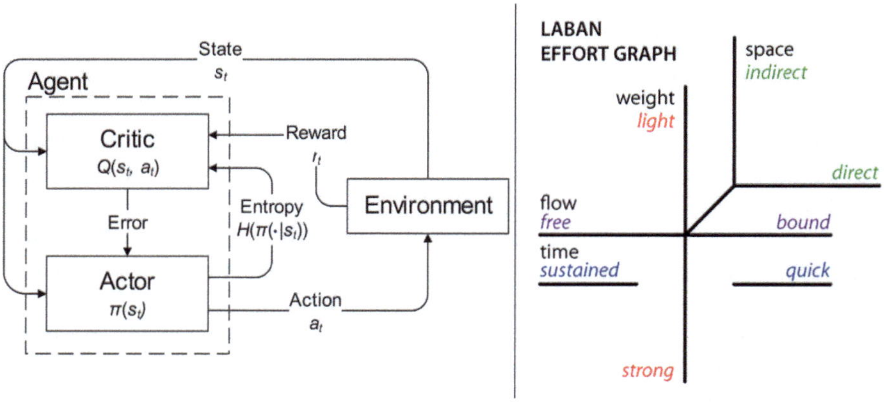

Fig. 12.11 Soft actor critic model architecture and Laban effort factors. The image on the left depicts the information flow between the main components of the SAC architecture. The following acronyms are used: t for time, s for observation state, a for action state, r for reward, H for entropy, Q for Action Value, π for policy. The image on the right shows a graph representation of the four Laban Effort Factors with their respective polarities. (Image Credits: Raphael Cottin, reprint with permission of the author)

and is further divided into four Effort Factors (EF): *Space*, *Weight*, *Time*, and *Flow* (see Fig. 12.11, right side). Each factor has two opposing polarities. The *Space* factor describes the directness of movement, which can either be *Direct* or *Indirect*. The *Weight* factor describes the strength of movement, which can either be *Light* or *Strong*. The *Time* factor describes the urgency of movement, which can either be *Sustained* or *Sudden*. The *Flow* factor describes the continuity of movement, which can either be *Free* or *Bound*. In *Expressive Aliens*, each EF is represented as a scalar value ranging from 0.0 to 1.0 to indicate the polarity of the factor. The EFs are derived from the agent's physical movement quantities using algorithms described by Larboulette and Gibet (2015). The numerical representation of the EFs is converted into rewards by computing the absolute difference between the current EF value and a given target value. The target values for the EFs are chosen at the beginning of a training run and affect the expressive characteristics of the resulting agent behavior.

12.3.3.2 Application

Expressive Aliens is meant to serve as a tool for studying the relationship between the physical characteristics of different morphologies and their capability to exhibit expressive movements. By combining a machine-learning and simulation-based approach, empirical experiments can be conducted. For the study presented here, three agent morphologies have been designed, which vary in terms of the number and shape of body parts, the number of joints, and the assignment of foot or hand roles to body parts. Body parts that don't serve as feet or hands will cause the agent to receive a negative reward upon ground contact. These morphological differences impact the type and level of stability that each morphology exhibits during the simulation. The morphology entitled *Biped* (see Fig. 12.12, left side) possesses two legs and maintains an upright posture through dynamic stability. The morphology

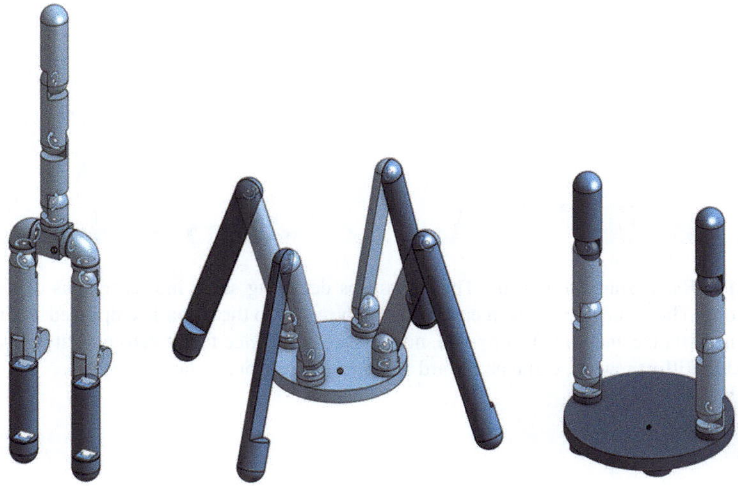

Fig. 12.12 Non-anthropomorphic morphologies. The images depict three different non-anthropomorphic morphologies that are trained to exhibit expressive movements. From left to right: *Biped*, *Quadruped*, *Legless*. The body parts rendered in dark blue have been assigned foot or hand status

entitled *Quadruped* (see Fig. 12.12, middle) possesses for legs and maintains an upright posture through static stability. This morphology entitled *Legless* (see Fig. 12.12, right side) possesses no legs but two arms and its circular torso forms the main contact with the ground.

To study how the relationship between morphology and movement expressivity changes under the influence of each Effort polarity, a set of training runs was devised in which either no or one polarity was rewarded at a time, with or without rewarding also for movement distance. After each training run, several inference runs were conducted during which reward values and simulation visualization were recorded. A selection of inference runs is depicted as filmstrips (see Figs. 12.13, 12.14, and 12.15). Corresponding animations are available online.[11,12,13,14,15,16,17,18,19,20,21,22]

[11] Expressive Aliens—Biped D0: https://vimeo.com/941214160

[12] Expressive Aliens—Biped D0S0: https://vimeo.com/941215962

[13] Expressive Aliens—Biped D1: https://vimeo.com/941216654

[14] Expressive Aliens—Biped D1S1: https://vimeo.com/941217631

[15] Expressive Aliens—Quadruped D0: https://vimeo.com/941218570

[16] Expressive Aliens—Quadruped D0F1: https://vimeo.com/941219819

[17] Expressive Aliens—Quadruped D1: https://vimeo.com/941220660

[18] Expressive Aliens—Quadruped D1W1: https://vimeo.com/941221447

[19] Expressive Aliens—Legless D0: https://vimeo.com/941222221

[20] Expressive Aliens—Legless D0S0: https://vimeo.com/941223003

[21] Expressive Aliens—Legless D1: https://vimeo.com/941223540

[22] Expressive Aliens—Legless D1F1: https://vimeo.com/941224283

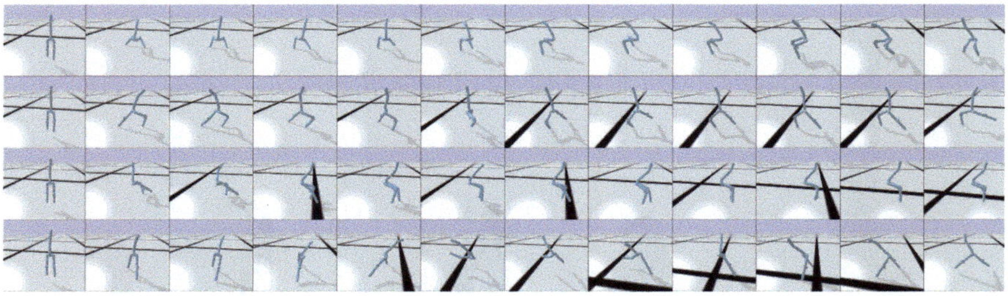

Fig. 12.13 *Biped* animation strips. The film strips depicting some inference runs of the *Biped* morphology. These runs differ from each other with regard to the rewards employed. From top to bottom: no distance and no Effort reward, no distance and *Space* factor zero polarity reward, distance and no Effort reward, distance reward and *Space* factor one polarity reward

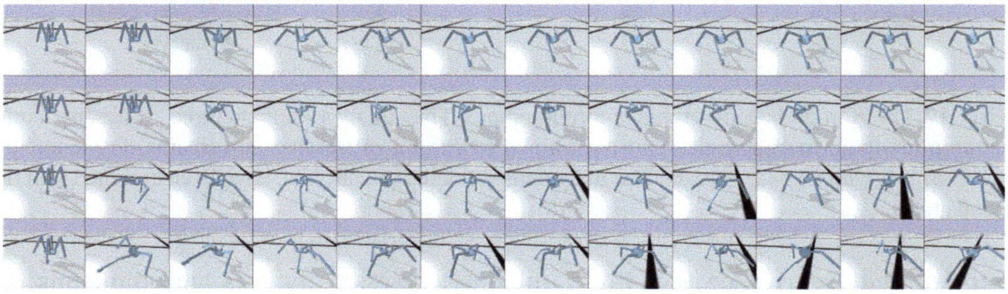

Fig. 12.14 *Quadruped* animation strips. The film strips depicting some inference runs of the *Quadruped* morphology. These runs differ from each other with regard to the rewards employed. From top to bottom: no distance and no Effort reward, no distance and *Flow* factor one polarity reward, distance and no Effort reward, distance reward and *Weight* factor one polarity reward

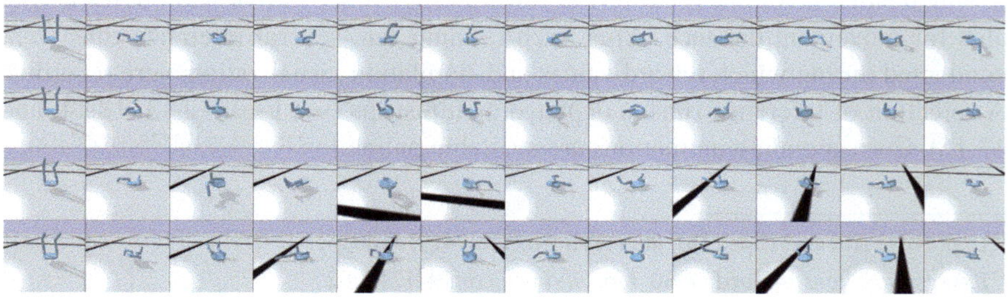

Fig. 12.15 *Legless* animation strips. The film strips depicting some inference runs of the *Legless* morphology. These runs differ from each other with regard to the rewards employed. From top to bottom: no distance and no Effort reward, no distance and *Space* factor zero polarity reward, distance and no Effort reward, distance reward and *Flow* factor one polarity reward

Some insights into the relationship between morphologies and Effort Factors can be gained by observing how the movements vary across different reward configurations for a single morphology and for identical reward configurations across different morphologies. It seems likely that the predispositions of morphologies for different forms of stability define a basic repertory of suitable poses and motions. The Effort rewards influence which of these poses and motions are chosen and how they are adjusted to match the given factor polarities. Vice versa, the appearance of Effort factors in an agent's movement is facilitated by certain morphologies and suppressed by other morphologies. Examples of this are the tendency of the *Biped* morphology to exhibit the *Bound* and *Sudden* polarities whereas the *Quadruped* morphology moves predomantly according to the *Free* and *Sustained* polarities. Another striking example is the *Strong* polarity that appears most pronounced for the *Legless* morphology whereas the *Light* polarity dominates for the *Quadruped* morphology.

Expressive Aliens also offers interesting possibilities for choreographic creation. It highlights how the focus of artistic creativity can be placed on the creation of morphologies rather than movements. In this setting, *Expressive Aliens* operates as a generative system that automatically creates movements for a given morphology through a combined process of training and simulation. Such an approach opens up new artistic possibilities and expands choreographic decision making beyond what is possible with human bodies.

12.3.4 RAMFEM

The acronym *RAMFEM* stands for Raw Music From Free Movement. *RAMFEM* is a system for translating the movements of a solo dancer into music. This system can be trained on motion capture and audio recordings of a dancer who improvises to music. *RAMFEM* operates directly on raw audio. For this reason, any recorded music can be used for training and subsequently recreated through movement sonification. *RAMFEM* exploits the idiosyncratic movement decisions a dancer makes during improvisation as mechanism for controlling music through movement.

12.3.4.1 Model

RAMFEM comprises three main components: an adversarial autoencoder (AAE), a sequence to sequence transducer (Seq2Seq), and an audio concatenation mechanism (see Fig. 12.16).

The AAE model is composed of four neural networks: an encoder, a decoder, a discriminator for the prior distribution of encodings, and a discriminator for audio waveforms. The autoencoder's primary function is to encode and decode short audio waveforms into and from latent vectors, reducing their dimensionality and temporal resolution. This, in turn, decreases the length of the audio sequence that must be learned by the Seq2Seq model. The autoencoder operates with audio at a sample rate of 32 kHz, waveforms that are 256 samples long, and a latent dimension of 32.

The Seq2Seq model is made up of five neural networks: a pose sequence encoder, a pose sequence decoder, an audio decoder, a discriminator for latent encodings, and a discriminator for audio waveforms. The audio decoder and audio waveform discriminator are reused from the AAE model. The sequence encoder and decoder

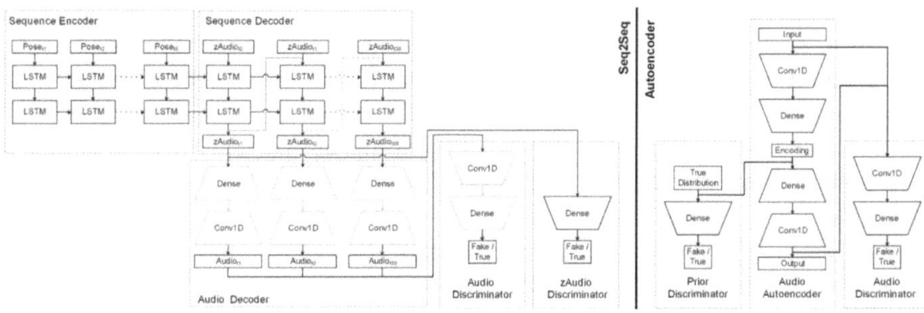

Fig. 12.16 Model architectures of a motion to audio sequence transducer. Neural network layers consisting of identical units are depicted as geometrical shapes: rectangles for layers in the sequence encoder and decoder, trapezoids for the layers in the audio encoder, decoder and discriminators. The labels used to identify the units are LSTM for Long-Short-Term-Memory and Dense for fully connected artificial neurons. Shapes with dark outlines refer to layers whose weights are changed during Seq2Seq training. Shapes with light outlines refer to layers with fixed weights

are deterministic and do not possess an attention mechanism. The Seq2Seq model receives a sequence of poses as input and translates them into a sequence of audio encodings. These encodings are passed to the audio decoder, which transforms them into waveforms.

The audio concatenation mechanism takes a sequence of waveforms, applies a Hanning window as an amplitude envelope to each of them, and then concatenates them with a 50% overlap, resulting in the final audio sequence.

The AAE and Seq2Seq models are trained separately. Initially, the AAE model is trained in alternation with its two discriminators using categorical cross-entropy for the discriminators and mean square error between audio waveforms for audio reconstruction loss. Subsequently, the Seq2Seq model is trained while keeping the weights of the pre-trained autoencoder and discriminators fixed. The sequence encoder and decoder are trained in alteration with its single discriminator on the following loss functions: categorical cross-entropy for adversarial loss, mean square error for audio reconstruction loss, and audio encoding reconstruction loss.

12.3.4.2 Application

RAMFEM is a system that captures the decisions made by a dancer when translating music into movement. It then imitates these decisions to create music from the dancer's movements. *RAMFEM* is trained on motion capture recordings of a dancer, allowing the dancer to use full body gestures to generate music. Although *RAMFEM* doesn't currently operate in real-time, once its performance has been improved, it can be used as a gestural interface for dancers to create music during live performances. *RAMFEM* allows a dancer to create a movement sonification system that is closely tied to the dancer's own approach of combining music and movement. The dancer can choose any musical material and movement vocabulary and employ any principle of responding through movement to music.

Two types of experiments have been conducted with *RAMFEM* so far. The first experiment verified the system's ability to learn and reproduce basic and strict correlations between movement and music. For this experiment, synthetic music was created from recorded movements by using a simple mapping between movement properties and sound synthesis parameters. The sound synthesis consists of two sine oscillators controlled by the dancer's hands. The height and velocity of each hand are mapped to the frequency and amplitude of each oscillator. The second experiment involved several professional dancers with a specialization in contemporary dance and improvisation. They were invited to freely improvise to a given music. Short excerpts of royalty-free music were selected, including experimental electronic music, free jazz, and contemporary classical music.

The movement and music recordings obtained from the two experiments were used to create two different datasets. The sonification dataset combines motion capture recordings with synthetically generated audio, while the improvisation dataset consists of the motion capture and audio recordings from the dancers' free improvisation sessions. These two datasets were used to train two different versions of *RAMFEM*. Figure 12.17 shows spectrograms of the original audio and the audio generated by the autoencoder and Seq2Seq components of *RAMFEM*.

A qualitative comparison of these spectrograms leads to several observations. The audio autoencoder can reconstruct the original audio for frequencies below 1000 Hz (see Fig. 12.17, left and middle columns). However, some spectral content is lost beyond this frequency range, which affects the music in the improvisation dataset but not the sonification dataset.

The results obtained from training the Seq2Seq model were more diverse. Two types of qualitative evaluations were conducted: a comparison between the original and predicted audio using pose sequences from the training dataset and a comparison between the predicted audio (see Fig. 12.17, left and right columns) and a visual rendering of the pose sequences that are not from the training dataset. Animations

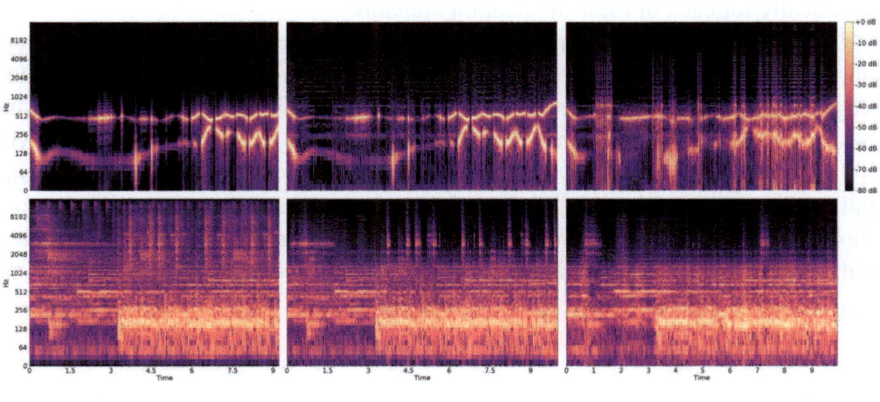

Fig. 12.17 Audio spectrogram comparison. The image depicts log-frequency power spectrograms of the original audio (left) and audio reconstructed by the audio autoencoder (middle) and the sequence to sequence transducer (right). The audio is part of the sonification (top) and improvisation (bottom) datatsets

showing these comparisons are available online.[23,24,25,26] The Seq2Seq model trained on the sonification dataset generated audio that is acoustically similar to the original audio for both audio generated from movements in the training set and for movements not in the training set. However, the quality of the generated audio is clearly inferior to the original audio. The generated audio includes noise, clicks, amplitude fluctuations, and dominant low frequencies. The results are similar for the Seq2Seq model trained on the improvisation dataset when using movements from the training dataset. The predicted audio is of low quality but acoustically similar to the original, and the correlation between movement and audio is preserved. When the Seq2Seq model was provided with movements that were not in the dataset, the results were more difficult to interpret. Some of the acoustic elements from the original audio are still present but appear to lack a clear temporal structure.

Based on these findings, it can be said that *RAMFEM* grasps some of the acoustic properties and correlations between movement and music, even with its current simple architecture. However, it is clear that the model architectures have to be improved to achieve higher audio quality and to fully capture the varied relations between movement and music that dancers work with when freely improvising to music.

12.4 Discussion and Outlook

The purpose of this article is to provide some insights into the role of generative approaches in the field of contemporary dance and the new possibilities that machine learning offers for this field. When developing machine learning-based systems for dance, it is crucial to take into account the creative principles that make this field unique, as well as the application domains that dancers and choreographers have shown the greatest interest in. This article focuses on four concrete examples of machine-learning-based generative systems that have been developed by the author specifically for each of these application domains.

Granular Dance, highlights how navigating the space of latent representations employed by a motion synthesis system provides an interesting approach for choreographic ideation. *Puppeteering AI*, demonstrates how machine learning can be used to create a semi-autonomous artificial dancer that can be interacted with via body gestures. *Expressive Aliens*, explores how experimentation with non-anthropomorphic entities can offer insights into the relationship between body morphologies and movement expressivity. *RAMFEM*, is a proof of concept system that demonstrates how the interactive and musical principles of a movement sonification system can be derived from a dancer's own creative decision-making when improvising to music.

[23] RAMFEM—Improvisation Dataset and Original Movement: https://vimeo.com/941234403

[24] RAMFEM—Sonification Dataset and Original Movement: https://vimeo.com/941237277

[25] RAMFEM—Improvisation Dataset and New Movement: https://vimeo.com/941238225

[26] RAMFEM—Sonification Dataset and New Movemen: https://vimeo.com/941239607

When creating tools for artists, it's essential for developers to consider potential applications beyond what was initially planned. Beyond choreographic ideation, *Granular Dance* could for example help choreographers gain insights into the characteristics of their movement vocabularies. Choreographers can identify movements that are over- or underrepresented in their vocabulary by locating regions that are particularly densely or sparsely populated in latent space. This observation can help choreographers to avoid habitual movements and look for novel movement material. Beyond the realisation of an artificial dancer, *Puppeteering AI* has for example potential applications for choreographic ideation, as it highlights how movement principles and latent space navigation can be related to each other. By following the same approach used for generating movements towards locations on a Kinesphere, different movement principles can be associated with their corresponding movement encodings. These encodings can be interpolated and subsequently decoded into movements that are representative of hybrid versions of the original movement principles. Beyond gaining insights into embodied creation, *Expressive Aliens* offers for example the possibility to realize artificial dancers that substantially deviate in their morphology from human dancers. Such entities prevent interpretations of their movements that are biased by preconceptions and anthropomorphic projections. By avoiding such interpretations, the audience is less likely to attribute meaning to the movement and instead appreciate the movement itself. Beyond its use for personalised forms of movement sonification, *RAMFEM* can for example be used for translating movement into other time-based media such as text and video. A particularly interesting application arises from the translation of different movements into each other. For this application, training data can be obtained by recording two dancers performing a duet. Once trained, *RAMFEM* can then take the role of an artificial dancer that responds to the movement of a human dancer by imitating the movement responses from the original duet.

Some of the four generative systems have been developed with specific target users in mind. These users are artists who possess technical skills and are active at the intersection between creative coding and contemporary dance. Accordingly, the systems are meant to be used by artists who are interested in adapting them to their specific creative needs.

The focus on these target users is reflected both in the system's technical implementation and the methods and contexts of their dissemination. The systems have been implemented by combining models and algorithms that are fairly common and for which documentations and tutorials exist that address non-academic audiences. Most systems make use of machine-learning-based and conventional generative methods, thereby catering both to artists who prefer to work in a data-driven manner and those who manually design their own algorithms. The machine-learning-based models possess relatively few parameters and can therefore be trained on datasets that are sufficiently small so that artists can use their own corpus of work for training. The code for the implementation of the systems is open-source and well-commented to facilitate modification by the artists. To disseminate the systems among the target users, an educational website[27] has been created by the author, providing introductory information about the methods used for acquiring the datasets and the technical principles of the machine learning architectures employed. The website also offers documentations and tutorials describing the functioning and use of each system. Additionally, the systems have been presented at workshops that are popular among the target users, including the so-called choreographic coding labs (CCL). CCLs are a recurring format for workshops that has originally been devised by MotionBank. These workshops specifically addresses artists who are proficient in coding and who are interested in drawing inspiration from Contemporary Dance. So far, the tools have been presented at three CCLS (Berlin 2022, Chatham 2022, Cologne 2023).

The author continues with the development of existing and new generative systems for contemporary dance. Currently, these development activities form part of the Horizon Europe project *Premiere*.[28] This project develops tools that make use of machine learning and extended reality to support digital workflows in dance and theater, covering all the stages of an artistic production from its inception, production, performance, dissemination, documentation, and archival.

The tools provided by *Premiere* are meant to cater to a wider community of artistic practitioners than those previously focused on by the author. The target users now also include performance artists who are interested in working with technology but have little technical expertise. To achieve this, the generative systems must be integrated into software that can run as standalone applications and are easy to use while still remaining flexible enough to be adapted for specific artistic needs. It also requires that the generative systems can be combined with other technical tools that artists commonly work with such as different sensing and recording hardware, audio and video applications, light and stage controls systems, and virtual reality authoring environments. But most importantly, it requires a more thorough and systematic evaluation of the interests and needs of performance artists than has been previously conducted.

This evaluation is currently ongoing and takes the form of guided group discussions, user surveys, and hands-on demonstrations with participants in dance and theater workshops or students at dance and theater academies. From these evaluations, the following preliminary insights have so far been gained: Users are mainly interested in generative systems that create output in real-time. Users would like to interact with generative systems through full-body or hand gestures. Users are interested in translating their movements into other media such as sound and images, either for the purpose of creating media-augmented performances or to increase the awareness of their movements. Users prefer forms of interaction that take qualitative aspects of their movements into account. Users would use generative systems that create movements for inspiration and ideation. Users find generated movements that are non-human-like more interesting than those that resemble humans. Users

[27] E2-Create Educational Website: https://wp.coventry.domains/e2edu/

[28] *Premiere*: https://premiere-project.eu/

are interested in working with artificial performers as partners for rehearsal and improvisation.

The feedback received is valuable but incomplete, as it is based on the opinions of performance artists who only briefly had the opportunity to experiment with the generative tools currently available or to think about the possible impact of future generative systems for their own creative work. To gain a broader understanding of the role generative systems play in Contemporary Dance, additional feedback will be collected from artists who work with the systems over an extended period. This will include both artists new to generative approaches and those with prior experience. By considering the interests and needs of a diverse group of artists, generative systems can be developed to have a more profound and sustained impact on the field of contemporary dance.

References

Akerly, J. (2015, August). Embodied flow in experiential media systems: a study of the dancer's lived experience in a responsive audio system. In *Proceedings of the 2nd International Workshop on Movement and Computing* (pp. 9–16).

Ali, H., Majeed, H., Usman, I., & Almejalli, K. A. (2021). Reducing entropy overestimation in soft actor critic using dual policy network. *Wireless Communications and Mobile Computing, 2021*, 1–13.

Antunes, R. F., & Leymarie, F. F. (2012, April). Generative Choreographie: Animieren von tanzenden Avataren in Echtzeit. In *International Conference on Evolutionary and Biologically Inspired Music and Art* (pp. 1–10). Springer.

Berman, A., & James, V. (2018, March). Learning as performance: Autoencoding and generating dance movements in real time. In *International Conference on Computational Intelligence in Music, Sound, Art and Design* (pp. 256–266). Springer.

Bisig, D. (2021). Granular dance. In *Ninth Conference on Computation, Communication, Aesthetics & X* (pp. 176–195). i2ADS.

Bisig, D. (2022a). Expressive Aliens-Laban effort factors for non-anthropomorphic morphologies. In *International Conference on Computational Intelligence in Music, Sound, Art and Design (Part of EvoStar)* (pp. 36–51). Springer.

Bisig, D. (2022b). Generativer Tanz—Taxonomie und Überblick. In *Proceedings of the 8th International Conference on Movement and Computing* (pp. 1–10).

Bisig, D., & Tatar, K. (2021). Rohe Musik aus freien Bewegungen: Early experiments in using machine learning to create raw audio from dance movements. In *Proceedings of the 2nd Conference on AI Music Creativity*.

Bisig, D., & Wegner, E. (2021). Puppenspielerische und KI-interaktive Steuerung einer maschinenlernbasierten künstlichen Tänzerin. In *Proceedings of the XXIII Conference on Generative Art. Rom, Italien* (pp. 315–332).

Bisig, D., & Wegner, E. (2022). Puppenspielerische KI-interaktive Steuerung eines künstlichen Tänzers. In *Proceedings of the Generative AI and H CI-CHI 2022 Workshop, New Orleans, USA*.

Blades, H. (2012, July). Creative computing und die Neukonfiguration der Tanzontologie. In *Electronic Visualisation and the Arts (EVA 2012)*. BCS Learning & Development.

Boden, M. A., & Edmonds, E. A. (2009). Was ist generative Kunst? *Digitale Kreativität, 20*(1–2), 21–46.

Bragato, M., Araújo, D., de Farias Filho, H. F., da Silva Costa, M., & Chellappa, T. (2023). *Das Becken als Brennpunkt: Marta Graham, eine Grammatik*. Seven Editora.

Christian, M. I. (2021). *Objects in Air: Kunstwerke und ihre Außenwelt um 1900*. University of Chicago Press.
Coton, A., Walker, K. S., Haddakin, L., & Cooper, M. (1975). *Schriften zum Tanz, 1938–68*. (No Title).
Crnkovic-Friis, L., & Crnkovic-Friis, L. (2016). Generative choreographie mit deep learning. *arXiv preprint arXiv:1605.06921*.
DeLahunta, S. (2002). Software für Tänzer: Formen kodieren. *Performance Research, 7*(2), 97–102.
DeLahunta, S. (2009). Der choreografische Sprachagent. In *Conference Proceedings of the 2008 World Dance Alliance Global Summit*.
Dorin, A., McCabe, J., McCormack, J., Monro, G., & Whitelaw, M. (2012). Ein Rahmen für das Verständnis generativer Kunst. *Digital Creativity, 23*(3–4), 239–259.
Downie, M. N. (2005). *Choreographing the extended agent: Performance graphics for dance theatre*. Dissertation, Massachusetts Institute of Technology, School of Architecture and Planning, Program in Media Arts and Sciences.
Dunlop, P., & Morthland, V. (1981). *The nature of the embodiment of choreutic units in contemporary choreography*. Goldsmiths, University of London (Vereinigtes Königreich).
Fiebrink, R., & Caramiaux, B. (2016). Der maschinelle Lernalgorithmus als kreatives musikalisches Werkzeug. *arXiv preprint arXiv:1611.00379*.
Fishwick, E. (2010). Wie durchdringen und prägen westeuropäische Vorstellungen von volkstümlichem/traditionellem Tanz historisch und im Kontext der Globalisierung den Bereich des Tanzes?
Galanter, P. (2003). Was ist generative Kunst? Die Komplexitätstheorie als Kontext für die Kunsttheorie. In *Proceedings of the 6th Generative Art Conference*.
Galili, D. F. (2015). Gaga: Jenseits der Technik mit Ohad Naharin im einundzwanzigsten Jahrhundert. *Dance Chronicle, 38*(3), 360–392.
Gilfillan, E. (2022). *Dance-making: Moment-to-moment collaboration in contemporary dance practices*. Dissertation, Macquarie University.
Gillies, M., Fiebrink, R., Tanaka, A., Garcia, J., Bevilacqua, F., Heloir, A., Nunnari, F., Mackay, W., Amershi, S., Lee, B., & et al. (2016). Human-centred machine learning. In *Proceedings of the 2016 CHI Conference Extended Abstracts on Human Factors in Computing Systems* (pp. 3558–3565).
Glynn, R. (2008). *Gesprächsumgebungen revisited*. Emerald Group Publishing Limited.
Gromala, D., & Sharir, Y. (1994). Tanzen mit dem virtuellen Derwisch: Virtuelle Körper. In *Virtual reality software and technology* (pp. 321–328).
Haarnoja, T., Zhou, A., Abbeel, P., & Levine, S. (2018, July). Soft actor-critic: Off-policy maximum entropy deep reinforcement learning with a stochastic actor. In *International Conference on Machine Learning* (pp. 1861–1870). PMLR.
Hardt, Y. (2011). Die Inszenierung des Ethnografischen der Tanzgeschichte: Zeitgenössischer Tanz und sein Spiel mit der Tradition. *Dance Research Journal, 43*(1), 27–42.
Hochreiter, S., & Schmidhuber, J. (1997). Langes Kurzzeitgedächtnis. *Neural Computation, 9*(8), 1735–1780.
Krug, M. (2015). Adaptive/responsive movement approach: Dance making as interactive system design. In *Proceedings of the 2nd International Workshop on Movement and Computing* (pp. 76–79).
Kuzian, E. (2021). Zeitgenössischer Tanz als modernistische Kunst. *Roczniki Kulturoznawcze, 12*(1), 51–63.
Laban, R. V. (1975). *Moderner pädagogischer Tanz*. McDonald & Evans, London, 3. Auflage.
Laban, R. V., & Ullmann, L. (1966). *Choreutics*. Macdonald and Evans.
Larboulette, C., & Gibet, S. (2015). A review of computable expressive descriptors of human motion. In *MOCO'15 Proceedings of the 2nd International Workshop on Movement and Computing* (pp. 21–28).
Leach, J., & Delahunta, S. (2017). Tanz wird Wissen: Gestaltung eines digitalen "Körpers". *Leonardo, 50*(5), 461–467.

Lee, H. Y., Yang, X., Liu, M. Y., Wang, T. C., Lu, Y. D., Yang, M. H., & Kautz, J. (2019). Dancing to music. *Advances in Neural Information Processing Systems, 32.*

Lepkoff, D. (1999). Was ist Befreiungstechnik. *Movement Research Performance Journal, 19,* 2–5.

Li, J., Yin, Y., Chu, H., Zhou, Y., Wang, T., Fidler, S., & Li, H. (2020). Lernen, verschiedene Tanzbewegungen mit Transformator zu erzeugen. *arXiv preprint arXiv:2008.08171.*

Maletic, V. (1987). *Körper, Raum, Ausdruck: Die Entwicklung der Bewegungs- und Tanzkonzepte von Rudolf Laban* (Bd. 75). Walter de Gruyter.

McCormick, J., Vincs, K., Nahavandi, S., Creighton, D., & Hutchison, S. (2014, June). Teaching a digital performing agent: Artificial neural network and hidden markov model for recognizing and performing dance movement. In *Proceedings of the 2014 International Workshop on Movement and Computing* (pp. 70–75).

Miller, R., & Wood, J. (2017). Merce Cunningham und Black Mountain: No fixed points. *Appalachian Journal, 44,* 494–511.

Morrison, J. D., & Adrien, J.-M. (1993). Mosaic: Ein Rahmenwerk für modale Synthese. *Computer Music Journal, 17*(1), 45–56.

Palazzi, M., Shaw, N. Z., Forsythe, W., Lewis, M., Albright, B., Andereck, M., Bhatawadekar, S., Ban, H., Calhoun, A., Drozd, J., et al. (2009). Synchrone Objekte für eine flache Sache, reproduziert. In *ACM SIGGRAPH 2009 Art Gallery* (p. 1).

Pettee, M., Shimmin, C., Duhaime, D., & Vidrin, I. (2019). Beyond imitation: Generative and variational choreography via machine learning. *arXiv preprint arXiv:1907.05297.*

Roads, C. (2004). *Microsound.* MIT Press.

Sagiv, T., Simons, T., & Drori, I. (2020). Die Konstruktion von Authentizität im kreativen Prozess: Lessons from choreographers of contemporary dance. *Organization Science, 31*(1), 23–46.

Sandvik, K. (2023). *Die vererbte Pädagogik der Humphrey-Limón-Tradition: Lineage and practitioners.* Dissertation, Wayne State University.

Schiphorst, T. (2013). Merce Cunningham: Making dances with the computer. In *Merce Cunningham* (pp. 79–98). Routledge.

Schwarz, D. et al. (2004). *Datengesteuerte konkatenative Klangsynthese.* Dissertation.

Shaw, N. Z. (2011). Synchrone Objekte, choreografische Objekte und die Übersetzung von Tanzideen. In *Emerging bodies* (pp. 207–224). transcript-Verlag.

Shoemake, K. (1985, July). Animating rotation with quaternion curves. In *Proceedings of the 12th Annual Conference on Computer Graphics and Interactive Techniques* (pp. 245–254).

Sicchio, K., Wang, Z., & Forbes, M. (2020). Live coding tools für Choreographie: Creating Terpsicode. In *Proceedings of the 2020 International Conference on Live Coding (ICLC2020). University of Limerick, Limerick, Irland* (pp. 87–94).

Sutil, N. S. (2015). *Bewegung und Darstellung: The language of human movement.* MIT Press.

Sutton, R. S., & Barto, A. G. (2018). *Reinforcement learning: An introduction.* MIT Press.

Tomczak, L. J. (2012). *GPU ray marching of distance fields* (p. 8). Technical University of Denmark.

Wang, Q., Artières, T., Chen, M., & Denoyer, L. (2020). Adversarial learning for modeling human motion. *The Visual Computer, 36*(1), 141–160.

Zannos, I., & Dimitrakopoulou, S. (2023). Telematischer Tanz und Live-Codierung: Ein kollaboratives Experiment in cyber-physischer Performance und Bildung. *Performance Philosophy, 8*(1), 5–5.

Zils, A., & Pachet, F. (2001). Musikalische Mosaikbildung. In *Digital audio effects (DAFx)* (Vol. 2, p. 135).

Automatic Annotation of Movements in Dance

13

Claudia Steinberg, Helena Miko, Julian Rogawski, David Rittershaus, Anton Koch, and Florian Jenett

Abstract

The article outlines research into the application of machine learning and deep learning in the field of dance. The implementation of automated systems for motion recognition and assessment is discussed, looking at various techniques such as pose estimation and human action recognition (HAR). It summarises existing research results and highlights the challenges and potentials using an application example from the #vortanz research project funded by the Federal Ministry of Education and Research (BMBF) from 2021 to 2024 (cf. Jenett et al., #vortanz—Automated pre-annotation in digital university dance education. https://vortanz.ai/#/de, 2021–2024).

13.1 Introduction

In recent years, the use and importance of computer-based analyses in the field of exercise and sport has increased significantly (Jayal et al., 2018; Siewert & Büning, 2023). This is largely due to technological advances in automated data collection, combined with the considerable public interest in major sporting events, which in turn mobilises considerable financial resources. However, one area that has not yet

C. Steinberg (✉) · H. Miko
German Sport University Cologne, Köln, Germany
e-mail: c.steinberg@dshs-koeln.de; h.miko@dshs-koeln.de

J. Rogawski
WizAI Soulution GmbH, Koblenz, Germany
e-mail: j.rogawski@wizai.com

D. Rittershaus · A. Koch · F. Jenett
Motion Bank, Hochschule Mainz—University of Applied Sciences, Mainz, Germany
e-mail: david.rittershaus@hs-mainz.de; anton.koch@hs-mainz.de; florian.jenett@hs-mainz.de

D. Memmert (ed.), *Artificial Intelligence and Machine Learning in Sports Science*,
https://doi.org/10.1007/978-3-662-70155-3_13

benefited from these advances is dance. Although dance offers great potential in the context of digitalisation (Steinberg et al., 2019, 2020, 2024), it has rarely been the subject of interdisciplinary research, particularly on issues of digitalisation and AI. Nevertheless, due to its immense complexity, dance has a long tradition and practice of observing, analysing and describing movement. It is almost impossible to capture individual movement qualities such as physical, spatial, temporal and dynamic progression and relations in a differentiated way in analogue and real-time, which is why they are usually recorded on video. In contemporary stage dance, initiatives have been built for the digital recording, transfer and communication of dance-specific knowledge as well as choreographic working methods and organisational principles (Bleeker, 2017). Video annotation has always been an important element in this, for example for the visualisation of choreographic concepts in William Forsythe's digital projects Synchronous Objects and Motion Bank (deLahunta et al., 2017; deLahunta & Jenett, 2016).

13.2 Annotation of Movements

Research into qualitative assessment methods for movement in sport and dance science began back in the 1980s with the aim of developing criteria for assessing movement quality. However, it encountered particular difficulties because the available data collection methods, which were based on characteristics such as speed, force and acceleration, could not adequately capture the complexity and holistic nature of movement quality. To date, the assessment of qualitative movement characteristics has mainly been based on expert judgement, which has led to limited interpersonal categorisation of movement phenomena and an increase in the subjective nature of assessment in this area (Leijen et al., 2008; McCutchen, 2006; Patterson-Price, 2020). As a result, various systems for the qualitative assessment of dance and movement have been developed. Many of these systems are mostly based on the principles of Laban Movement Analysis (LMA) (Fernandes, 2015), which have been modified and transformed into highly specialised movement analysis systems, such as the BAST system (Lausberg, 1998; Lausberg et al., 1996).

13.2.1 Morphological Methods of Movement Analysis in Dance Using the Example of Laban Movement Analysis

A complete formalisation of dance is difficult because contemporary dance in particular (in comparison to classical ballet) is no longer based on standardised movement vocabulary and is characterised by a heterogeneity of approaches that are subject to an ongoing artistic-discursive transformation process (Traub, 2001). This is also reflected in the failure of standardised dance notation systems over the course of the twentieth century (Pouillaude, 2017), meaning that a unified, standardised notation system for dance has not been established, as it has for example in music.

However, the Laban Movement Analysis System (LMA) provides useful approaches for the present project.

With the aim of making dance easier to communicate and discuss, Rudolf von Laban (1879–1958) developed an extremely detailed system of analysis and notation based on his dance practice at the beginning of the twentieth century (Kennedy, 2014; Wahl, 2019). Laban broke down the complex phenomenon of dance in a methodical and structured way and formed movement-specific basic parameters. It should be noted that Laban did not consider one specific dance technique or vocabulary only, but took the moving body as a general starting point—be it in dance, in sport or at work in a Taylorist-Fordist context (Herrmann, 2002). Laban Movement Analysis (LMA) does not only look at dance analytically, but also offers a framework and methods for teaching and organising movement. It is used not only in dance practice, movement theory and research, but also in the therapeutic field in an interdisciplinary manner (Kennedy, 2014; Lausberg, 1998). The movement analysis system was further developed and specified by Laban's students. Irmgard Bartenieff played a key role in the specification and dissemination of the LMA. For this reason, Laban's concept is often referred to as Laban/Bartenieff Movement Analysis (Kennedy, 2014). Bartenieff expanded Laban's basic principles and categories to include her expertise from the fields of dance and physiotherapy and thus developed her system of Bartenieff Fundamentals, which is used internationally in teaching and therapy as a training method or corrective bodywork (cf. Kennedy, 2014, p. 125). The LMA comprises of six movement analytical categories—body, space, effort, form, phrasing and relationship—which can be further differentiated in their facets resulting in around 60 movement parameters (Kennedy, 2014).

Laban Movement Analysis (LMA) serves as a basis in various research areas and projects for the development of evaluation categories and systems for the analysis of movement behaviour or for the classification of movements for data analysis (e.g. Burton et al., 2016; Büning et al., 2023; Garcia & Ronfard, 2020). In addition, LMA is also used in the fields of human-computer interaction, computer animation, machine learning of movement parameters and movement detection and evaluation in other work (Rallis et al., 2020). When using the LMA system as an observational tool, it is crucial to assess interrater reliability before the system is recognised as a method for motion analysis. Initial studies and literature reviews show the importance of comparative assessments, especially when comparing human assessments with marker-based assessments and AI-supported assessments (cf. Meng, 2022; Miko et al., 2022).

13.2.2 Video-Based Possibilities for Annotating Movement in Dance

In dance, annotation can build on the fact that video has established itself as a recording standard (Franko, 2011, p. 331) and media deficits are compensated for by dancers and an increasingly widespread movement literacy (Karreman, 2013, p. 121). Annotation is now seen as an additional way of adding dance knowledge to videos (Stancliffe, 2018, p. 203) or further layers of information and modalities and linking them semantically (Rittershaus et al., 2022). The annotation process also trains the analytical eye (Stancliffe, 2018) and allows expertise to be developed for choreographic works (deLahunta et al., 2017; Jenett et al., 2021–2024). The time-based annotations also open up the hermetic video format, making it more deeply referenceable and searchable.

Since 2016, the Motion Bank research group at Mainz University of Applied Sciences has been bringing together software prototypes for annotating dance and audio-visual media (dance recordings) in a web-based, modular open source software, the Motion Bank System. Annotations are stored in accordance with the W3C Web Annotation Model and are therefore standards-compliant and interoperable with other linked data collections. Nevertheless, the software interface is designed to be low-threshold and intuitive, so that dancers, students and researchers can work with the software without having to go through a long learning process. The easy accessibility is also supported by the fact that the annotations—unlike with some other annotation and coding tools—are freely customisable, i.e. no categories or labels have to be created in advance, but videos can be annotated descriptively. Thanks to its design as a web application, the Motion Bank System can also be used collaboratively in different constellations and groups. For this project, the software was used as a starting point and expanded to include automated video analysis processes (machine learning pipeline, see below).

13.3 Automated (Pre-)annotation

The Open Source Computer Vision Library (OpenCV) is a widely used software library that provides a broad collection of implementations of classic and modern machine vision methods. It offers numerous methods for segmenting, analysing and processing image data and has thus become an essential tool in digital image processing. In the past, this library was used to implement simple motion detection using visual markers and is today part of the basis of many newly developed methods for analysing image and sensor data using machine learning, often referred to synonymously as AI. Pose recognition, both in 2D and 3D, is now possible efficiently using standard hardware thanks to the further development of machine learning methods. OpenPose has established itself as a centralised and extremely efficient real-time system for recognising bodies, faces and joints in multi-person scenes. The system not only contains software based on TensorFlow, but also offers a range of pre-trained models.

The use of automated systems to generate feedback in movement learning processes is continuously increasing (Turmo Vidal et al., 2023). More and more research approaches and studies based on deep learning methods such as human pose estimation and human action recognition are being pursued, particularly in the field of movement learning.

Human action recognition (HAR) is based on video data and is a method for recognising and distinguishing human movement. HAR makes it possible to recognise various human activities in videos, from simple gestures to complex motion sequences. This technology utilises machine learning and computer vision and is used in various fields such as surveillance systems, sports analysis, healthcare, virtual reality and human-computer interaction. By automating processes, it helps to improve the user experience in various applications.

These approaches aim to support test subjects in identifying errors and offer direct correction options via the visualisation of the body and movement using skeletal models or avatars (Kang et al., 2023; Steinberg et al., 2024; Tharatipyakul & Pongnumkul, 2023).

Tharatipyakul and Pongnumkul (2023) define pose estimation as a computer vision technique that uses 2D or 3D images to analyse human posture by determining the position and orientation of individual body parts. This makes it possible to use the analysis of movements to generate feedback. With the help of deep learning, a recent development in the field of machine learning, 2D video data can be analysed efficiently and posture data can then be provided in both 2D and 3D. In this way, automated feedback can be generated in a very short time without further post-processing of the video material (Voulodimos et al., 2018).

Selected technological developments for the visualisation and reflection of movement, its tracking and analysis, will now be presented in more detail in order to determine the potential of the application of pose estimation and human action recognition and to understand how these findings can be used in sport and dance science.

13.3.1 Overview of Procedures and Methods for System Utilisation and Model Development

The current state of research shows that video and image material often form the basis for the automatic processing of movements using pose estimation and action recognition methods (see Table 13.1). Researchers mainly use camera-based, markerless tracking systems to record movements. This also shows that new methods that rely on person and pose recognition using image analysis can manage with conventional camera systems and uncontrolled recording situations. They are better suited for the use in teaching as well as in artistic contexts than complex motion capture systems, which often have to be adjusted and calibrated first.

In the work of Baş et al. (2023), Kang et al., 2023), Liu et al. (2021) and Aristidou et al. (2019) shown in Table 13.1, pose data and information are used on the one hand for training and optimising systems and models to improve tracking functions and pose recognition and on the other hand to generate and output classifications and information on movement executions based on the data. Visual feedback is provided via literal annotations or comments, stick figures or avatars. The studies by Liu et al. (2021) and Aristidou et al. (2019) deal with the development of models for improved processing and analysis of movement data. These models are based on joint angle analyses in different variations. In the application of these analyses, the researchers draw on data sets from open-source libraries and use open-source

Table 13.1 Current studies on the use of machine learning/deep learning for motion annotation

No.	Authors (year), country Title	Study design/research activity	Technology and functions	System output
1	Baş et al. (2023), Turkey DancÆR: Efficient and Accurate Dance Choreography Learning by Feedback Through Pose Classification	Comparative study (in planning): 200 students from Hisar + similar high schools—3 randomised groups: 2 control groups, 1 experimental group, different learning settings • Survey by questionnaire	Technology: • Augmented reality application for step-by-step choreography learning Function: • AI feedback on accuracy—position and timing, rhythm-based segmentation, speed reduction, step-by-step movement introduction, three main interfaces: 1. Home page–choice of dance sessions, 2. AR dance session, 3. Self-evaluation	• Display of incorrect poses in optimised position (movement marker/ low opacity avatar) • Accuracy score and general advice by AI-Instructor • 2D avatar option/ anonymisation • Annotation option
2	Kang et al. (2023), Korea Dancing on the inside: A qualitative study on online dance learning with teacher-AI cooperation	Preliminary study: 4 dance teachers (3 f/1 m; 31–40 years): Semi-structured interviews—qualitative content analysis Two-week field study: 11 dance beginners (6 f/5 m; 21–27 years): • Quantitative data analysis—login times and scores, focused, semi-structured interviews—qualitative analysis	Technology: • Online dance learning system and platform Functions: • Video comparison and evaluation • 4 modules: Warm up + tutorial, recording of practice, AI tutor report, human teacher report	• AI report: Overall similarity score, timing evaluation, summary on body parts (correct/ incorrect + Emoji) • Teachers Report: written comment • Duet video display • Naming phrases by annotation
3	Kimm et al. (2021a, 2021b), USA MoveFeel: Expressive Dance Movement Determination Through Video Analysis	Feasibility study: • Video evaluation: 10 videos (5 religious dance/hip hop, 5 modern/interpretative dance) • Quantitative evaluation	Technology: • MoveFeel—computing framework for the assessment of expressive dance Function: • Categorisation of dances and movements, Laban Efforts recognition	• Automatic categorisation (happy or sad) of dances and movements

(continued)

Table 13.1 (continued)

No.	Authors (year), country Title	Study design/research activity	Technology and functions	System output
4	Liu et al. (2021), China Learning 3-D Human Pose Estimation from Catadioptric Videos	Exploration: New ways of obtaining 3D human pose data without manual annotation via catadioptric systems. • Proposition of large-scale dance-before-mirror (DBM) dataset, development of novel algorithm, cross-data transfer experiments with other datasets	Technology: • Novel algorithm that extracts human 3D pose from catadioptric videos Function: • 3D Keypoint estimation and 3D pose collection	• Pose data animated as a stick figure skeleton • 3D human pose estimation
5	El Raheb et al. (2020) Greece Symbolising Space: From notation to movement interaction	Interdisciplinary, experimental, qualitative study, research through design: 30 participants: 5 dance and movement experts, 5 technology experts, 5 from wider public (children and teenager), 15 digital natives (8–15 years, no information about gender) some with ballet background, no experiences with Laban Movement Analysis (LMA) • Tryouts with several user groups • Observation → Survey by questionnaire and interviews	Technology: • Interactive XR-learning environment, prototypes: A. Augmented mirror version B. VR setting Functions: • Movement tracking • Interaction with interface	• A: Visualisation in augmented mirror • B: Visualisation in VR • A + B: Laban notation signs displayed in relation to execution. • Sound and colour change to indicate correct or incorrect execution • Summarising score
6	Tharatipyakul, Choo and Perrault (2020), USA Pose Estimation for Facilitating Movement Learning from Online Videos	User study: 12 participants (7 f/5 m; 22–37 years): 11 with no previous experience, 1 with experience in Tai Chi • Quantitative analysis	Technology: • Web based learning application Function: • Visualisation of movement execution	• Skeleton visualisation in four different versions

(continued)

Table 13.1 (continued)

No.	Authors (year), country Title	Study design/research activity	Technology and functions	System output
7	Aristidou et al. (2019), Israel Cypers Digital Dance Ethnography: Organising Large Dance Collections	Development of a computer supported analysis system • Establishing a systematic database of 72 dances • Cluster analysis • Experiments to evaluate method • Interrater test with 5 dance experts	Technology: • Computer supported analysis system Function: • Enriching existing data base and design of a meta-database • Catalogue, categorize and identify similarities	• Automated analysis, classification in style—motion and style signatures • Figural representation of sequences and relation to other dance styles

models for programming. A frequently used example of this is OpenPose (Kang et al., 2023; Kimm et al., 2021a, 2021b). In addition, proprietary tracking systems such as Kinect are used and own, supplementary data sets are created (Aristidou et al., 2019).

The study by Liu et al. (2021), for example, describes the development and application of different models for analysing the dance technique used in each case. Liu et al. (2021) use videos that record movement sequences in front of a mirror to generate high-quality 3D human keypoints using an algorithm without the need for a complex camera system. These videos are known as catadioptric videos. By using this recording setup, 3D human keypoints can be extracted from the videos. The advantage is that the recording via the mirror offers an additional perspective on the moving person. This means that pose data from two perspectives of the same movements can be obtained from a single recording. This helps to avoid depth distortions and inaccuracies without the need for additional effort. The extracted pose data is then animated as stick figures to train a model. Existing online databases are used to develop this model and the freely available videos are inserted into the model.

13.3.2 Automated Pre-annotation

Due to the frequent further developments in the field of deep learning, a modular structure of a pipeline for video processing makes sense, which corresponds to the modular overall concept of the Motion Bank Software System.

The term "pipeline" comes from computer science and engineering and originally refers to a series of processing steps through which data flows continuously. In software development and data processing, the concept is used to describe the flow and processing of data in a system, similar to how fluids flow through a physical pipeline. In this context, a pipeline describes the sequential process in which raw data is automatically processed and evaluated based on learnt patterns and relationships.

With the modular structure, it is possible to respond to new releases of more powerful and completely new models in order to keep the pipeline fit for the future and at the cutting edge of technology. The data builds on each other and more complex properties can be determined from the videos through the flexible combination of different methods.

Due to the focus on dance, the choice was primarily based on benchmarks of various models in this area.

Benchmarks are standardised tests and data sets that are used to objectively measure and compare the performance of AI models and ensure high standards in AI research and development. They serve as reference points to objectively measure the effectiveness and efficiency of different algorithms and systems. Benchmarks often use well-known and well-curated data sets that are accepted by the research and development community. Examples include Common Objects in Context (COCO) for object and image segmentation.

DanceTrack is particularly relevant in this project (Sun et al., 2022). The movement patterns here, for example, differ greatly from other video recordings in everyday situations such as in a pedestrian zone and are ideally suited for the "#vortanz" project due to their focus on dance performances.

The simplified illustration of the pipeline (Fig. 13.1) shows the relationship between the individual methods. In the first step, YOLOX (https://arxiv.org/abs/2107.08430) is used to recognise people (Ge et al., 2021). This recognises areas in the image that depict a person for each individual frame of the video. To establish a connection between the data of individual images, OC-SORT (https://arxiv.org/abs/2203.14360) is used as a tracking method in order to recognise people in the individual images of the video as realistically as possible and to be able to correctly assign the areas that these people occupy in the image (Cao et al., 2022). This basic data enables more complex analyses, but can also be used in its own right, for example by allowing people to be selected in the video image and annotations to be assigned to individual people. In addition to visual data, the audio data is also analysed and automatically transcribed with the help of Whisper (https://arxiv.org/abs/2212.04356) in order to quickly create written annotations from speech in the video (Gao et al., 2023).

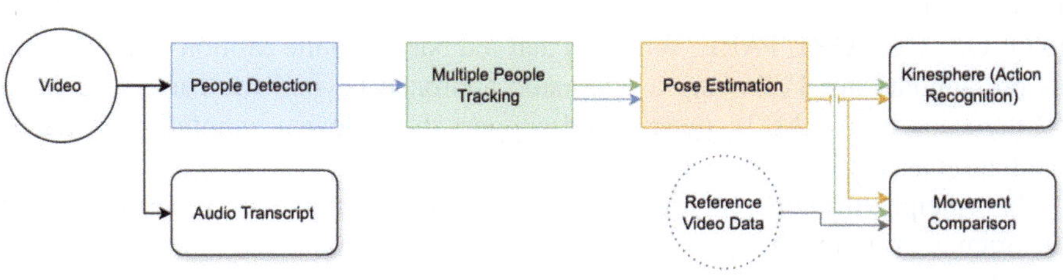

Fig. 13.1 Simplified representation of the pipeline for the #vortanz project (own illustration)

13.3.3 Human Action Recognition/Pose Estimation Procedure

A sequence of image areas of individual, specific persons is created through the data of individual images, their temporal relationship and coupling to identified persons. This allows PARE (https://arxiv.org/abs/2104.08527) to generate pose data that quantitatively maps the position and orientation of the persons' bodies (Kocabas et al., 2021). This results in data points that describe the position of different anchor points of the body over time. These values can in turn be used to determine movement characteristics. The AI model architecture of the AGCN from the field of human action recognition uses this pose data and classifies movements that are represented by the pose data in certain time periods (Shi et al., 2019). As part of the #vortanz project and presented here, a data set was created to describe the kinesphere (e.g. Lleshi, 2021; Steinberg et al., 2024). On the one hand in relation to the entire body, on the other hand in relation to individual limbs. On this data set, several AI models could be trained with the AGCN architecture, each of which annotates the movement characteristics of individual persons in different time ranges according to the respective kinesphere (see below). This is an example of a movement feature that can be automatically annotated with this method and thus shows the potential for many other possibilities. Further analyses of quantitative pose or movement data that allow statements to be made about movement qualities, such as those already used with older computer vision technology (Fdili Alaoui et al., 2017; Volpe, 2003), are obvious.

13.4 Application Using the Example of Motion Bank System

In the present research project #vortanz (cf. Jenett et al., 2021–2024), machine-learning-supported software was implemented as an intelligent assistance system to support the learning process within selected courses and modules of dance education and analysed with regard to learning success. For this purpose, the Motion Bank software system was sustainably integrated into teaching and learning at three locations in Germany (Cologne, Frankfurt, Berlin) at different levels of study (BA, MA) and with a profile-related focus (dance, choreography, movement culture). By connecting an analysis pipeline to the system, additional data could be generated from the video material with the help of established, previously described methods (YOLOX, OC-SORT, PARE, Whisper) and fed back into the Motion Bank System as pre-annotations. Following a pragmatic approach beyond the current AI hype (Rittershaus et al., 2023), these could support the reflexive annotation of dance mediation as application-oriented functions.

In principle, the concrete implementation of the ML-supported functions should be orientated as much as possible towards the local conditions and the subject-specific didactic requirements that were determined in the first project phase together with the teachers (subject-specific didactics and university didactics) at the training locations. This should ensure that they actually and sustainably contribute to (facilitating) the achievement of learning objectives. The functions are based on the following basic concepts that extend video annotation:

- Visual anchors in the video (people, poses) created through image analysis serve as reference points for free text annotation
- Pose data enables automatic recognition and annotation of movement characteristics (kinesphere)
- Automated transcriptions of spoken words and language analyses (e.g. term frequency) support the analysis of verbal communication in combination with non-verbal communication and movement

The following application examples describe pilot tests from the project that were developed in particular for didactic use in dance teaching modules at the German Sport University Cologne. The so-called use cases are independent of dance style and offer reflection and feedback options for students (among themselves) and teachers, as well as individual options for analysing movements and comparison options—mainly with a focus on movement quality and accuracy.

13.4.1 Use Case 1: People Tracking

The use case “People Tracking” refers to the application of computer vision techniques to track people in a dance group or a teaching situation in a video and to mark them visually with the help of bounding boxes. The process starts with finding people in the video using object recognition algorithms. These algorithms automatically identify and localise the positions of all people in the video and provide data that can be used to display bounding boxes around the people. Each bounding box represents a person in the dance group. In the setting presented here, a dance video showing a group of dancers performing a choreography is used as input (see Fig. 13.2). The aim of the “people tracking” function is to track each person in the dance group over the duration of the video in order to be able to provide individual, person-specific feedback via the input window of the annotation interface. This feedback appears on the left-hand side as an annotation. In the so-called “swimlane” view below the video window, the annotations also appear as visual markers on the video timeline.

The use of bounding boxes facilitates the visual tracking of individuals as they are clearly marked and highlighted in colour when individual annotations are selected. This enables researchers, facilitators or choreographers to analyse specific movements or patterns within groups, provide feedback or improve certain aspects of the choreography. In the future, these functions can also be used for archives and media libraries, which can use them to mark and link people related content in databases to videos.

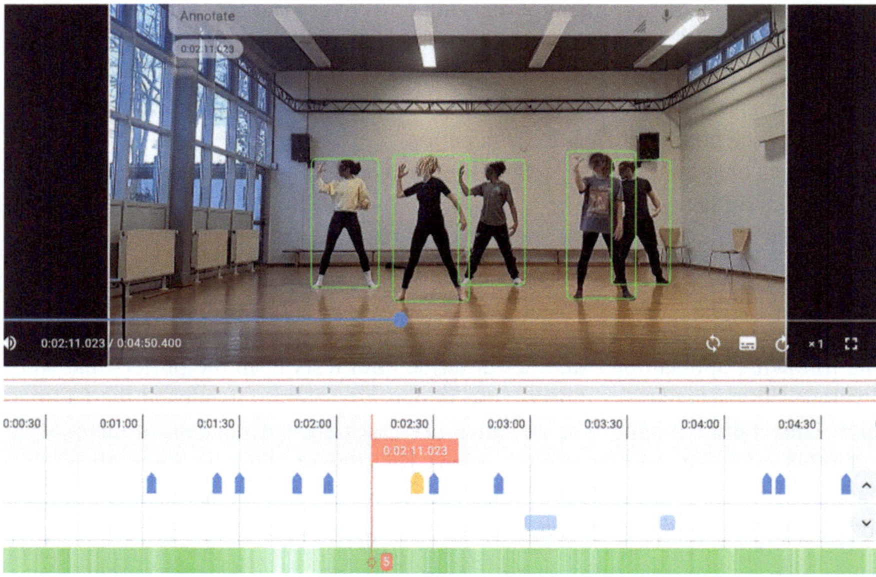

Fig. 13.2 "People tracking" use case in the #vortanz project (own illustration)

13.4.2 Use Case 2: Pose Estimation

The "Pose Estimation" use case is used in the context of feedback on movement execution and refers to the recording and analysis of dancers' postures and movements in order to give them constructive feedback on the quality of their movement performance, for example. In this scenario, the pose estimation algorithms analyse the captured images to determine the positions of the dancers' body parts in space. This typically includes the positions of extremities and joints such as arms, head, shoulders, legs and especially knees and lower legs (see Fig. 13.3). In the Motion Bank System, the body parts can be marked and commented on specifically. The recorded video can then be analysed in order to give the dancers detailed feedback on their individual movement execution. The free form of the commentary or annotation function allows the feedback to be customised to individual needs and goals of the dancers and can include emphasising certain aspects of a movement, tips on improving techniques or recommendations for specific exercises.

13.4.3 Use Case 3: Kinesphere Recognition

In relation to the Laban Movement Analysis (see above), the kinesphere describes the personal movement space and its use. The range of movement is differentiated into the dimensions of narrow, medium and wide: Does someone move close along their own body, medium with strongly flexed joints or wide, stretched away from the centre of the body (Miko et al., 2022). These categories were defined on the

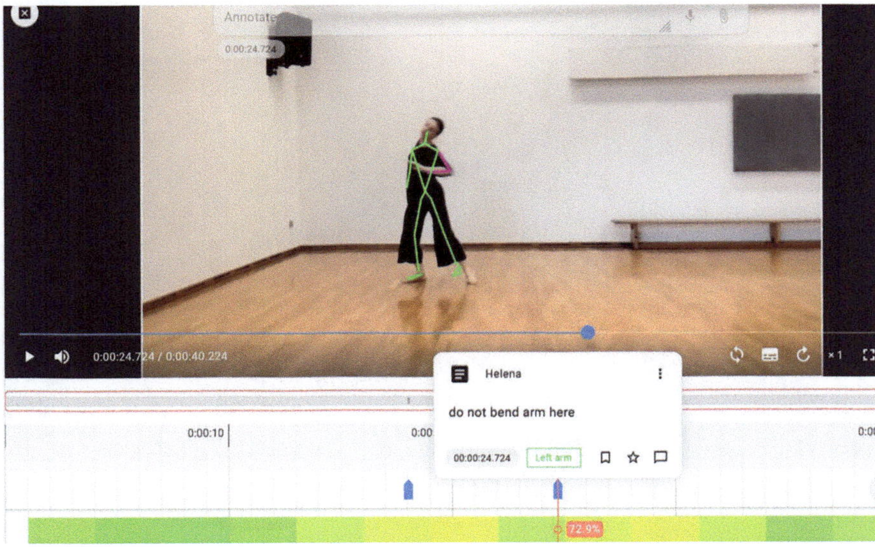

Fig. 13.3 Use case "Pose Estimation" in the #vortanz project (own illustration)

basis of the principles of Laban Movement Analysis (cf. Kennedy, 2014) and supplemented by exceptional or borderline cases. A catalogue of criteria was developed to classify and categorise the movements for the training of ML-supported movement recognition. Category-specific movement recordings were created in collaboration with students from the German Sport University Cologne and from the field of dance training. To this end, 60 students with different levels of dance experience and professional dancers were asked to move in the individual categories, which was recorded on video. Of the 240 videos recorded, around 120 were annotated in detail. The human action recognition (HAR) models were trained with the kinesphere material and then evaluated with different dance sequences to test their robustness (cf. Miko et al., 2022). The automated feedback on movement execution (narrow, medium or wide) is visualised in the interface of the Motion Bank System on the bounding box as well as in colour in the swimlane view (see Fig. 13.4). This feedback can help students to become aware of their range of movement within dance choreographies. It can also be used to check the maximum or minimum range of movement in order to train body and movement awareness.

13.4.4 Use Case 4: Comparison of Dance Movements

The "Comparison of dance movements" use case is used to analyse and improve dance technique and performance. This use case relates to the recording and comparison of dance movements in order to identify differences and similarities between different versions of dance sequences (see Fig. 13.5). On the other hand, it can also be used to visualise the attempt to achieve maximum variation in the movement. In

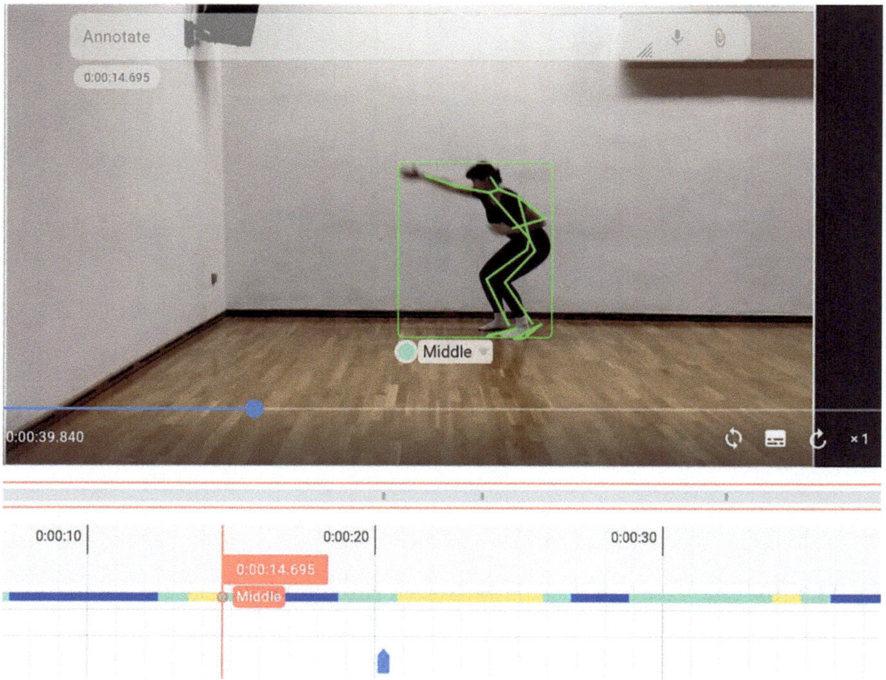

Fig. 13.4 Use case "Kinesphere Recognition" in the #vortanz project (own illustration)

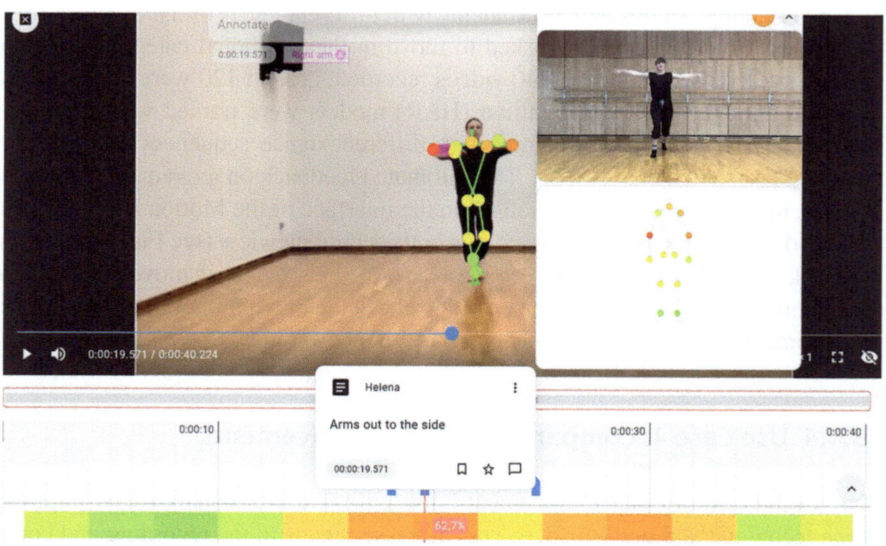

Fig. 13.5 Use case "comparison of dance movements" in the #vortanz project (own illustration)

this scenario, pose estimation algorithms are also used to analyse the dance movements. The algorithms capture the positions and angles of the dancers' body parts in space, including extremities and joints such as arms, head, shoulders, legs and especially knees and lower legs. The movement data extracted from the video is then compared with that from a template to identify differences and similarities in movement execution. The feedback is based on the results of the comparison of the joint angle positions and speed of the movements.

13.5 Conclusion

Research into the application of machine learning and deep learning in the field of dance has become increasingly important in recent years, although there has been little research into the application context of higher education to date. The annotation of movements has played an important role in dance for some time in various application contexts. Automated pre-annotation using systems such as pose estimation and HAR enables efficient, mobile and practical analysis, visualisation and reflection of dance movement. This article outlines the use of automated methods for recognising and evaluating movement in dance and the benefits of digital feedback against the background of the university didactic setting of the German Sport University Cologne, in which the application scenarios are used. Research results from the BMBF project #vortanz on specific challenges and potentials were presented using application examples. The integration of new techniques and technologies into the Motion Bank System (Mainz University of Applied Science) enables a wide range of possibilities for analyzing dance movements and contributes to research in the field.

References

Aristidou, A., Shamir, A., & Chrysanthou, Y. (2019). Digital dance ethnography: Organising large dance collections. *ACM Journal on Computing and Cultural Heritage, 12*(4), Article 29, 27 pages. https://doi.org/10.1145/3344383

Baş, I., Alp, D., Ergenç, L. C., Koçak, A. E., & Yalçın, S. (2023). DancÆR: Efficient and accurate dance choreography learning by feedback through pose classification. In *International Conference on Artificial Intelligence in Education* (pp. 747–755). Springer.

Bleeker, M. (Ed.). (2017). *Transmission in motion: The technologising of dance*. Routledge.

Büning, C., Steinberg, C., & Lausberg, H. (2023). Impact of divergent learning experiences and diversification on movement self-perception and movement performance in university sport students. *Research Quarterly for Exercise and Sport*. https://doi.org/10.1080/02701367.2023.2174489

Burton, S. J., Samadani, A. A., Gorbet, R., & Kulić, D. (2016). Laban movement analysis and affective movement generation for robots and other near-living creatures. In *Dance notations and robot motion* (pp. 25–48).

Cao, J., Weng, X., Khirodkar, R., Pang, J., & Kitani, K. (2022). Observation-centric SORT: Rethinking SORT for robust multi-object tracking. In *Proceedings of the IEEE/CVF Conference on Computer Vision and Pattern Recognition* (pp. 9686–9696). https://arxiv.org/abs/2203.14360

deLahunta, S., & Jenett, F. (2016). Making digital choreographic objects interrelate: A focus on coding practices. In M. Leeker, I. Schipper, & T. Beyes (Eds.), *Performing the digital. Performance studies and performances in digital cultures* (pp. 63–79). Transcript.

deLahunta, S., Cramer, F. A., & Jenett, F. (2017). The digital and back again—A conversation on Motion Bank. In B. Büscher & F. A. Cramer (Eds.), *Fluid access: Archiving performance based arts* (pp. 211–224). Olms.

El Raheb, K., Stergiou, M., Katifori, A., & Ioannidis, Y. (2020). Symbolising space: From notation to movement interaction. In *Proceedings of the International Conference on Technologies for Music Notation and Representation-TENOR* (Vol. 20, p. 21).

Fdili Alaoui, S., Françoise, J., Schiphorst, T., Studd, K., & Bevilacqua, F. (2017). Seeing, sensing and recognising laban movement qualities. In *Proceedings of the 2017 CHI Conference on Human Factors in Computing Systems.*

Fernandes, C. (2015). *Moving researcher: Laban/Batenieff movement analysis in performing arts education and creative arts therapies*. Jessica Kingsley Publishers.

Franko, M. (2011). Writing for the body: Notation, reconstruction, and reinvention in dance. *Common Knowledge, 17*(2), 321–334.

Gao, D., Xu, H., Raj, D., Perera, L. P. G., Povey, D., & Khudanpur, S. (2023). Learning from flawed data: Weakly supervised automatic speech recognition. In *2023 IEEE Automatic Speech Recognition and Understanding Workshop (ASRU)* (pp. 1–8). IEEE. https://arxiv.org/abs/2212.0435

Garcia, M., & Ronfard R. (2020). Recognition of Laban Effort Qualities from Hand Motion. 7th International Conference on Movement and Computing (MOCO '20), Jul 2020, Jersey City/Virtual, United States. hal-02899999. https://doi.org/10.1145/3401956.3404227

Ge, Z., Liu, S., Wang, F., Li, Z., & Sun, J. (2021). Yolox: Exceeding yolo series in 2021. *arXiv preprint arXiv:2107.08430.*

Herrmann, H.-C. (2002). Movement writings. On the scientific and media-historical context of Rudolf von Laban's kinetography around 1930. In S. Dinkla & M. Leeker (Eds.), *Tanz und Technologie/dance and technology. Auf dem Weg zu medialen Inszenierungen/moving towards media productions* (pp. 134–161). Alexander-Verlag.

Jayal, A., McRobert, A., Oatley, G., & O'Donoghue, P. (2018). *Sports analytics*. Routledge. https://doi.org/10.4324/9781315222783

Jenett, F., Furbach, U. & Steinberg, C. (2021–2024). *#vortanz—Automated pre-annotation in digital university dance education*. Funded by the Federal Ministry of Education and Research (BMBF) in the funding line "Digital Higher Education (Big Data and AI)". Retrieved December 21, 2022, from https://vortanz.ai/#/de

Kang, J., Kang, C., Yoon, J., Ji, H., Li, T., Moon, H., Ko, M., & Han, J. (2023). Dancing on the inside: A qualitative study on online dance learning with teacher-AI cooperation. *Education and Information Technologies, 28*. https://doi.org/10.1007/s10639-023-11649-0

Karreman, L. (2013). The dance without the dancer: Writing dance in digital scores. In J. Fletcher & R. Allsopp (Eds.). *Performance Research, 18*(5), 120–128.

Kennedy, A. (Ed.). (2014). *Moving knowledge. Understanding and experiencing Laban/Bartenieff movement studies*. Logos Publishers.

Kimm, H., Sullivan, A. Y., & Jain, S. (2021a). MoveFeel: Expressive dance movement determination through video analysis. In *ACM Workshop on Body Centric Computing Systems (BodySys 21), Virtual, WI, USA*. ACM (6 pages). https://doi.org/10.1145/3469260.3469668

Kimm, H., Sullivan, A. Y., & Jain, S. (2021b, June). MoveFeel: Expressive dance movement determination through video analysis. In *Proceedings of the Workshop on Body-Centric Computing Systems* (pp. 18–23).

Kocabas, M., Huang, C. P., Hilliges, O., & Black, M. J. (2021). PARE: Part attention regressor for 3D human body estimation. In *2021 IEEE/CVF International Conference on Computer Vision (ICCV)* (pp. 11107–11117). https://arxiv.org/abs/2104.08527

Lausberg, H. (1998). Movement diagnostic test with rating scales for diagnostics and therapy evaluation in dance therapy. *Journal for Dance Therapy, 7*, 35–42.

Lausberg, H., von Wietersheim, J., & Feiereis, H. (1996). Movement behaviour of patients with eating disorders and inflammatory bowel disease: A controlled study. *Psychotherapy and Psychosomatics, 65*(5), 272–276. https://doi.org/10.1159/000289087

Leijen, Ä., Admiraal, W. F., Wildschut, L., & Simons, P. R. J. (2008). Pedagogy before technology: What should an ICT intervention facilitate in practical dance classes? *Teaching in Higher Education, 13*(2), 219–231.

Liu, C., Li, Y., Ma, K., Zhang, D., Bao, P., & Mu, Y. (2021). Learning 3-D human pose estimation from catadioptric videos. In *IJCAI* (pp. 852–859).

Lleshi, R. (2021). *Video based human action recognition using deep learning*. Master Thesis, University of Koblenz-Landau.

McCutchen, B. P. (2006). *Teaching dance as art in education*. Human Kinetics.

Meng, Y. (2022). Design of automatic motion capture algorithm for Yao's long drum dance based on multieye machine vision. *Scientific Programming, 2022*, 1126020. https://doi.org/10.1155/2022/1126020

Miko, H., Siewert, K., Büning, C., Rittershaus, D., Jenett, F., & Steinberg, C. (2022). Possibilities and limits of an AI-supported annotation system in the context of movement-analytical and choreographic reflection processes. In *Tradition—Transition—Vision, Laban EUROLAB Conference*: 25–27 November 2022, , Loheland, Fulda, Abstract volume (pp. 14–16). https://www.laban-eurolab.org/konferenz-2022/

Patterson-Price, J. (2020). Youth dance fundamental movement skills and assessment. In S.-H. Lee, M. L. Morris, & S. V. Nicosia (Eds.), *Perspectives in performing arts medicine practice* (pp. 347–370). Springer.

Pouillaude, F. (2017). *Unworking choreography: The notion of the work in dance*. Oxford University Press.

Rallis, I., Voulodimos, A., Bakalos, N., Protopapadakis, E., Doulamis, N., & Doulamis, A. (2020). Machine learning for intangible cultural heritage: A review of techniques on dance analysis. In F. Liarokapis, A. Voulodimos, & N. Doulamis (Eds.), *Springer series on cultural computing. Visual computing for cultural heritage* (pp. 103–119). Springer. https://doi.org/10.1007/978-3-030-37191-3_6

Rittershaus, D., Koch, A., deLahunta, S., & Jenett, F. (2022). Recording "Effect": A case study in technical, practical and critical perspectives on dance data creation. In C. Fernandes, V. Evola, & C. Ribeiro (Eds.), *Dance data, cognition, and multimodal communication* (pp. 71–88). Routledge. https://doi.org/10.4324/9781003106401-7

Rittershaus, D., Otto, L., Miko, H., & Stange, J. P. (2023). After the hype: Routines of artificial intelligence in dance, choreography and mediation. In K. Schneider (Ed.), *Routines in dance: Artistic practices between stabilisation and destabilisation* (pp. 255–259). Jahrbuch TanzForschung 2022. transcript.

Shi, L., Zhang, Y., Cheng, J., & Lu, H. (2019). Two-stream adaptive graph convolutional networks for skeleton-based action recognition. In *Proceedings of the IEEE/CVF Conference on Computer Vision and Pattern Recognition* (pp. 12026–12035).

Siewert, K., & Büning, C. (2023). Avatar-based feedback in movement-related educational programmes at university level. *Physical Education, 72*(2), 59–64.

Stancliffe, R. (2018). *Video annotation for the articulation and transmission of dance and movement knowledge*. Unpublished doctoral thesis. Coventry University

Steinberg, C., Jenett, F., Bindel, T., Zühlke, M., Koch, A., & Rittershaus, D. (2019). #digitanz—The question of digital support for creative processes. In B. Jörissen, S. Kröner, & L. Unterberg (Eds.), *Cultural education and digitality: Vol. 1. Research on digitalisation in cultural education* (pp. 49–62). kopaed.

Steinberg, C., Zühlke, M., Bindel, T., & Jenett, F. (2020). Aesthetic education revised: A contribution to mobile learning in physical education. *German Journal of Exercise and Sport Research, 50*(1), 92–101. https://doi.org/10.1007/s12662-019-00627-9

Steinberg, C., Lleshi, R., & Miko, H. (2024). Video-based human action recognition and its application in dance research. In D. Memmert (Ed.), *Sports technology*. Springer Spektrum. https://doi.org/10.1007/978-3-662-68703-1_14

Sun, P., Cao, J., Jiang, Y., Yuan, Z., Bai, S., Kitani, K., & Luo, P. (2022). Dancetrack: Multi-object tracking in uniform appearance and diverse motion. In *Proceedings of the IEEE/CVF Conference on Computer Vision and Pattern Recognition* (pp. 20993–21002). https://arxiv.org/abs/2111.14690

Tharatipyakul, A., Choo, K. & Perrault, S.T. (2020). Pose Estimation for Facilitating Movement Learning from Online Videos Proceedings of the 2020 International Conference on Advanced Visual Interfaces, 2020. https://doi.org/10.48550/arXiv.2004.03209

Tharatipyakul, A., & Pongnumkul, S. (2023). Deep learning-based pose estimation in providing feedback for physical movement: A review. *Preprint, 2023*, 2023060395. https://doi.org/10.20944/preprints202306.0395.v1

Traub, S. (2001). Contemporary dance. In S. Dahms, C. Jeschke, & M. Woitas (Eds.), *Dance* (pp. 181–188). Bärenreiter, Metzler.

Turmo Vidal, L., Márquez Segura, E., & Waern, A. (2023). Intercorporeal biofeedback for movement learning. *ACM Transactions on Computer-Human Interaction, 30*(3), 1–40. https://doi.org/10.1145/3582428

Volpe, G. (2003). *Computational models of expressive gesture in computational models of expressive gesture in multimedia systems*. PhD thesis. InfoMus Lab.

Voulodimos, A., Doulamis, N., Doulamis, A., & Protopapadakis, E. (2018). Deep learning for computer vision: A brief review. *Computational Intelligence and Neuroscience, 2018*, 7068349, 13 pages. https://doi.org/10.1155/2018/7068349

Wahl, C. (2019). *Laban/Bartenieff movement studies. Contemporary applications*. Human Kinetics.

14 Profiling of High-Performance Skateboarders from Anthro-Fitness Variables: A Random Forest-Based-Machine Learning Analysis

Rabiu Muazu Musa, Aina Munirah Ab Rasid, Che Nadia Che Samsudin, Anwar P. P. Abdul Majeed, and Mohamad Razali Abdullah

Abstract

The advent of machine learning techniques has paved the way for comprehensive athlete profiling, which is instrumental in identifying excellence in sports activities. The present investigation endeavours to investigate the correlation between Anthro-fitness variables and skateboarding performance. A total of 45 skateboarders were subjected to a comprehensive evaluation that included various skateboarding maneuvers, along with anthropometric and fitness assessments. The performances of the skateboarders were categorized, and a Random Forest (RF) classification model was developed to predict the grouping of the skateboarders. The cluster analysis revealed two distinct groups: High Anthro-fit Skateboarders (HAS) and Low Anthro-fit Skateboarders (LAS). The RF model demonstrated a high degree of accuracy, achieving 97% for training and 90% for testing. Further diagnostics of the model revealed that the HAS group exhibited superior performance in all the measured variables, except for abdominal circumference, compared to the LAS group. The integration of machine learning provides coaches and stakeholders with valuable insights, aiding in informed

R. M. Musa (✉) · A. M. Ab Rasid · C. N. C. Samsudin
Centre for Fundamental and Continuing Education, Universiti Malaysia Terengganu, Kuala Nerus, Malaysia
e-mail: rabiu.muazu@umt.edu.my; chenadia@umt.edu.my

A. P. P. Abdul Majeed
School of Engineering and Technology (SET), Sunway University, Selangor Darul Ehsan, Malaysia

M. R. Abdullah
Faculty of Applied Social Sciences, Universiti Sultan Zainal Abidin, Kuala Nerus, Terengganu, Malaysia

D. Memmert (ed.), *Artificial Intelligence and Machine Learning in Sports Science*,
https://doi.org/10.1007/978-3-662-70155-3_14

decision-making for athlete development and thereby enhancing competitiveness.

14.1 Introduction

Skateboarding, an extreme sport that has gained immense popularity among the youth, carries an inherent risk of injury. However, its appeal lies in its ability to foster self-expression, nurture creativity, and embrace a freestyle approach without rigid rules (Reuters, 2021). Furthermore, the inclusion of skateboarding as an Olympic sport at the 2020 Tokyo Summer Games aimed not only to attract a younger audience but also to provide skateboarders with a productive platform to showcase their talents (Johnston et al., 2018). The global popularity of skateboarding, with its 85 million participants and a market value projected to grow at a compound annual growth rate of 5.32% from 2023 to 2028, reflects the sport's potential for fostering a new generation of athletes who can elevate a nation's standing in the competitive arena (Thakur, 2023). This highlights the significance of developing and supporting aspiring skateboarders, enabling them to reach their full potential and contribute to their country's success in this sport (Borden, 2016).

The rapid development of technology in the areas of data gathering, storage, and analysis has significantly accelerated the adoption of data-driven decision-making in various fields, including sports and physical activity (Bunker & Susnjak, 2022; Majumdar et al., 2022). This progress is facilitated by the use of wearable sensors, camera capture methods, and questionnaires, which collectively contribute to the creation of comprehensive information systems. These systems empower coaches, stakeholders, and practitioners with an in-depth understanding of athletes' physical, technical, and psychological characteristics (Haiyun & Yizhe, 2020). Such datasets are crucial in identifying and selecting top-performing athletes based on their distinct attributes and skills (Weiwei, 2022). However, despite the availability of these systems, coaches often still depend on observation-based decision-making, using their expertise and experience to identify athletes they believe have the potential to excel in sports (Karakaya et al., 2022). However, such intuitive decision-making processes are prone to errors, including biased selections influenced by personal preferences, and may not always produce the best results (Lath et al., 2021; Ofoghi et al., 2013; Roberts et al., 2019).

A multitude of studies have highlighted the importance of motor skills and fitness-related factors in enhancing athletic performance and reducing the risk of injuries. For example, a recent study revealed that athletes with superior motor skills and fitness capabilities tend to exhibit better performance on the field and increased endurance, thereby reducing fatigue (García-Pinillos et al., 2017). Skateboarding athletes, specifically, rely heavily on their lower body limbs to execute tricks with precision and ease. Consequently, augmenting lower and core body strength, along with static and dynamic balance controls, could be pivotal elements in enhancing skateboarding performance.

Moreover, research suggests that an athlete's fitness and motor skills significantly contribute to the learning process of new sports skills, facilitating the efficient acquisition of new techniques. In general, it can be deduced that the integration of motor skills and fitness-related factors can serve as a valuable tool for talent identification, performance improvement, and injury prevention in various sports, including skateboarding. Therefore, this paper aims to explore the potential of machine learning in profiling and predicting the grouping of skateboarding athletes based on their Anthro-fitness performance variables.

14.2 Materials and Methods

The following steps were adhered to achieve the purpose of the current investigation.

14.2.1 Participants

A group of 45 male skateboarders were randomly chosen from various skate parks located on the East Coast of Malaysia. They were briefly introduced to the experimental procedures and willingly agreed to fully cooperate throughout the entire process. It's important to note that these skateboarders participated in the activity purely for leisure, and none of them were affiliated with any governmental or private skateboarding clubs. It important to note that prior to the initiation of the study, informed consent was secured from the skateboarder. Additionally, ethical approval was granted by the University Human Research Ethics Committee (UMT/JKEPHMK/2021/53). This ensures that the study was conducted following established ethical guidelines and standards, safeguarding the rights and welfare of the participants.

14.2.2 Fitness Evaluation

The participants were subjected to specific Anthro-fitness evaluations designed to measure their physical strengths and body characteristics in relation to skateboarding requirements. The Anthro-fitness evaluation protocol included seven standard tests: weight, height, abdominal circumference, leg length, stork balance, star excursion balance test (SEBT), vertical jump, single-leg wall sits, standing broad jump, plank, and sit-up tests. Before the start of the assessments, participants were given detailed explanations of the experiment and underwent health screenings to ensure they were in good physical health and free from any current or past injuries. To minimize the risk of injuries, participants were advised to perform 3 min of warm-up and stretching exercises before the tests.

14.2.3 Skateboarding Performance Test

After the Anthro-fitness assessments were completed, the athletes were given additional rest time before their skateboarding skills were evaluated. The study focused on assessing five basic skateboarding tricks: Ollie, Nollie, Frontside 180, Pop-Shuvit, and Kickflip, all of which have been thoroughly examined in previous studies (Abdullah et al., 2020; Borden, 2016; Corrêa et al., 2017; Frederick et al., 2006; Rasid et al., 2021; Shapiee et al., 2020). Each participant was given three attempts to perform each trick under the close supervision of the researchers. A performance was considered successful if an athlete executed the trick or landed securely on the skateboard without stumbling. On the other hand, if an athlete failed to perform the trick or land successfully on the skateboard, the performance was deemed unsuccessful. The evaluation criteria included five general elements: difficulty and variety of performed tricks, quality of execution, use of the course and individual obstacles, flow and consistency, and repetition. The performance scores for each trick were determined by the percentage of successful tricks out of the total attempts (tricks' performance %) (Rasid et al., 2020).

14.2.4 Clustering

K-means clustering is an unsupervised algorithm that divides datasets into k distinct clusters. It starts by selecting k centroids and iteratively assigns each data point to the nearest centroid. The centroid is then updated to reflect the average of its assigned points, and this continues until the centroid stabilizes (Eswaramoorthi et al., 2018; Razali et al., 2017; Taha et al., 2009). The goal is to minimize the total distance between data points and their centroids. In this study, k-means clustering was used to create k non-overlapping clusters, aiming to increase intra-cluster similarity and maintain inter-cluster differences. The Euclidean distance metric was used for cluster formation.

14.2.5 Formulation of Random Forest Model

In the study, a Random Forest (RF) model was formulated for the classification task. The RF model is an ensemble learning method that operates by constructing multiple decision trees during training and outputs the class which is the mode of the classes for classification (Iranzad & Liu, 2024). RF model is preferred for its ability to handle a large number of input variables without variable deletion, providing an internal unbiased estimate of the generalization error, and offering insight into the importance of each predictor. Moreover, the RF model was chosen for the classification task in this study due to its robustness and high predictive accuracy as it is capable of handling large datasets with high dimensionality and can model complex interactions between predictors (Palczewska et al., 2014).

The model was trained and validated using a five-fold cross-validation technique (Muazu Musa et al., 2022; Taha et al., 2018a). This technique involves partitioning the original sample into five equal-sized subsamples or folds. Of the five subsamples, a single subsample is retained as the validation data for testing the model, and the remaining four subsamples are used as training data. The cross-validation process is then repeated five times, with each of the five subsamples used exactly once as the validation data (Musa et al., 2018; Taha et al., 2018b). The data was split into a ratio of 70:30 for training and testing respectively (Musa et al., 2019). This is considered a suitable practice in machine learning as it offers a good trade-off between training and testing. The Jamovi Statistical package version 2.4.14 as well as Spyder IDE version 5.4.3 integrated with Python 3.9 were utilized as a programming environment for the current study.

14.3 Results and Discussion

Figure 14.1 presents the results of the k-means clustering analysis, which classifies each skateboarder based on their performance levels. This clustering process is based on the evaluation of the skateboarding performance previously determined. The visualization of the clustering outcomes clearly shows the formation of two distinct groups: high-Anthro fitness skaters (HAS) and low-Anthro-fitness skaters (LAS), differentiated by their skateboarding performance scores. According to our analysis, 31 out of the total 45 participants are classified as HAS, while the remaining 14 participants are categorized as LAS.

Figure 14.2 provides a comprehensive profile of the skateboarders based on the Anthro-fitness variables examined. It is observed that High-Anthro-fitness skateboarders (HAS) exhibited superior performance in all the measured variables, except for abdominal circumference, compared to Low-Anthro-fitness

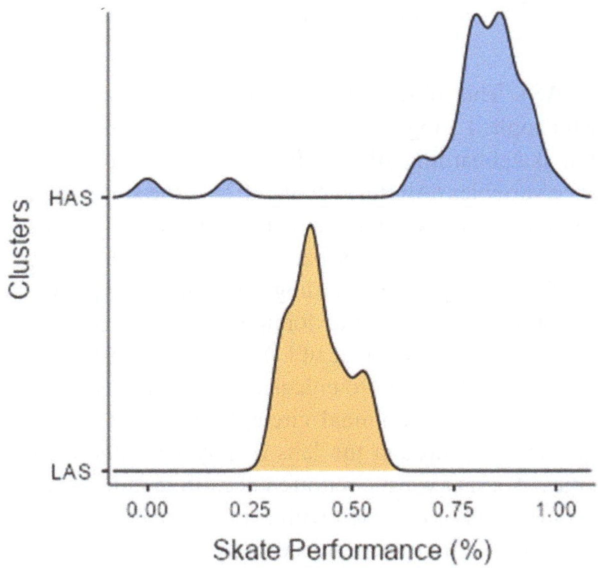

Fig. 14.1 Grouping of the skateboarders based on k-Means clustering analysis

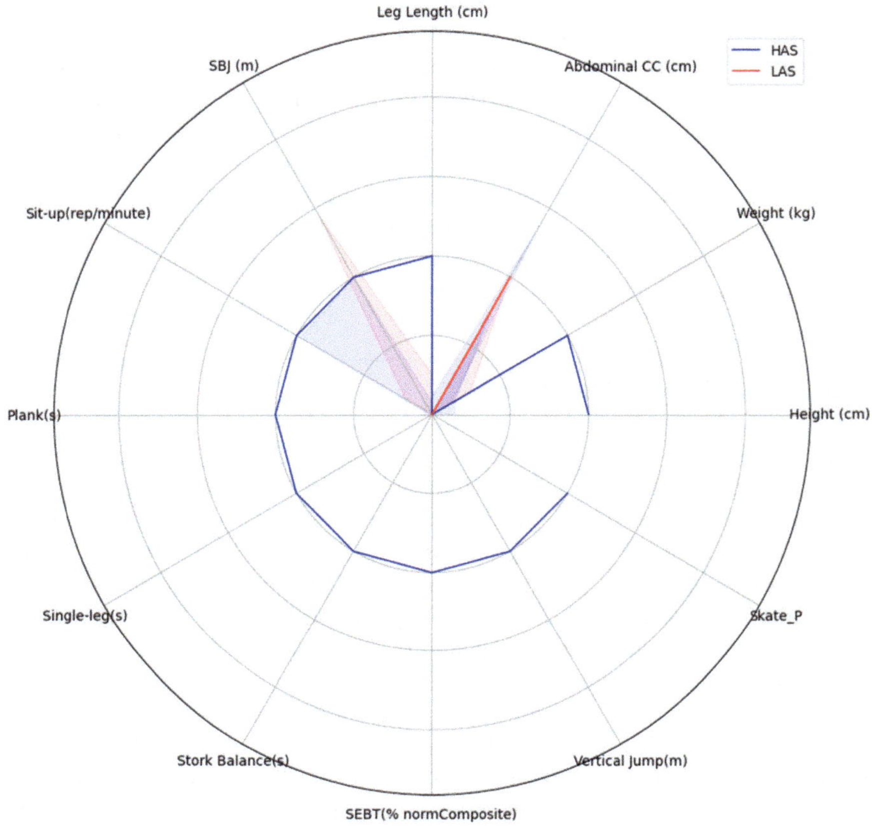

Fig. 14.2 Profiles of the skateboarders based on the Anthro-fitness evaluated

skateboarders (LAS). This observation aligns with the findings of the previous investigators, who applied a logistic regression-based machine learning pipeline to identify potential skateboarding athletes from a combination of fitness and motor skills performance variables (Ab Rasid et al., 2024). The study found that static and dynamic balance, lower body strength, and endurance were the most important factors contributing to the model's performance. These factors are therefore essential for successful performance in skateboarding. The superior performance of HAS in all the measured variables, except for abdominal circumference, could be attributed to these factors. The lower performance of LAS in these variables suggests that they may need to improve in these areas to enhance their skateboarding performance.

The association between abdominal circumference and skateboarding performance can be understood through the lens of physical fitness and its impact on athletic capabilities. A larger waist circumference, often indicative of a higher body fat percentage, is negatively associated with explosive leg strength, which is a critical component of skateboarding performance (Fitzpatrick et al., 2012). Consequently,

Table 14.1 Model performance parameters

Model parameters	Accuracy	Precision	Recall	F1 score
Training	0.97	0.98	0.99	0.99
Test	0.90	0.91	0.89	0.89

it is not surprising that the study findings indicated higher abdominal circumference attributed to the LAS group.

Table 14.1 provides a summary of the Random Forest (RF) model's performance in predicting the two distinct groups of skateboarders. The RF model achieved an accuracy of 97% for training and 90% for testing, indicating an excellent prediction of the skateboarder classes. Furthermore, the precision, recall, and F1 scores demonstrate high values, reflecting a perfect model fit. This suggests that the specific anthro-fitness variables examined in the study are crucial for excellent skateboarding performance.

This finding aligns with the preceding researchers, who applied a logistic regression-based machine learning pipeline to identify potential skateboarding athletes from a combination of fitness and motor skills performance variables (Ab Rasid et al., 2024). The study found that static and dynamic balance, lower body strength, and endurance were the most important factors contributing to the model's performance. These factors have also been previously reported as essential for successful performance in skateboarding (Henriques-Neto et al., 2020).

Moreover, the RF model is one of the most popular machine-learning models used for both classification and regression problems (Dudek, 2022). As an ensemble model, it demonstrates high predictive accuracy and low variance, while being easy to learn and optimize. This makes it an excellent tool for predicting skateboarder classes based on the anthro-fitness variables examined.

Figure 14.3 depicts the confusion matrix of the developed model after the completion of cross-validation. This analytical technique was used to evaluate the model's ability in accurately classifying the skateboarder's categories. The results from the confusion matrix show that the model correctly classified both the (HAS) and (LAS) without a single misclassification during the training phase. However, during the testing stage of the model, only one misclassification occurred for each group. This result further emphasizes the capability of the RF model in profiling and predicting the classes of skateboarders.

A confusion matrix is a powerful tool used in machine learning to assess the performance of a classification model. It provides a comprehensive summary of the model's predictions by comparing them to the actual values from the labelled dataset (Fabris et al., 2017). The confusion matrix displays the four possible outcomes of a binary classification problem: true positives (TP), true negatives (TN), false positives (FP), and false negatives (FN). These outcomes are organized in a tabular format, making it easier to visualize and interpret the model's performance.

The high accuracy of the RF model in this study aligns with the findings of the previous study which applied cluster analysis in conjunction with an artificial neural network to identify potential skateboarding athletes from a combination of

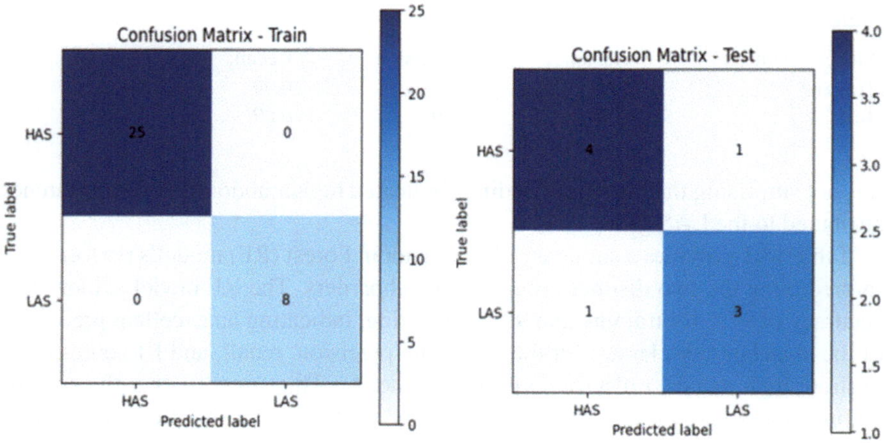

Fig. 14.3 Profiles of the skateboarders based on the Anthro-fitness evaluated

bio-fitness performance variables (Ab Rasid et al., 2023). The study found that static and dynamic balance, lower body strength, and endurance were the most important factors contributing to the model's performance of 98% accuracy. These factors are therefore essential for successful performance in skateboarding.

14.4 Conclusion

The correlation between Anthro-fitness variables and skateboarding performance was investigated in the current study. The findings revealed that a combination of specific anthropometrics and fitness-related variables contribute to the successful performance of skateboarding maneuvers. In conclusion, the use of machine learning models like the RF model, in conjunction with anthro-fitness variables, can provide valuable insights into the factors that contribute to superior performance in skateboarding. This can assist coaches and other relevant stakeholders in making informed decisions regarding athlete performance and training programs. This further accentuates the potential of machine learning in revolutionizing athlete training and performance assessment.

References

Ab Rasid, A. M., Suhaimi, M. Z., Abdul Majeed, A. P. P., Mohd Razman, M. A., Hassan, M. H. A., Najmi, N., Abu Osman, N. A., & Musa, R. M. (2023). A cluster analysis and artificial neural network of identifying skateboarding talents based on bio-fitness indicators. In *Innovation and Technology in Sports: Proceedings of the International Conference on Innovation and Technology in Sports, (ICITS) 2022, Malaysia* (pp. 47–56).

Ab Rasid, A. M., Muazu Musa, R., Abdul Majeed, A. P. P., Musawi Maliki, A. B. H., Abdullah, M. R., Mohd Razmaan, M. A., & Abu Osman, N. A. (2024). Physical fitness and motor ability parameters as predictors for skateboarding performance: A logistic regression modelling analysis. *PLoS One, 19*(2), e0296467.

Abdullah, M. A., Ibrahim, M. A. R., Shapiee, M. N. A. B., Mohd Razman, M. A., Musa, R. M., & Abdul Majeed, A. P. P. (2020). The classification of skateboarding trick manoeuvres through the integration of IMU and machine learning. In Z. Jamaludin & M. N. Ali Mokhtar (Eds.), *Intelligent Manufacturing and Mechatronics. SympoSIMM 2019. Lecture Notes in Mechanical Engineering*. Springer. https://doi.org/10.1007/978-981-13-9539-0_7

Borden, I. (2016, August). *Ollies at the Olympics: Why having skateboarding at Tokyo 2020 is a winning move.*

Bunker, R., & Susnjak, T. (2022). The application of machine learning techniques for predicting match results in team sport: A review. *Journal of Artificial Intelligence Research, 73*, 1285–1322. https://doi.org/10.1613/jair.1.13509

Corrêa, N. K., de Lima, J. C. M., Russomano, T., & dos Santos, M. A. (2017). Development of a skateboarding trick classifier using accelerometry and machine learning. *Research on Biomedical Engineering, 33*(4), 362–369.

Dudek, G. (2022). A comprehensive study of random forest for short-term load forecasting. *Energies, 15*(20), 7547.

Eswaramoorthi, V., Abdullah, M. R., Musa, R. M., Maliki, A. B. H. M., Kosni, N. A., Raj, N. B., Alias, N., Azahari, H., Mat-Rashid, S. M., & Juahir, H. (2018). A multivariate analysis of cardiopulmonary parameters in archery performance. *Human Movement, 19*(4), 35–41. https://doi.org/10.5114/hm.2018.77322

Fabris, F., de Magalhães, J. P., & Freitas, A. A. (2017). A review of supervised machine learning applied to ageing research. *Biogerontology, 18*(2), 171–188. https://doi.org/10.1007/s10522-017-9683-y

Fitzpatrick, C., Pagani, L. S., & Barnett, T. A. (2012). Early childhood television viewing predicts explosive leg strength and waist circumference by middle childhood. *International Journal of Behavioral Nutrition and Physical Activity, 9*, 1–6.

Frederick, E. C., Determan, J. J., Whittlesey, S. N., & Hamill, J. (2006). Biomechanics of skateboarding: Kinetics of the Ollie. *Journal of Applied Biomechanics, 22*(1), 33–40.

García-Pinillos, F., Cámara-Pérez, J. C., Soto-Hermoso, V. M., & Latorre-Román, P. Á. (2017). A high intensity interval training (HIIT)-based running plan improves athletic performance by improving muscle power. *The Journal of Strength & Conditioning Research, 31*(1), 146–153. https://doi.org/10.1519/JSC.0000000000001473

Haiyun, Z., & Yizhe, X. (2020, September). Sports performance prediction model based on integrated learning algorithm and cloud computing Hadoop platform. *Microprocessors and Microsystems, 79*, 103322. https://doi.org/10.1016/j.micpro.2020.103322

Henriques-Neto, D., Minderico, C., Peralta, M., Marques, A., & Sardinha, L. B. (2020). Test–retest reliability of physical fitness tests among young athletes: The FITescola® battery. *Clinical Physiology and Functional Imaging, 40*(3), 173–182. https://doi.org/10.1111/cpf.12624

Iranzad, R., & Liu, X. (2024). A review of random forest-based feature selection methods for data science education and applications. *International Journal of Data Science and Analytics*. https://doi.org/10.1007/s41060-024-00509-w

Johnston, K., Wattie, N., Schorer, J., & Baker, J. (2018). Talent identification in sport: A systematic review. *Sports Medicine, 48*(1), 97–109. https://doi.org/10.1007/s40279-017-0803-2

Karakaya, A., Ulu, A., & Akleylek, S. (2022). GOALALERT: A novel real-time technical team alert approach using machine learning on an IoT-based system in sports. *Microprocessors and Microsystems, 93*, 104606.

Lath, F., Koopmann, T., Faber, I., Baker, J., & Schorer, J. (2021). Focusing on the coach's eye; Towards a working model of coach decision-making in talent selection. *Psychology of Sport and Exercise, 56*, 102011. https://doi.org/10.1016/j.psychsport.2021.102011

Majumdar, A., Bakirov, R., Hodges, D., Scott, S., & Rees, T. (2022). Machine learning for understanding and predicting injuries in football. *Sports Medicine—Open, 8*(1), 73. https://doi.org/10.1186/s40798-022-00465-4

Muazu Musa, R., Abdul Majeed, A. P. P., Abdullah, M. R., Kuan, G., & Mohd Razman, M. A. (2022). Current trend of analysis in high-performance sport and the recent updates in data mining and machine learning application in sports. In *Data mining and machine learning in high-performance sport* (pp. 1–11). Springer.

Musa, R. M., Abdullah, M. R., Maliki, A. B. H. M., Kosni, N. A., Mat-Rasid, S. M., Adnan, A., & Juahir, H. (2018). Supervised pattern recognition of archers' relative psychological coping skills as a component for a better archery performance. *Journal of Fundamental and Applied Sciences, 10*(1S), 467–484. https://www.ajol.info/index.php/jfas/article/view/168316

Musa, R. M., Majeed, A. P. P. A., Taha, Z., Abdullah, M. R., Maliki, A. B. H. M., & Kosni, N. A. (2019). The application of artificial neural network and k-nearest neighbour classification models in the scouting of high-performance archers from a selected fitness and motor skill performance parameters. *Science & Sports, 34*(4), e241–e249.

Ofoghi, B., Zeleznikow, J., Macmahon, C., & Dwyer, D. (2013). Supporting athlete selection and strategic planning in track cycling omnium: A statistical and machine learning approach. *Information Sciences, 233*, 200–213. https://doi.org/10.1016/j.ins.2012.12.050

Palczewska, A., Palczewski, J., Marchese Robinson, R., & Neagu, D. (2014). Interpreting random forest classification models using a feature contribution method. *Integration of Reusable Systems*, 193–218.

Rasid, A. M. A., Kamarudin, N. A., Abdullah, M. A., Ibrahim, M. A. R., Shapiee, M. N., Bin, A., Razman, M. A. M., Abdul Majeed, P. P., Abdullah, M. R. A., & Musa, R. M. (2020). Development of skill performance test for talent identification in amateur skateboarding sport. In *International Conference on Innovative Technology, Engineering and Science* (pp. 385–390).

Rasid, A. M. A., Kamarudin, N. A., Abdullah, M. A., Ibrahim, M. A. R., Shapiee, M. N. A. B., Razman, M. A. M., Abdul Majeed, A. P. P., Abdullah, M. R., & Musa, R. M. (2021). Development of skill performance test for talent identification in amateur skateboarding sport. In J. A. Mat Jizat, I. M. Khairuddin, M. A. Mohd Razman, A. F. A. Nasir, M. S. Abdul Karim, A. A. Jaafar, L. W. Hong, A. P. Abdul Majeed, P. Liu, H. Myung, H.-L. Choi, & G.-A. Susto (Eds.), *Advances in robotics, automation and data analytics—Selected papers from iCITES 2020. (Advances in Intelligent Systems and Computing; AISC)* (Vol. 1350, pp. 385–390). Springer. https://doi.org/10.1007/978-3-030-70917-4_37

Razali, M. R., Alias, N., Maliki, A., Musa, R. M., Kosni, L. A., & Juahir, H. (2017). Unsupervised pattern recognition of physical fitness related performance parameters among Terengganu youth female field hockey players. *International Journal on Advanced Science, Engineering and Information Technology, 7*(1), 100–105.

Reuters, T. (2021). *Skateboarding legend Tony Hawk shreds new Olympic venue, calling it "surreal"*. CBC Sports Canada.

Roberts, A. H., Greenwood, D. A., Stanley, M., Humberstone, C., Iredale, F., & Raynor, A. (2019). Coach knowledge in talent identification: A systematic review and meta-synthesis. *Journal of Science and Medicine in Sport, 22*(10), 1163–1172. https://doi.org/10.1016/j.jsams.2019.05.008

Shapiee, M. N. A., Ibrahim, M. A. R., Mohd Razman, M. A., Abdullah, M. A., Musa, R. M., Hassan, M. H. A., & Abdul Majeed, A. P. P. (2020). The classification of skateboarding trick manoeuvres through the integration of image processing techniques and machine learning. *Lecture Notes in Electrical Engineering, 632*, 347–356. https://doi.org/10.1007/978-981-15-2317-5_29

Taha, Z., Haque, M., Musa, R. M., Abdullah, M. R., Maliki, A., Alias, N., & Kosni, N. A. (2009). Intelligent prediction of suitable physical characteristics toward archery performance using multivariate techniques. *Journal of Global Pharma Technology, 9*(7), 44–52.

Taha, Z., Musa, R. M., Abdul Majeed, A. P. P., Abdullah, M. R., Zakaria, M. A., Alim, M. M., Jizat, J. A. M., & Ibrahim, M. F. (2018a). The identification of high potential archers based on relative psychological coping skills variables: A support vector machine approach. *IOP Conference Series: Materials Science and Engineering, 319*(1). https://doi.org/10.1088/1757-899X/319/1/012027

Taha, Z., Musa, R. M., Abdul Majeed, A. P. P., Alim, M. M., & Abdullah, M. R. (2018b). The identification of high potential archers based on fitness and motor ability variables: A support vector machine approach. *Human Movement Science, 57*, 184–193. https://doi.org/10.1016/j.humov.2017.12.008

Thakur, A. (2023). *Global skateboard market overview, 2023–28*. September 2021.

Weiwei, H. (2022). Classification of sport actions using principal component analysis and random forest based on three-dimensional data. *Displays, 72*, 102135. https://doi.org/10.1016/j.displa.2021.102135

Part IV

Applications in Sports Science

Defining Player Dominant Region in Soccer: Background, Data-Driven Models Advances, and Scientific Challenges

15

Felipe Arruda Moura, Fabio Giuliano Caetano, and Ricardo da Silva Torres

Abstract

Soccer players constantly move during the match to occupy different regions of the pitch to generate opportunities for their team and to prevent offensive actions from the opponents. The dominant regions are defined as regions of the pitch where players can reach before the others. Over the last two decades, several models were developed to describe player-dominant regions allowing insights into pitch control, and invasion strategies. This chapter provides an overview of the development and interpretation of these models and the recent advances based on data-driven models. The concepts, strengths, and limitations of traditional metrics are presented, such as the Voronoi Diagram and models derived probabilistic inferences from players' kinematic data. Finally, new perspectives and scientific challenges to the improvement of the models are presented.

15.1 Introduction

During soccer matches, players experience a decision-making process in which they perform actions influenced by constraints from the environment (teammates or opponents), the task (rules of the game), and the individual capabilities (technical skill, for instance) (Araujo et al., 2014). Team sports players make their movement decisions based on information about the positioning or direction of movement of other players in this system (Davids et al., 2005). Therefore, these concepts can

F. A. Moura (✉) · F. G. Caetano
Sport Sciences Department, Universidade Estadual de Londrina, Londrina, Brazil
e-mail: felipemoura@uel.br

R. da Silva Torres
Artificial Intelligence Group, Wageningen University and Research, Wageningen, Netherlands
e-mail: ricardo.dasilvatorres@wur.nl

D. Memmert (ed.), *Artificial Intelligence and Machine Learning in Sports Science*,
https://doi.org/10.1007/978-3-662-70155-3_15

elucidate the tactical behavior of players or teams in their complex interactions, such as their spatiotemporal relationships.

These dynamic relationships can be elucidated from the investigation of interactions between players based on analyses of dyads (Araujo et al., 2014). The interaction between players in soccer has been the subject of studies that often use dyads of the closest players, as well as the distance between them as the main parameter to analyze it (Caetano et al., 2019, 2020; Clemente et al., 2013; Duarte et al., 2010; Ric et al., 2017). The interactions between the closest players and consequently the distance between them are relevant concepts for analyzing the spatial exploration of teams in soccer because they are commonly used to compute the players' dominant region (Rico-González et al., 2021).

A player's dominant region is defined as the region of the field that he/she can reach before any other player (Taki & Hasegawa, 2000). Traditionally, the dominant region associated with players is determined using the Voronoi diagram (Rico-González et al., 2021). Voronoi diagram allows a space to be decomposed into cells associated with each point which, in the case of soccer, are respectively represented by the playing field, the dominant regions, and the players, taking into consideration only the position of the players and the distance between them and the closest players (Kim, 2004; Okabe et al., 2009). This approach has already been used in soccer to determine the teams' dominant region in the last third of the field (Memmert et al., 2019; Rein et al., 2017), showing that greater control of space is related to a greater probability of winning the match (Rein et al., 2017).

From the Voronoi Diagram model, other proposals were presented which involved maximization of the area of Voronoi regions of a set of points placed inside a circle (Bhattacharya, 2010; Filetti et al., 2017). However, Voronoi diagram approaches in which determined dominant regions are associated with players only because they are the closest to them, promoting a static perspective that can only be representative for the instant of time in question. To solve this problem, data-driven approaches were developed to improve the representation of the Voronoi Diagram, considering players' kinematic data and/or probabilistic models (Brefeld et al., 2019; Caetano et al., 2021; Fujimura & Sugihara, 2005; Taki & Hasegawa, 2000). Those methods explore statistical and machine learning methods to determine the dominant regions associated with each player based on trends and patterns found in the spatiotemporal positional data available.

The purpose of this chapter is to present the general concepts of the different models presented in literature, as well as their strengths, and weaknesses. Finally, some new perspectives and challenges to future studies are addressed.

15.2 Voronoi Diagram and Adapted Models

Initially, the determination of the players' dominant regions was proposed using the Voronoi diagram. This technique makes it possible to decompose a space into cells that are associated with each point (Okabe et al., 2009). The Voronoi cell, or Voronoi region, associated with the given site is the set of all points in whose distance to it is not greater than their distance to all the other sites. In the context of soccer, space represents the pitch, cells are the dominant regions and players are represented by the sites. In this method, the dominant regions are determined considering only the position of the players. These regions are delimited by calculating the distance between each player and neighboring players, with the limit defined as half the distance between neighboring players, as illustrated in Fig. 15.1 (Kim, 2004).

Another approach to the Voronoi diagram focused on maximizing the Voronoi regions in a set of points within a circle that results in regions with rounded characteristics (Bhattacharya, 2010). Filetti and colleagues (2017) used this approach to determine the dominant regions of players, which they called "K-regions," relating these regions to the technical performance of the teams in passing actions. The study considered a positive technical-tactical behavior when a given player presented a dominant region that provided an opportunity for his team to pass correctly in the offensive context. From a defensive perspective, positive behavior was considered when the player's dominant region limited the opposing team's passing opportunities. Furthermore, an index of the teams' technical-tactical behavior was computed and demonstrated that greater differences between teams in this index were related to greater probabilities of team victory during matches.

Voronoi diagram was also explored to determine the dominant region of the teams in the last third of the field (Rein et al., 2017). The team's space control relative to the area of this region of the field was identified, with the value of space control at the moment the player performed the pass subtracted from the value of space control at the moment a teammate received the ball. In this proposal, the

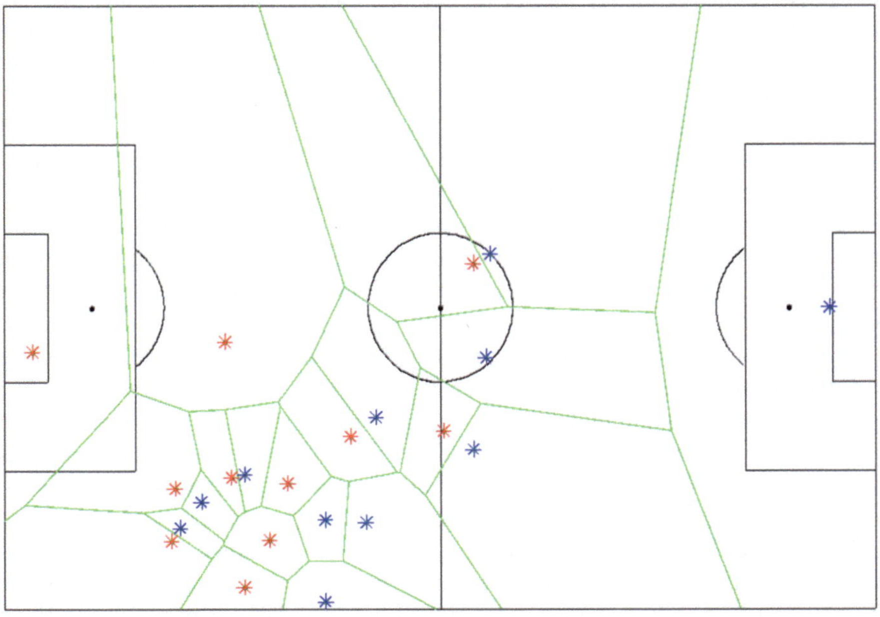

Fig. 15.1 Representation of Voronoi Diagram from players' position

authors suggest that negative values of space control represent that attackers dominate a smaller region when the pass is completed. It has also been reported that greater space control is related to a greater probability of teams winning the match. Memmert and colleagues (2019) also applied this method during simulated matches. The behavior of this variable was compared between situations in which teams used different playing styles, with no significant differences between them. Their approach limits the application of the Voronoi diagram to the last third of the field. They argue that employing this method across the entire field may result in the formation of extensive dominant regions for players positioned far from the area of offensive play.

The limitation of the Voronoi diagram is its static nature because the dominant regions are determined considering only the position of the players and the distance between them. Therefore, several data-driven models were proposed in the literature, using movement models defined from displacement data, such as direction, velocity, and acceleration, aiming to improve the determination of the dominant region, making it more compatible with the players' displacement capabilities.

15.3 Data-Driven Models

Data-driven models within the realm of Artificial Intelligence (AI) pertain to methodologies that heavily lean on data to inform decisions, glean insights, and enhance system efficacy. These techniques use vast datasets to train machine learning algorithms, uncover patterns, and formulate forecasts or choices grounded in the data collected. Considering the dominant region problem, data-driven methods are promising because other information that directly influences the 'reachable locations' of the players may be provided to the models. In this sense, kinematic data are essential, once displacement direction, velocity, and acceleration determine the probable regions of the players within a given time window. With the proposal of these models, the parameter called future time-window consequently arises, representing the temporal variation used to generate the different models.

The first proposal presented by Taki and Hasegawa (2000) determined a sphere of influence for each player, replacing the distance function in the Voronoi region with a time function. The players 'dominant regions change according to the vectors of the direction of displacement and velocity at every time-stamp, while the acceleration vector was considered to be of equal magnitude for all directions, suffering a reduction in the direction of displacement according to the velocity increases. The proposed concept understands that players can accelerate with the same magnitude in all directions when they are stationary or moving at low velocities. However, as velocity increases, it becomes more difficult to accelerate in the direction of the displacement. Then, the spheres of influence are used to determine the parameter called 'shortest time pattern' which represents the time required for each player to move to each x position. Finally, the dominant regions are determined using the Voronoi diagram with the replacement of the Euclidean distance parameter between players by the "shortest time pattern" parameter proposed by them. The authors also presented applications based on the players' dominant regions. For example, they determined the teams' dominant region area as the sum of the areas of the players' dominant regions belonging to each team. The ratio between the areas of the teams' dominant regions over time was used to evaluate the teams' collective space generation behavior in defending or attacking actions. Experimental results showed an increase in the attacking team's dominant region throughout an attack attempt that resulted in a goal, as well as the pressure work on the opponent was analyzed using a metric called "spatial pressure" which was determined considering the distribution of dominant regions around the player, distance from the goal line and the ball. Furthermore, passing actions were evaluated as successful when the player performed a pass and the ball reached an area dominated by a teammate.

Fujimura and Sugihara (2005) proposed a new movement model by adding a 'resistive force' in order to avoid a limitation of the model proposed by Taki and Hasegawa (2000) which generates infinite acceleration caused by constant acceleration. In the model proposed by the authors, the force resistive to acceleration was considered proportionally opposite to the velocity of the players. The movement model generated was evaluated from an experiment with the displacement of three players in a future time window of 2 s, resulting in an acceleration of 1.3 m/s^2 and a maximum velocity of 7.8 m/s. Subsequently, this movement model was used to determine the dominant region of the players using the Voronoi diagram. Furthermore, the researchers applied this analysis in a real hockey match to evaluate each player's contribution to teamwork. They first determined the dominant regions of only the defensive players, assuming that each defensive player marks an opposing player. Then, they calculated the area of the players' dominant region and suggested weighting this area according to the distance from the region to the goal and the ball. These two indices are used to represent the importance of positioning players in a better way compared to the simple dominant area, suggesting that regions far from the ball or goal have a low density of players and result in larger dominant areas, but without the same relevance to the game. The authors also proposed measuring the degree of pressure by calculating the minimum time that the defensive player takes to reach the current region of the offensive player, as well as using the dominant region of a player as a region of viability to pass it on to your team.

A previous study used the data-driven model proposed by Fujimura and Sugihara (2005) to determine the players' dominant region, focused on identifying the teams' dominant region and analyzing its relationship with their offensive performance (Ueda et al., 2014). The dominant region of the teams was determined in a delimited area that was referred to as the teams' occupation area. This occupation area was defined as a quadrilateral formed by the position of the most extreme players on the pitch (excluding goalkeepers), that is, the vertices defined by the player closest to their own goal line, the player closest to the opponent's goal line and the players closest to each sideline. The authors analyzed the teams with and without possession of the ball according to where possession of the ball was acquired. Then, it was analyzed the teams in relation to the outcome of the offensive attempt, i.e., an effective action was considered when the teams were successful. Finally, the areas of the teams' dominant regions were determined at the moment they gained the ball possession. In this study, no statistically significant differences were found in the values of the dominant region of the teams when there was an exchange of ball possession in the defending and attacking thirds. However, differences were observed in midfield third, with higher values of the dominant area of the teams in possession of the ball during successful actions. The researchers suggested that the exchange of ball possession in the central third of the field presents a causal relationship between the teams' dominant area and their offensive performance, in which, from an offensive perspective, greater values of the dominant area are related to success.

More recently, a probabilistic movement model based on positional data as a correction parameter for the Voronoi diagram to determine the dominant regions, called the zones of control, was proposed (Brefeld et al., 2019). The proposed probabilistic movement model considers players' positions, displacement angle, and velocity to determine zones of control in a 1-s future time window. The estimation of the probabilistic movement model based on the players' positions used a set of parameters that was called a triplet. The triplet was composed of the initial position and final position data to represent the trajectory of displacement, as well as the points that players can reach on the field considering their origin and velocity. Subsequently, the triplet was translated so that its origin was at zero in coordinates (x, y) and the angle was rotated to be aligned with the horizontal. The velocities used to generate the movement models were separated into five different velocity ranges: stopped (<1 km/h), walking (1–7 km/h), jogging (7–14 km/h), running (14–20 km/h), and sprint (>20 km/h). Furthermore, movement models were generated considering the athletes' playing positions, divided into goalkeepers, defenders, and attacking midfielders. Finally, the probabilities of players reaching the different regions over time were smoothed through the two-dimensional kernel density estimation function and thus the zones of control were determined. In this study, the researchers determined the players' dominant regions at the same instant of time using the simple Voronoi diagram, the movement model proposed by Taki and Hasegawa (2000), the movement model proposed by Fujimura and Sugihara (2005), as well as the probabilistic model they proposed for visual comparison of the resulting dominant regions. In their study, they suggested that the probabilistic model promotes more realistic control zones when compared with the other proposals. The simple Voronoi diagram was identified as the method that generates the most overestimation of the dominant regions. They also highlight that proposals based on movement models contain simplified assumptions and only the probabilistic movement model presents results that are aligned with the players' movements and distances. Furthermore, the authors highlight that the different positions present particularities regarding the characteristics of the movements and consequently specific needs to generate movement models. Even though, satisfactory results are obtained in generating movement models for outfield players with positional data obtained in just one match. However, goalkeepers are generally moving at low velocity during the match, which results in a few samples of higher values of initial velocity. Brefeld et al. (2019) suggest that more realistic motion models can contribute to dominant region analyses, as well as to the assessment of the influence of players' dominant region on regions controlled by their team. In their study, players' position likelihoods are computed based on kernel density estimation (KDE) and assumptions on conditions related to velocity and time horizon intervals (see Fig. 15.2). Finally, Fadel and colleagues (2023) proposed to model players' movements using conditional normalizing flows. Normalizing flows has been demonstrated to be an effective framework for density learning based on invertible deep neural networks. The goal is to compute simple distributions based on an invertible chain of transformations applied to complex data distributions. In their formulation, distributions are computed based on a triplet that encodes the current, past, and future positions. Normalizing flows is then used to map from this distribution to a known base distribution (e.g., standard Gaussian).

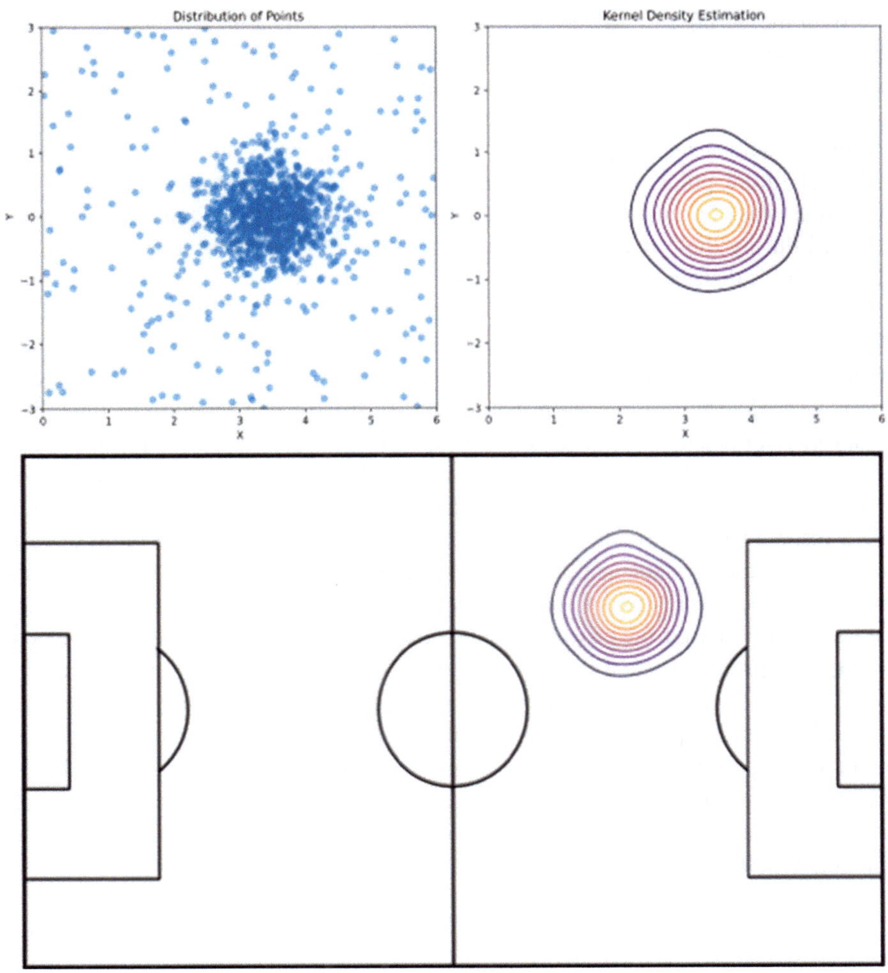

Fig. 15.2 Kernel density estimation (KDE) based on positional data (top). Dominant region on the field for a given player (bottom)

From this overview, it is possible to visualize the valuable contributions of these studies regarding tactical analysis in soccer based on the players' dominant regions, as well as the limitations that make it possible to identify possible contributions to be made on this topic. Thus, an approach to creating motion models considering the kinematic variables of the players, including instantaneous acceleration, can generate several motion models representing the distinct characteristics of displacements that have different accelerations. For example, movements with positive or negative accelerations cannot be included in the same movement model, because players can reach different regions of the field according to this parameter. Furthermore, a method for determining the dominant region of players that considers as free spaces the regions of the field where players are not able to reach in a specific time interval can promote dominant regions that are more compatible with human physical capacity and make it possible to visualize free spaces on the pitch.

With the purpose of addressing these limitations, a data-driven model based on players' positions, displacements, velocities, and instantaneous acceleration data, as well as the future time window determined from data collected from official matches, was proposed recently (Caetano et al., 2021). The positions, displacements, velocities, and accelerations at each timestamp were identified and the positions of future points up to 3 s were recorded. The future time window of the positions was chosen from an experiment with a dataset of technical actions performed during three matches, resulting in almost 3000 passing actions analyzed. Based on the distribution of the time windows of these passes, the maximum future time window of 3 s was determined, representing approximately 95% of the data. Considering the information of more than 15 million players' positions, the displacement angles in relation to the longitudinal axis, the velocities, and accelerations were determined based on the players' positions as a function of time. The displacement angles were rotated to 90° and the origin of the future positions was translated to coordinate 0 on the x and y axes. The displacements were classified into the same combinations of velocity and acceleration ranges and then stored in their respective matrix cells, generating an accumulation of future positions of various displacements that constitute a movement model. Also, the displacements were mirrored around the y-axis to balance the curvatures of the motion models on both sides. Finally, based on possible future positions from the respective movement model, the probabilities of the player reaching each region of the field were calculated using a histogram function with normalization by a probability density function. The matrix with the probability values was smoothed using a 2-D Gaussian filter to reduce the discrete characteristics of the histogram results, and values below 0.001 were rounded to 0 to avoid an association with the very low probability regions with the players' dominant regions. Figure 15.3 shows an example of this model, with the dominant areas of all players.

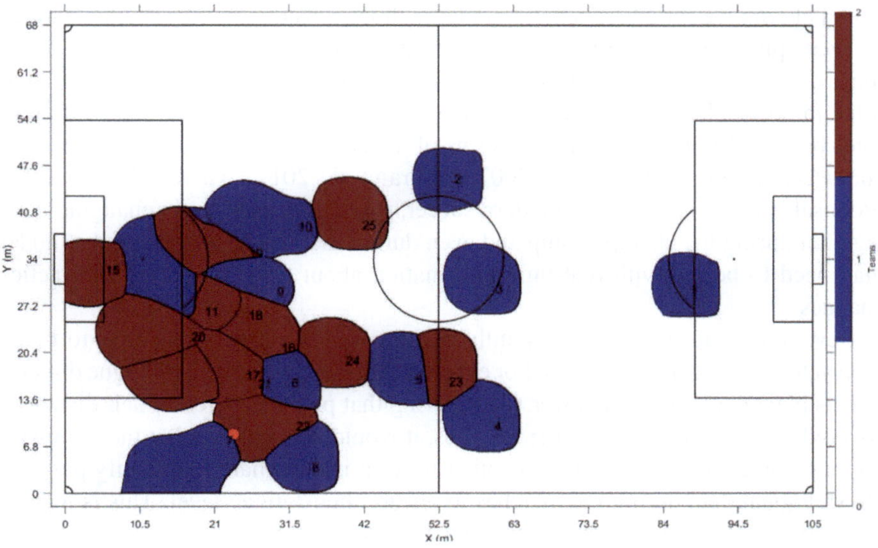

Fig. 15.3 Representation of players' dominant regions using a data-driven model proposed by Caetano et al. (2021)

15.4 New Perspectives and Scientific Challenges for Dominant Regions' Representation

Because recent advances in modeling the players' dominant regions rely on kinematic data, the player's ability to move as fast as possible, to accelerate, and to change running direction clearly affects the position the player can reach in a given time window. However, data-driven models presented in literature take into consideration the kinematic data of several players from different positions, during several matches against different opponents. Thus, the models generated are generic representations from a wide range of players with individual capabilities. Although this is a positive aspect considering that the dominant regions defined by these models are grounded by probabilistic inferences, i.e., it is expected that the predictions from these models are accurate enough for a large percentage of the population, one can argue whether individual models would provide a more realistic representation. For instance, if a given player is an outlier and is able to reach very high velocities and accelerations, it is clear that this player will present a greater dominant area than the generic model. On the other hand, slower and less powerful players probably will have their dominant areas overestimated. An important investigation for the scientific community is to evaluate the impact of generic models compared to individual models on the accuracy of the dominant region estimate. The challenge beyond this investigation relies on the need for enough data from every player to make a good prediction. In other words, considering that it is rare for players to reach high velocities and accelerations during the match, modeling a single player would require data collected from several matches to provide enough data.

An alternative for a less generic model without considering the specific features of every player would be to model dominant regions according to players' roles (e.g., defensive or forward player). It is well-known in the literature that different roles in soccer lead to different demands during matches. Players with different roles present different outcomes in terms of distances covered, range of velocities, and accelerations (Barros et al., 2007; Moura et al., 2017). The challenge of this proposal is to consider that in modern soccer, players are constantly changing their position during the championship and even during the match. Therefore, the model may need to be fed with real-time information about team formation and tactics changes.

Several circumstances directly influence the player's ability to move, to reach high velocities and accelerations. For example, we may consider that fatigue directly affects player velocity and power. Considering that players decrease their distances covered and velocities during the matches, it would be expected that the dominant regions change accordingly. On the other hand, it is quite hard to quantify players' physiological data during the matches to ensure their fatigue level. This is a challenge to the scientific community and may require previous research with validation of indirect measures (such as kinematic data or behavioral profiles) to estimate players' physiological conditions. Again, the investigation of online learning algorithms that can adjust predictions based on real-time collected seems to be a promising research direction for future studies.

Finally, there is a concern about the relevance of the dominant areas in different parts of the pitch. One can argue that, as a pitch control inference, areas that are far from the goals or from the ball have less importance. Further investigations may discuss solutions for implementing weighted models that take into consideration players and ball position, or even xG (expected goal) values associated with each athlete. The challenge is to define clear criteria for proper weights considering that, eventually, players far away from the ball may have great relevance in terms of pitch control, such as set-play situations (corner kicks, free kicks, goal kicks, etc). Table 15.1 summarizes these new perspectives and the scientific challenges for improving the representation of the players' dominant areas.

15.5 Conclusion

As an invasion sport, the representation of dominant regions is promising for soccer analytics approaches. From the quantification of the players' dominant regions, insights about pitch control, space creation, and team organization on the pitch are possible. Traditionally, Voronoi diagram-derived models were presented and explored in literature considering different parts of the pitch. Because of its static representation and unrealistic huge areas, data-driven models were presented considering the probable regions each player can reach given their kinematic data. These advances allowed a better representation of space sharing and free spaces to be explored by teammates and opponents. Modeling players´ dominant region is still challenging considering all the variables that may directly influence players'

Table 15.1 Overview of current modeling

Approach	Research problem	Required analyzes	Scientific challenges
Individual x generic models	Are there differences between generic and individual models on the dominant region prediction?	Evaluation of the advantages of individual modeling on the dominant region prediction compared to generic models	Modeling every player would require data collection from several matches to provide good predictions
Modeling dominant regions by position	Do modeling dominant regions by position provide more accurate predictions?	Evaluation of the advantages of modeling groups of players with the same role on the dominant region prediction compared to generic models	Players are constantly changing their position during the match/ championship, requiring real-time information about team tactics
Modeling dominant regions sensitive to performance changes during the match	Is it necessary to model the players' dominant region dependent on physical/physiological indicators collected during the match?	Evaluation of the advantages of modeling players' dominant region as a function of fatigue and physical performance	To quantify players' physiological data during the matches to ensure their fatigue level or to validate indirect real-time measures to estimate players' physical condition
Modeling weighted dominant regions	How to model players' dominant regions weighted by the relevance of the player location?	Development and validation of models dependent on player distance from the ball and the goal	To establish objective criteria for proper weights considering different scenarios, including set-plays

ability to reach high velocities and acceleration, as well as changes in displacement direction. The scientific community needs to consider constantly improving the available machine learning methods to provide more accurate and realistic representations that consider individual and group profiles, performance variation during the match, and the relevance of every region to teams successfully achieving their objective during attacking and defending actions.

References

Araujo, D., Diniz, A., Passos, P., & Davids, K. (2014). Decision making in social neurobiological systems modeled as transitions in dynamic pattern formation. *Adaptive Behavior, 22*(1), 21–30.

Barros, R. M. L., Misuta, M. S., Menezes, R. P., Figueroa, P. J., Moura, F. A., Cunha, S. A., et al. (2007). Analysis of the distances covered by first division Brazilian soccer players obtained with an automatic tracking method. *Journal of Sports Science and Medicine, 6*, 10.

Bhattacharya, B. B. (2010). Maximizing Voronoi regions of a set of points enclosed in a circle with applications to facility location. *Journal of Mathematical Modelling and Algorithms, 9*, 375–392.

Brefeld, U., Lasek, J., & Mair, S. (2019). Probabilistic movement models and zones of control. *Machine Learning, 108*(1), 127–147.

Caetano, F. G., da Silva, V. P., da Silva Torres, R., de Oliveira Anido, R., Cunha, S. A., & Moura, F. A. (2019). Analysis of match dynamics of different soccer competition levels based on the player dyads. *Journal of Human Kinetics, 70*(1), 173–182.

Caetano, F. G., de Souza, N. M., Bueno, M. J. D. O., Cunha, S. A., & Moura, F. A. (2020). Interpersonal interaction during official soccer matches considering the coupling of different playing positions. *International Journal of Performance Analysis in Sport, 20*(4), 646–658.

Caetano, F. G., Barbon Junior, S., Torres, R. D. S., Cunha, S. A., Ruffino, P. R. C., Martins, L. E. B., & Moura, F. A. (2021). Football player dominant region determined by a novel model based on instantaneous kinematics variables. *Scientific Reports, 11*(1), 18209.

Clemente, F. M., Couceiro, M. S., Martins, F. M. L., Dias, G., & Mendes, R. (2013). Interpersonal dynamics: 1v1 sub-phase at sub-18 football players. *Journal of Human Kinetics, 36*(1), 179–189.

Davids, K., Araújo, D., & Shuttleworth, R. (2005). Applications of dynamical systems theory to football. In T. Reilly, J. Cabri, & D. Araújo (Eds.), *Science and football V* (pp. 537–550). Routledge.

Duarte, R., Araújo, D., Gazimba, V., Fernandes, O., Folgado, H., Marmeleira, J., & Davids, K. (2010). The ecological dynamics of 1v1 sub-phases in association football. *The Open Sports Sciences Journal, 3*, 16–18.

Fadel, S. G., Mair, S., da Silva Torres, R., & Brefeld, U. (2023). Contextual movement models based on normalizing flows. *AStA Advances in Statistical Analysis, 107*(1), 51–72.

Filetti, C., Ruscello, B., D'Ottavio, S., & Fanelli, V. (2017). A study of relationships among technical, tactical, physical parameters and final outcomes in elite soccer matches as analyzed by a semiautomatic video tracking system. *Perceptual and Motor Skills, 124*(3), 601–620.

Fujimura, A., & Sugihara, K. (2005). Geometric analysis and quantitative evaluation of sport teamwork. *Systems and Computers in Japan, 36*(6), 49–58.

Kim, S. (2004). Voronoi analysis of a soccer game. *Nonlinear Anal-Model, 9*(3), 233–240.

Memmert, D., Raabe, D., Schwab, S., & Rein, R. (2019). A tactical comparison of the 4-2-3-1 and 3-5-2 formation in soccer: A theory-oriented, experimental approach based on positional data in an 11 vs. 11 game set-up. *PLoS One, 14*(1), e0210191.

Moura, F. A., Marche, A. L., Caetano, F. G., Torres, R. D., Martins, L. E., & Cunha, S. A. (2017). Analysis of high-intensity efforts in Brazilian professional soccer players. *Human Movement Special Issues, 2017*(5), 55–62.

Okabe, A., Boots, B., Sugihara, K., & Chiu, S. N. (2009). *Spatial tessellations: Concepts and applications of Voronoi diagrams*. Wiley.

Rein, R., Raabe, D., & Memmert, D. (2017). "Which pass is better?" Novel approaches to assess passing effectiveness in elite soccer. *Human Movement Science, 55*, 172–181.

Ric, A., Torrents, C., Gonçalves, B., Torres-Ronda, L., Sampaio, J., & Hristovski, R. (2017). Dynamics of tactical behaviour in association football when manipulating players' space of interaction. *PLoS One, 12*(7), e0180773. https://doi.org/10.1371/journal.pone.0180773

Rico-González, M., Ortega, J. P., Nakamura, F. Y., Moura, F. A., & Los Arcos, A. (2021). Identification, computational examination, critical assessment and future considerations of spatial tactical variables to assess the use of space in team sports by positional data: A systematic review. *Journal of Human Kinetics, 77*(1), 205–221.

Taki, T., & Hasegawa, J. (2000). *Visualization of dominant region in team games and its application to teamwork analysis*. Paper presented at the Proceedings Computer Graphics International 2000.

Ueda, F., Masaaki, H., & Hiroyuki, H. (2014). The causal relationship between dominant region and offense-defense performance–focusing on the time of ball acquisition. *Football Science, 11*, 1–17.

Applications of Computer Vision in Sports Sciences

16

Thomas Koller and Solange Emmenegger

Abstract

In the following we are going to present the working principle of two computer vision concepts, which are of great value for sports science: Tracking and human pose estimation. We are going to present these complex methods and their underlying working principles. We also provide the tools, that sport scientists are able select the most effective algorithm or implementation for their specific case. Additionally, we give real world examples of how to make use of these concepts for sport science and address the most common challenges. As these algorithms often were engineered for everyday tasks the adaption to the domain specific tasks is of high importance. As this process often requires the collection of sports data, we show the most important points that need to be considered.

16.1 Introduction

The broad availability of video data as well as the advancement of computer vision has revolutionized numerous industries, and sports science is no exception. Two key technologies, namely tracking and monocular human pose estimation, offer profound benefits across various types of sports. Tracking technology enables real-time monitoring of an athlete's position, speed and trajectory. This is invaluable for team sport like soccer, basketball and hockey, where player positioning and movement

T. Koller
Fakultät für Informatik und Informationstechnologie, Hochschule Luzern, Rotkreuz, Switzerland
e-mail: thomas.koller@hslu.ch

S. Emmenegger (✉)
Thalwil, Switzerland
e-mail: solange@uem.com

D. Memmert (ed.), *Artificial Intelligence and Machine Learning in Sports Science*,
https://doi.org/10.1007/978-3-662-70155-3_16

are critical to team strategies. Concurrently, the well-being of athletes can be monitored, and injuries prevented. In contrast, pose estimation is particularly advantageous in the context of compositional or technical sports such as gymnastics, dance and martial arts, where precision and form are key.

The advantages of incorporating computer vision in sports are numerous. It offers objective feedback, eliminating the biases and inconsistencies inherent in human judgment. Therefore, the integration of computer vision into sports science provides a powerful toolkit to analyze a large amount of data. It not only offers precise, objective and scalable insights but is also readily available. This chapter will explore these applications in depth and at the same time look at challenges which can occur.

16.2 Tracking

Prior to an examination of the general working principle of tracking, it is necessary to define how tracking is understood in computer vision.

The term tracking refers to the process of following one or more objects or people in an image sequence. There are two main types of tracking, single object tracking (SOT) and multiple object tracking (MOT), where the latter is more challenging as new objects must be detected when they enter or leave a frame.

There are many applications of tracking in sport. For example, the tracking of player locations in a soccer match allows for the calculation of distances covered by each player or a higher-level analysis enables the assessment of the effectiveness of the defense. Additionally, the ball's trajectory can be monitored to determine the accuracy and velocity of goal kicks. However, this approach is not straightforward, as merely tracking the ball's location does not directly provide the information to find its velocity and potential trajectory. To illustrate this, we consider the example to determine the speed and accuracy of a volleyball serve and show how to solve it in Sect. 16.2.3.

Multi-object tracking must solve several issues.

- **Detection/Initialization**: How are the objects that should be tracked detected in the image?
- **Data Association:** How are the currently tracked objects associated with detections in the next frame?
- **End of tracking**: When can tracking be stopped as an object will no longer appear in subsequent frames?
- **Prediction**: How can we predict likely object position or calculate attributes like speed?
- **Occlusion**: How do we handle object that are temporarily occluded?

The tracking of objects in sports presents additional difficulties due to the possibility of objects with high velocities, for example a ball or a shuttle, or for the similar appearance of players from the same team. Furthermore, the tracking process requires the derivation of additional state information, rather than merely the location of the object.

16.2.1 Tracking-by-Detection

The most common method of tracking is known as tracking-by-detection, in which individual objects are identified and located independently in each frame.

Object detection has a long history in computer vision (CV). Early attempts involved the use of a sliding window to scan all potential object locations in an image after which a classifier, such as a support vector machine (SVM), was run on features calculated from the pixels in the window.

The advent of modern convolutional neural networks (CNNs) has rendered the necessity for hand-crafted features and manual classification obsolete. However, the problem persists in finding regions suitable for evaluation. Two-step detection approaches, such as Faster-RCNN (Ren et al., 2015) employ a region proposal network (RPN) to calculate potential bounding boxes of objects. A second network subsequently classifies the region, while also updating the bounding boxes more tightly around the objects. By sharing layers between the networks, the time required for calculations can be reduced.

Another approach to object detection is to utilize a single pass through the neural network, with the objective of predicting both the class and location of objects in one step. This is known as a single-shot detector. A popular implementation is YOLO (Redmon et al., 2015) which divides the image into a grid and predicts bounding boxes and object classes for each grid cell.

Pretrained versions of those networks can be employed directly, for instance to detect people. Alternatively, the networks can be fine-tuned to align with the task at hand. One strategy for integrating tracking with a two-step detector is to substitute the region proposals with the prediction of the tracker and subsequently only run the second stage to find more accurate positions (Bergmann et al., 2019).

The majority of tracking is conducted in two dimensions (2D), utilizing a single camera. However, the integration of supplementary cameras allows for the tracking of players in three dimensions (3D) in world coordinates. This enables the precise determination of their location within a given environment, for example by the mapping the location in an image frame to the position on a soccer field. An example how this could look like is given in Fig. 16.1.

16.2.2 Data Association

After detecting likely object positions in a new frame, the tracks and positions must be matched. This is known as data association or reidentification (re-ID). To illustrate, the detector will identify all players in a soccer match and must now associate one of those detections with the specific tracked player with number 10.

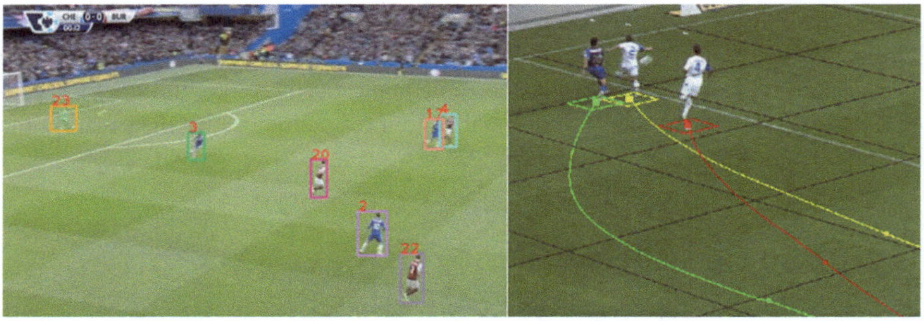

Fig. 16.1 Left: Example of tracking result of a soccer game showing some of the challenges in tracking in sports. Players are blurred and have low contrast (23), players have very similar appearance (20, 22) or occlude each other (4, 17). Right: Result of a 3D tracking approach where multiple cameras are used to track the players at their correct positions on the soccer field, as can be seen by the boxes projected on the ground. This approach allows to derive additional, valuable information

Similarities between the previous and new appearances of the tracked objects can be used, for instance by extracting features and comparing them. In earlier approaches, those features were often hand-crafted to the task, for example by comparing color and/or gradient histograms between the regions. As CNNs learn to extract features from their input in the early layers of the network, they can be used instead, for example in Wang et al. (2019) the CNN is trained to perform both detection and to extract feature embeddings for the comparison. Another approach is to train an additional network to do the feature comparison. A common architecture for such an approach is the Siamese network (Chicco, 2021).

Online tracking refers to the problem of tracking an object in real-time, whenever a new frame becomes available. In contrast, offline tracking processes the whole image sequence to find the optimal tracks over all frames.

If we consider the potential matches of an object between the individual frames as the edges of a graph, weighted by their similarities, the problem of finding the optimal tracks in offline tracking can be solved using graph-based methods (Fig. 16.2).

16.2.3 Kalman Filtering

In the context of our application example, our objective is to determine the speed of a serve in volleyball and the accuracy of the serve. However, tracking in the images will only provide us with the positions. A straightforward approach would be to measure the differences between the locations found in two frames and calculate the velocity by dividing the result by the time difference between the frames. However, this approach does not yield accurate results due to potential inaccuracies in the locations and the resulting error would be unknown.

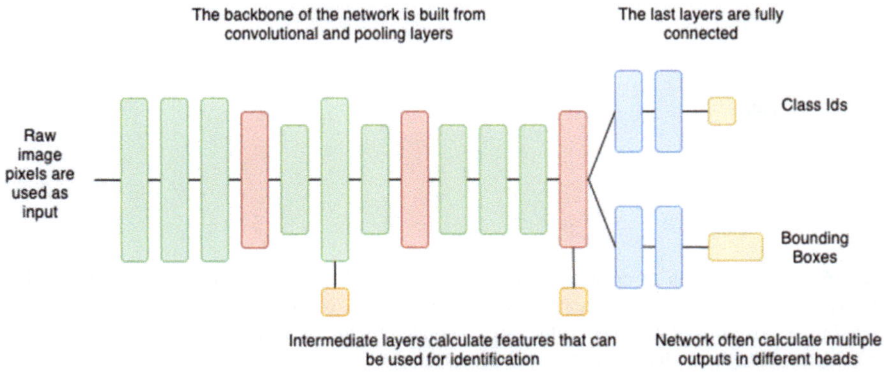

Fig. 16.2 Overview of neural network for tracking-by-detection. Different architectures exist for the backbone of the network, the fully connected layers at the end calculate the output, such as the class ids and bounding box coordinates for 2-step detectors

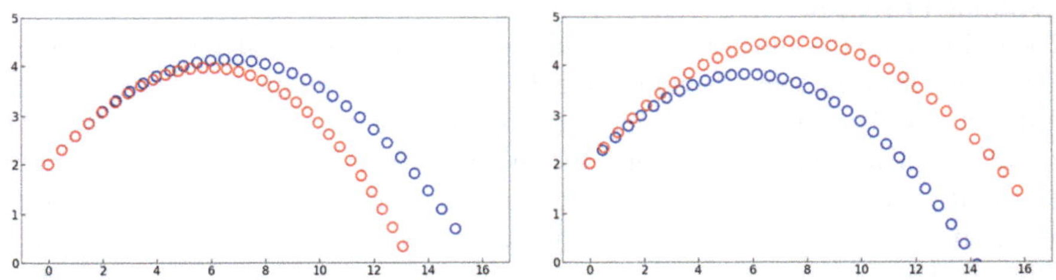

Fig. 16.3 Visualization of error terms in the prediction of movement. On the left, the differences in predicted positions between the low (blue) and high (red) error margin of the velocity are shown. Right displays a possible error with a simplified prediction model (red) is used, but the object moves as shown in blue

> There is a more principled way to determine the state of the object, in this case the velocity, from measurements. This is the Kalman filter. The Kalman filter uses a linear model of the system to calculate predictions of future state and a measurement model to link them with measured values. The approach also involves dealing with several errors.

16.2.3.1 Errors

Any estimate will contain a margin of error, so prediction will vary for the lower and higher bounds of the margins. The system model for calculating the prediction might be too simple and not capture the ground truth exactly. The predictions in Fig. 16.3 shown in blue are calculated using a simple dynamical model with constant acceleration, but the volleyball would follow the curve in red, if air resistance and for example spin is incorporated (Fig. 16.4).

On the measurement side, it might not be possible to measure the location exactly, for example due to blurring or occlusion. And lastly the found position in

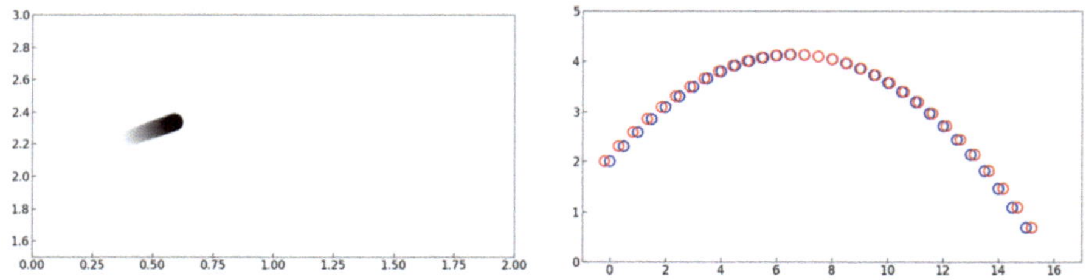

Fig. 16.4 Visualization of error terms in measurements. Left: The object might not be visible clearly in a frame, so the exact location of the measurement is difficult to obtain. Right: The predicted pixel positions of an object (blue) are calculated by the measurement model, but might not be fully accurate due to lens distortion or other artefacts

the frame, might not be the correct location calculated by the measurement model as additional factors like lens distortion could affect the result.

16.2.3.2 Principle

The Kalman filter assumes a linear model for the state change and a linear model for linking the measurements with the state. If all the errors are modelled by Gaussian distributions, then the filter calculates the optimal solution. A prediction step first calculates the new state, for example the speed and location of the volleyball in the next time step, and then corrects these predictions with the measurement results of that frame. It also updates the error of the state in form of a covariance matrix (Figs. 16.5 and 16.6).

16.2.3.3 Non-linear Models

Particle filtering is an alternative to the Kalman filter that can be used if the models are non-linear or multiple hypotheses need to be tracked, as then, the error is non-Gaussian. The object-state distribution is described by a weighted set of particles which are then propagated and updated by the measurement model (Breitenstein et al., 2009).

16.2.4 Metrics for Tracking

To evaluate and compare different tracking approaches, several metrics have been proposed. MOTA (accuracy of tracking) is the most commonly used and is considered to match human assessment of tracking results. It is calculated using the following equation

$$\mathrm{MOTA} = 1 - \frac{\sum_t \left(m_t + \mathrm{fp}_t + \mathrm{mme}_t \right)}{\sum_t g_t}$$

where m_t is the number of missed tracks, fp_t is the number of false positives, mme_t is the number of mismatches and g_t is the total number of ground truth objects.

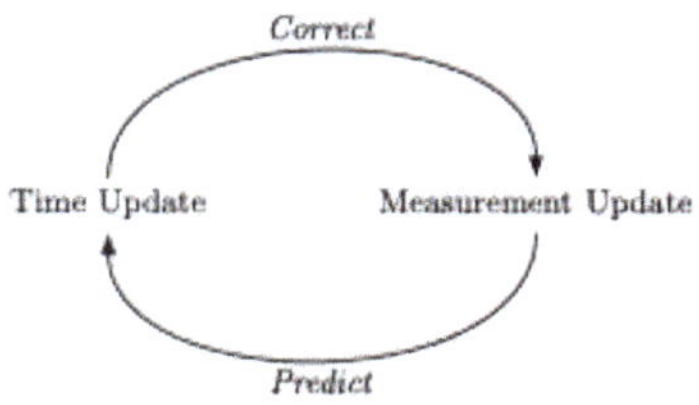

Fig. 16.5 The discrete Kalman filter: The prediction calculates the estimate of the state at the next time step and the measurement update adjusts the prediction by the actual measurement at this time

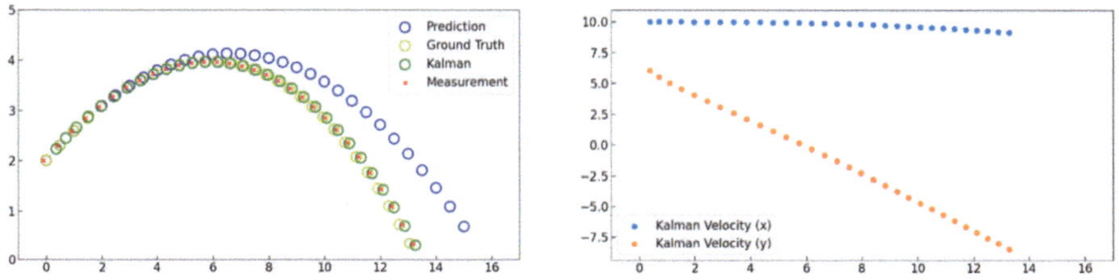

Fig. 16.6 Left: Result of applying a Kalman filter. Blue shows the prediction due to the incorrect system model without drag from the initial position. Red are the measurements and yellow shows the ground truth. The predicted position by the Kalman filter (green) is between the measurement and the prediction. Right: The predicted velocity calculated by the Kalman filter

General tracking algorithms are usually evaluated on benchmark data sets. The newest edition is MOTA 20 (Dendorfer et al., 2020) where the best approaches currently reach MOTA = 0.67. There is a large benchmark dataset for tracking in sports images with 240 sequences in Cui et al. (2023) where the best score is currently MOTA = 0.96.

16.3 Monocular Human Pose Estimation

A complex computer vision task which is beneficial to sport science problems is monocular human pose estimation (MHPE).

MHPE describes the task, where based on an input image or video, the two- or three-dimensional coordinates of human body keypoints are inferred. These body keypoints coincide with the anatomical joints of a human and are also connected accordingly.

An example of such a set of keypoints calculated with mmpose (Sengupta et al., 2019) can be seen in Fig. 16.7.

In standard computer vision MHPE finds further usage in tasks such as human action recognition, person reidentification or even human—computer interaction.

There are some obvious advantages to the use of pose estimation algorithms to retrieve positions of the joints: The alternative to computer vision algorithms is the use of wearable markers to calculate the positions. Especially in competitions this additional gear is forbidden and thus it is not possible to record data. With the

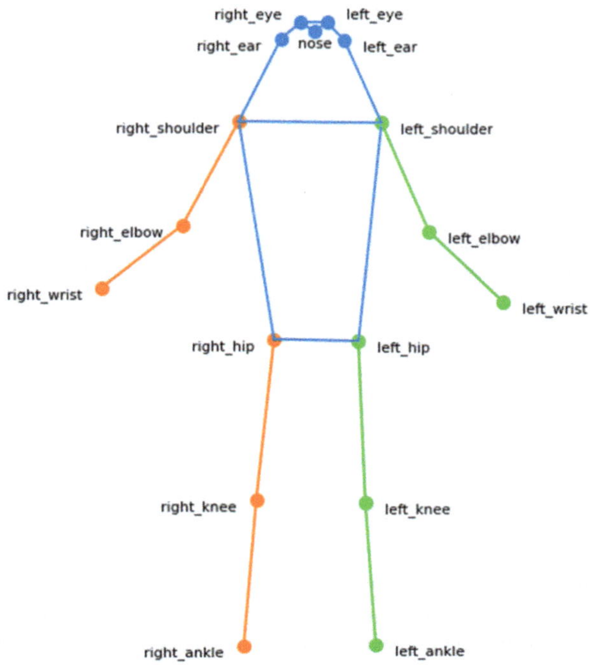

Fig. 16.7 Example of keypoints that are the result of MHPE. These keypoints were calculated with mmpose. The number of keypoints, their names as well as their interconnections depend on the pose estimation algorithm used

computer vision approach though, there is no need of wearable markers and even competition data can be evaluated. With MPHE keypoints evaluations are no longer restricted to the laboratory setting but can also leverage studies in field settings. The only requirement is a camera with a sufficient high speed to capture the dynamics of the sport under investigation.

As we are focusing on monocular human pose estimation there is no need for a complicated multicamera setting. One camera is enough to retrieve the desired data.

Video recordings of sports events are now widely available, creating a vast database that can be readily accessed. With MHPE these data can even be evaluated even though there was no original intention to do so.

It is important to note that most of the MHPE algorithms research has been conducted in the context of everyday life scenarios, such as pedestrians crossing a street. We need to bear in mind that these algorithms must first be evaluated and transferred to the intended application in the sports domain, as the detected poses can differ significantly. Before we dive into adapting the MHPE algorithms to our applications in sports in Sect. 16.3.2, we will provide an overview of the vast selection of algorithms and their fundamental working principles in the next Sect. 16.3.1.

16.3.1 Classification and Working Principles

In this section we will give an overview of the different general working principles of pose estimation. MHPE can be subdivided into two different tasks. Namely the reconstruction of 2D and 3D keypoints. First, we will explore the 2D case, specifically the task of single person pose estimation where only one person is featured in the RGB image which serves as input source. In the recent years it became state-of-the-art to use deep learning to execute this task. At an abstract level single person pose estimation works as depicted in Fig. 16.8.

> ▶ The input of the pose estimator is an RGB image, which is then fed into the pose decoder. The pose decoder acts as a feature extractor, meaning that through a network architecture which is narrowing down towards the output layer, high-level features can be extracted and encoded from the image. The resulting representation of the RGB image in the latent space is then supplied to the second part of the pose estimator, which is called pose decoder. The pose decoder then provides the 2D coordinates of the keypoints as output.

There are two main methods by which the output can be calculated either by a regression method or by a detection-based method where heatmaps are calculated.

The pose-encoder is typically a ResNet architecture pretrained on a large Image dataset such as ImageNet/COCO and can extract the main features of the RGB image. For state-of-the-art research it is much more interesting to tune the architecture of the decoder part.

When moving on to multiperson pose estimation, where the objective is to reconstruct multiple poses from an input image, the concept of the pose encoder-decoder architecture is still of value. There are two approaches for multiperson pose estimation.

In the top-down approach a person detector is employed to outputs bounding boxes around each identified individual. A similar encoder-decoder architecture then calculates the poses from every of the ingoing bounding boxes.

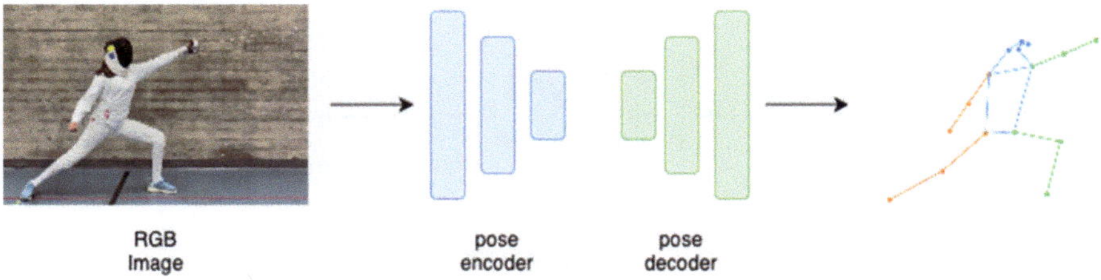

Fig. 16.8 General working principle of a single person MHPE algorithm. A RGB is first processed by the pose encoder resulting in latent representation which is reduced in dimensionality. This representation will then be transformed into keypoints by the pose decoder

The calculation of the pose estimation for every bounding box, explains why this approach is rather slow as the runtime is proportional to the number of people in an image. However, the top-down approach has the advantage that it is relatively easy to implement.

In the bottom-up approach. Here the encoder-decoder architecture is directly applied to the RGB image, which contains multiple persons. In a second step the calculated keypoints of all the individuals are assembled and assigned to the different individuals.

In this approach mainly the association part is challenging to achieve good results. The working principle of the two approaches is visualized in Fig. 16.9.

For all the 2D approaches the networks are trained in a supervised manner—meaning that RGB images as well as annotated 2D keypoints need to be provided. These annotations imply very tedious manual labelling tasks, which is a significant factor limiting the number of datasets on which these algorithms can be trained from scratch. Therefore, good 2D pose estimation datasets are sparse. It needs to be noted that these datasets mainly contain everyday task and are not directly related to sports activities.

> One of the main problem is, that sport specific poses tend to be more extreme (e.g. Yoga, gymnastics) than everyday activities. Additionally, occlusions, caused by either team members or oneself due to rotations tend to occur more often. Therefore, algorithms need to be adapted to the sports domain by the use of specific sport-related datasets.

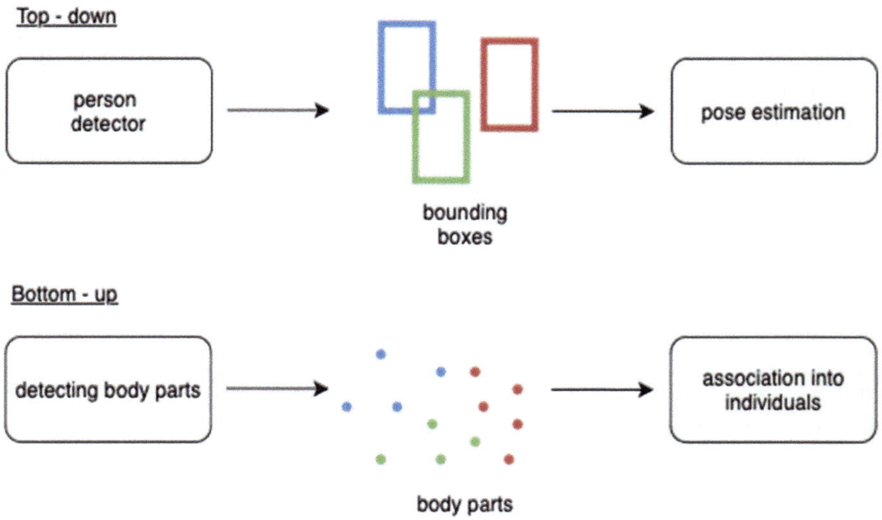

Fig. 16.9 Visualization of the top-down and the bottom—up approach for multiperson pose estimation

Table 16.1 Overview of publicly available sports datasets for 2D pose estimation

Dataset	Characteristics	Sports
Leeds Sports Pose (LSP) (Johnson & Everingham, 2010)	2000 images	Athletics, badminton, baseball, gymnastics, parkour, soccer, tennis, volleyball
Penn Action (Zhang et al., 2013)	2326 videos	15 different sports actions
KTH Multiview Football dataset (Kazemi et al., 2013)	5907 images	Soccer
Ski 2D Pose (Bachmann et al., 2019)	1982 images	Ski (downhill, Super-G, giant slalom, slalom)
Kanoya dataset (Kitamura et al., 2022)	1500 frames	Acrobatic movements (e.g. backflips)

Such public sports datasets are even more rare, partially due to the reason that a lot of the collected data is kept private. In Table 16.1 a list of public sports related datasets and their characteristics is given.

These domain specific datasets can then be used to transport the power of the pose estimation algorithms to your sport-specific domain.

When moving on to 3D pose estimation, we are facing two principal challenges. Firstly, the reconstruction of 3D keypoints from 2D input is inherently ambiguous due to the additional dimension. Secondly curating and annotating a dataset is more complex and time-consuming as also depth information is required. This frequently necessitates the use of some sort of sensor for motion capture which often restrains the data collection to indoor situations with very limited background and overall variety in the dataset. An overview of the few publicly available 3D sports datasets is shown in Table 16.2.

We first investigate basic 3D single person reconstruction which can be divided into three paradigms. An approach which is analogous to the 2D method is generating volumetric heatmaps directly from the RGB inputs. These heatmaps containing Gaussian distribution for every keypoint can then be postprocessed and transformed into 3D coordinates. A second approach is to use the much more stable 2D pose estimation as an intermediate step. The task of the architecture is then simply to perform a regression from 2D to the 3D coordinates which is learned on paired 2D/3D training data. Lastly the 2D pose estimation can be combined with images features which are then fed in a regression model which can predict the 3D poses.

It also needs to be mentioned that pose estimation is still an active field of research where new contributions are made every year. Thus, it is worth reviewing the newest advancements in the field to find the most suitable algorithm/framework for a given problem.

16.3.2 Sports Science Applications

Once poses have been extracted with the HPE algorithm which is most suitable for the task or an adaption of it there are many ways how the keypoints can be made use of in sports science and many examples can be found in literature. The most straight forward application is to use the extracted poses to calculate features or metrics,

Table 16.2 Overview of publicly available sports datasets for 3D pose estimation

Dataset	Characteristics	Sports
Martial Arts, Dancing and Spots (MADS)	53,000 frames	Tai-chi, Karate, Jazz dance, Hip-hop dance and different sports
KTH Multiview Football dataset II (3D) (Kazemi et al., 2013)	800 frames	Soccer
SportsPose (Ingwersen et al., 2023)	176,000 poses	Soccer, volleyball, jump, baseball pitch, tennis
Australian Sports Pose Dataset (ASPset) (Nibali et al., 2021)	100,000 poses	Outdoor activities (jumping, running, throwing, hitting)

such as angles of joints or forces and velocity of body parts. Very often in sport science these are key performance indicators, especially in very technical sports.

Example

A good example of such a case is ski jumping (Ludwig et al., 2020), where the angles of the body in take-off and during the flight phase are directly decisive for the quality of the jump. As additionally the angles between the body and the skis are relevant, the algorithm based on a Mask R-CNN (Girshick et al., 2013) architecture was trained to correctly predict the position of the ski tips and tails. For this purpose, the parameters of the chosen algorithm must be trained on annotated data. In this case around 10,000 frames from approximately 300 jumps were manually labelled to serve as training data. An additional 3400 frames from 100 jumps were used as test set. ◀

When conducting such data taking campaign to train a pose estimation algorithm several points need to be considered. First, it is important to provide a sufficient amount of data. Here it is decisive whether the algorithm needs to be trained from scratch or a finetuning or repurposing is enough. In the first case significantly more data needs to be collected than in the later.

▶ During data taking we need to include many different actions with different backgrounds, lighting conditions and athletes. This is to make sure that the trained model will later generalize well to all sorts of conditions and out of distribution samples.

With a finetuning of the algorithm on the self-produced dataset, the relevant angles of the ski jumper could be reconstructed with a high accuracy.

Secondly the reconstructed poses or series of poses can be used to be further processed by other deep learning algorithms.

Example

We here showcase the example of fencing (Emmenegger et al., 2023), where pose estimation was used to do action recognition but also classify error patterns in the movements. The poses of more than 1200 videos of four different fencing movements and up to six corresponding error patterns were reconstructed with Openpose (Cao et al., 2018). As fencing takes place on a 2D plane called the fencing strip it is sufficient to work with 2D poses.

The time series of reconstructed keypoints was then used in further deep learning architecture such as BiLSTM, CNN and attention-based models to classify them into the different actions and error pattern. The whole pipeline showed a good accuracy and also the capability to generalize on in the wild found fencing videos. ◀

Such an application can then further be used to generate statistics of matches to generate tactical profiles, to help monitoring athletes or to generate a self-assessment tool where athletes no longer only rely on the coaches' feedback.

Overall we can see that there are a broad range of applications for MHPE in sports science which have not been exploited yet.

16.4 Conclusion

In this chapter we introduced the working principles of tracking as well as human pose estimation, which are sophisticated state-of-the art computer vision tools relevant to sport science. Even though there are many ready-to-use implementations of these two methods many are not yet tailored to the use in sports science. Either because they make high demands on performance, such as very fast-moving objects, dealing with occlusion or because additional finetuning of the algorithms to the sports domain are needed. Sports science can contribute to the research of this field by curating and publicly providing datasets for various task. Thanks to these advancements, sport scientists can in the future obtain needed data conveniently from image or video material.

References

Bachmann, R., Spörri, J., Fua, P., & Rhodin, H. (2019). Motion capture from pan-tilt cameras with unknown orientation. In *2019 International Conference on 3D Vision (3DV)* (pp. 308–317). IEEE.

Bergmann, P., Meinhardt, T., & Leal-Taixé, L. (2019, March). Tracking without bells and whistles. In *IEEE International Conference on Computer Vision* (pp. 941–951).

Breitenstein, M. D., Reichlin, F., Leibe, B., Koller-Meier, E., & Van Gool, L. (2009). Robust tracking-by-detection using a detector confidence particle filter. In *2009 IEEE 12th International Conference on Computer Vision* (pp. 1515–1522). IEEE.

Cao, Z., Hidalgo, G., Simon, T., Wei, S.-E., & Sheikh, Y. (2018). OpenPose: Realtime multi-person 2D pose estimation using part affinity fields. *ArXiv [Cs.CV]*. arXiv. http://arxiv.org/abs/1812.08008

Chicco, D. (2021). Siamese neural networks: An overview. *Methods in Molecular Biology, 2190*, 73–94.

Cui, Y., Zeng, C., Zhao, X., Yang, Y., Wu, G., & Wang, L. (2023, April). SportsMOT: A large multi-object tracking dataset in multiple sports scenes. *IEEE International Conference on Computer Vision* (pp. 9887–9897).

Dendorfer, P., Rezatofighi, H., Milan, A., Shi, J., Cremers, D., Reid, I., Roth, S., Schindler, K., & Leal-Taixé, L. (2020). MOT20: A benchmark for multi object tracking in crowded scenes. *ArXiv [Cs.CV]*. arXiv. http://arxiv.org/abs/2003.09003.

Emmenegger, S., Egli, M., & Pouly, M. (2023). Mastering fencing techniques with machine learning: A video-based classification and correction system. In *2023 10th IEEE Swiss Conference on Data Science (SDS)* (pp. 120–127). IEEE.

Girshick, R., Donahue, J., Darrell, T., & Malik, J. (2013). Rich feature hierarchies for accurate object detection and semantic segmentation. *ArXiv [Cs.CV]*. arXiv. http://arxiv.org/abs/1311.2524

Ingwersen, C. K., Mikkelstrup, C., Jensen, J. N., Hannemose, M. R., & Dahl, A. B. (2023). SportsPose—A dynamic 3D sports pose dataset. *ArXiv [Cs.CV]*. arXiv. http://arxiv.org/abs/2304.01865

Johnson, S., & Everingham, M. (2010). *Clustered pose and nonlinear appearance models for human pose estimation*. https://doi.org/10.5244/C.24.12

Kazemi, V., Burenius, M., Azizpour, H., & Sullivan, J. (2013). Multi-view body part recognition with random forests. In *2013 24th British Machine Vision Conference, BMVC 2013; Bristol; United Kingdom; 9 September 2013 through 13 September 2013*. British Machine Vision Association. https://doi.org/10.5244/C.27.48

Kitamura, T., Teshima, H., Thomas, D., & Kawasaki, H. (2022). Refining OpenPose with a new sports dataset for robust 2D pose estimation. In *2022 IEEE/CVF Winter Conference on Applications of Computer Vision Workshops (WACVW)* (pp. 672–681). IEEE.

Ludwig, K., Einfalt, M., & Lienhart, R. (2020). Robust estimation of flight parameters for SKI jumpers. In *2020 IEEE International Conference on Multimedia & Expo Workshops (ICMEW)* (pp. 1–6). IEEE.

Nibali, A., Millward, J., He, Z., & Morgan, S. (2021, July). ASPset: An outdoor sports pose video dataset with 3D keypoint annotations. *Image and Vision Computing, 111*, 104196.

Redmon, J., Divvala, S., Girshick, R., & Farhadi, A. (2015). You only look once: Unified, real-time object detection. *ArXiv [Cs.CV]*. arXiv. http://arxiv.org/abs/1506.02640

Ren, S., He, K., Girshick, R., & Sun, J. (2015). Faster R-CNN: Towards real-time object detection with region proposal networks. In C. Cortes, N. Lawrence, D. Lee, M. Sugiyama, & R. Garnett (Eds.), *Advances in neural information processing systems* (Vol. 28). Curran Associates, Inc. https://proceedings.neurips.cc/paper_files/paper/2015/file/14bfa6bb14875e45bba028a21ed38046-Paper.pdf

Sengupta, A., Jin, F., Zhang, R., & Cao, S. (2019). Mm-Pose: Real-time human skeletal posture estimation using MmWave radars and CNNs. *ArXiv [Eess.SP]*. arXiv. http://arxiv.org/abs/1911.09592

Wang, Z., Zheng, L., Liu, Y., Li, Y., & Wang, S. (2019). Towards real-time multi-object tracking. *ArXiv [Cs.CV]*. arXiv. http://arxiv.org/abs/1909.12605

Zhang, W., Zhu, M., & Derpanis, K. G. (2013). From actemes to action: A strongly-supervised representation for detailed action understanding. In *2013 IEEE International Conference on Computer Vision* (pp. 2248–2255). IEEE.

Machine Learning-Based Analysis of Multi-Agent Trajectories in Basketball

17

Keisuke Fujii, Kazuhiro Yamada, Rikako Kono, Ziyi Zhang, and Rory Bunker

Abstract

This chapter introduces machine learning-based analysis of multi-agent trajectories in basketball. It addresses the challenge of comprehending multi-agent behaviors, which are typical in team sports, and characterized by intricate interactions and cognitive processes. The chapter outlines how machine learning methods can effectively analyze such behaviors, despite the inherent difficulty in modeling the behaviors and interpreting these non-linear models. We highlight two primary strategies: learning-based feature and rule extraction, and generating and controlling behaviors from models. The first involves visualizing representations and identifying underlying structures, while the second allows for the simulation and control of potential and hypothetical scenarios. The chapter concludes by discussing the practical implications of these methods, emphasizing their utility in enhancing our understanding of multi-agent behaviors in basketball.

17.1 Introduction

The emergence of advanced measurement technologies such as camera-based tracking systems has significantly enhanced the field of sports analytics, for example, in basketball (Yao, 2021). These tools have made it possible to precisely measure and analyze player movements, offering new insights into the dynamics of team sports.

K. Fujii (✉) · K. Yamada · Z. Zhang · R. Bunker
Graduate School of Informatics, Nagoya University, Nagoya, Aichi, Japan
e-mail: fujii@i.nagoya-u.ac.jp; yamada.kazuhiro@g.sp.m.is.nagoya-u.ac.jp; zhang.ziyi@g.sp.m.is.nagoya-u.ac.jp

R. Kono
Research School of Physics, The Australian National University, Canberra, Australia
e-mail: rikako.kono@anu.edu.au

D. Memmert (ed.), *Artificial Intelligence and Machine Learning in Sports Science*, https://doi.org/10.1007/978-3-662-70155-3_17

Notably, the application of these technologies in basketball enables a deeper understanding of the principles behind team sports behaviors, which is a key challenge in sports analytics. The interactions in team sports, such as basketball are complex and often characterized by high-order social dynamics, cognition, and body movements. Traditional mathematical and statistical models in team sports (Supola et al., 2022; Wu et al., 2022) have been useful for understanding behavior, but they usually consider simple assumptions while the rules underlying team sports behaviors are often largely unknown (Fujii et al., 2018). Therefore, to obtain a better understanding of these behaviors, a data-driven and model-free (or equation-free) approach is needed.

This chapter targets researchers who are interested in the application of machine learning in sports analytics, particularly those focusing on basketball. We introduce the usage of machine learning models in sports analytics, and the trade-off that exists between interpretability and predictive performance. We highlight two primary strategies: learning-based feature and rule extraction, and generating and controlling behaviors from models (Fujii, 2021). Several surveys have explored the spatiotemporal elements of this topic (Gudmundsson & Horton, 2017) and the use of machine learning for sports analytics in sports including basketball (Fujii, 2021; Sarlis & Tjortjis, 2020; Terner & Franks, 2021). Specifically, applications of such predictive approaches using machine learning cover areas such as match outcome prediction, tactical decision-making, player investments, fantasy sports, and injury forecasting. This chapter aims to systematically organize recent progress in machine learning-based analysis, with multi-agent trajectories in basketball as a core focus.

17.2 Data Sources

Multi-agent trajectories in basketball have been analyzed in recent studies, but data sources are limited. Most studies have used the only open-source dataset from the National Basketball Association (NBA) 2015–2016 season, preprocessed by the STATS[1] SportVU system. The dataset is available at https://github.com/rajshah4/BasketballData. It includes the trajectories of basketball players and the ball (25 frames per second) from approximately 630 games. The positional data contains the (x,y) coordinates of each player on the court and the (x,y,z) coordinates of the ball. Even in 2023, the dataset is still used based on the following reasons. First, an alternative open-source dataset has not appeared yet, which might be related to the tracking data provider changing firstly to Second Spectrum, and then (2023-) Sony's Hawk-Eye. Second, the size of the dataset may be also sufficient for most machine learning models. However, the dataset will become outdated as time passes, thus the availability of additional large trajectory datasets is expected in the future. We will also discuss how computer vision and other sensor-based technologies will solve this problem. Although not considering trajectory data, there have been studies that have analyzed games using movement annotations, for example, in the FIBA Basketball Champions League (Stavropoulos et al., 2021) and the Liga ACB Spanish Championship (Lamas et al., 2015).

[1] STATS is now STATS Perform (Northbrook, IL, USA)

17.3 Learning-Based Feature and Rule Extraction

This section provides an overview of learning-based methods for feature and rule extraction from data, with minimal prior knowledge. These encompass unsupervised, supervised, semi-supervised, and self-supervised learning methods. These methods help interpret sports data through feature extraction and rule discovery. Traditional non-learning methods rely on researchers' expertise and theories to analyze multi-agent behaviors. For instance, researchers assess athletes' distances, action frequencies, and phases, leading to hypothesis testing often via statistical analysis (e.g., Bourbousson et al., 2010).

Advanced mathematical methods have also been employed for more complex analyses, such as visual analytics and movement evaluations in sports (Supola et al., 2022, 2023; Wu et al., 2022). Nonetheless, flexible modeling is essential for the detailed representation of interactions. Data-driven techniques, including machine learning, offer solutions for automatic feature extraction, classification, and regression. This section further delves into unsupervised and supervised learning, highlighting their applications in interpreting team sports data using player positions (cf. Chap. 2).

17.3.1 Unsupervised Learning

Unsupervised learning, a key machine learning technique, uncovers hidden structures from unlabeled data. It is crucial for knowledge discovery without predefined hypotheses. Common techniques include dimensionality reduction and clustering.

Dimensionality reduction transforms high-dimensional data into lower-dimensional, meaningful formats. Techniques such as t-distributed stochastic neighbor embedding (t-SNE) (van der Maaten & Hinton, 2008), non-negative matrix factorization (NMF) (Miller et al., 2014), and topic modeling (Miller & Bornn, 2017) can summarize complex sports behaviors. These methods, however, sometimes overlook dynamic properties due to assumptions about sample independence. Thus, methods recognizing the dynamic aspects of data, for example, neural network-based image processing (Wang & Zemel, 2016) and dynamic mode decomposition (DMD) (Fujii et al., 2018, 2020), are needed.

Clustering organizes similar objects into groups. For example, hierarchical clustering based on object connectivity, centroid-based clustering such as k-means, and distribution-based clustering such as Gaussian mixture models are considered. In the context of sports, researchers have applied similarity methods such as dynamic time warping (DTW) (Sha et al., 2016) and Gaussian mixture model clustering (Perše et al., 2009).

Multi-agent trajectory analysis faces challenges, such as player permutation. Rule-based (Fujii et al., 2018) and data-driven methods like the Hungarian algorithm (Sha et al., 2016, 2017), DMD, as permutation-invariant analysis (Fujii et al., 2020), and graph neural networks (Yeh et al., 2019) address this issue. While unsupervised learning does not rely on labeled data, combining it with supervised learning can enhance validation and effectiveness.

17.3.2 Supervised Learning

Supervised learning is a machine learning that involves inferring infers functions from labeled data. It includes classification for discrete data (e.g., play types) and regression for continuous data (e.g., scores). This involves using static features in models for tasks including score prediction (Cervone et al., 2014) and play classification (Hojo et al., 2018; McQueen et al., 2014; Tian et al., 2019). Incorporating time-series structures is essential, especially for analyzing complex behaviors. Dynamic features from unsupervised learning can aid in classifying tactics (Fujii et al., 2020) and predicting outcomes (Fujii et al., 2018).

End-to-end approaches, such as neural networks, integrate feature extraction and prediction. They have been used in classifying plays (Wang & Zemel, 2016) and analyzing outcomes (Sicilia et al., 2019), but may lack interpretability. Balancing predictability and interpretability, researchers have used matrix (Yue et al., 2014) and tensor (Zheng et al., 2018) models, and Poisson point processes (Mortensen & Bornn, 2019). Trajectory mining (Bunker et al., 2024) and interpretable models with attention mechanisms (Zhang et al., 2023) have also been employed.

17.3.3 Semi-Supervised and Self-Supervised Learning

Supervised learning is a powerful approach given a sufficient amount of labeled data, but, in fact, it is labor-intensive and the amount of labeled data is often limited. Semi-supervised learning is an approach that combines supervised and unsupervised learning by combining both labeled and unlabeled data for training machine learning models. In semi-supervised learning, a small amount of labeled data is used along with a larger amount of unlabeled data to achieve better performance. Even though the true labels of the unlabeled data are unknown, the data is expected to encode useful statistical patterns. In team sports, semi-supervised methods have been used to classify basketball screens (Zhang et al., 2022) and to predict annotated trajectories using generative models (Fassmeyer et al., 2022).

Self-supervised learning is also a solution to tackle the problem of annotation cost, and is a machine learning approach that uses unlabeled data and derives supervisory signals through different preprocessing techniques. It creates proxy labels from patterns within the raw data itself. The representative examples are transformer-based approaches. For example, in basketball, methods have been proposed to recognize group behaviors (Hauri & Vucetic, 2022) and to learn trajectory representations (Alcorn & Nguyen, 2021). There have been relatively few studies about semi- and self-supervised learning in basketball, but such methods will become more necessary in the future if a larger amount of (unlabeled) data is accessible.

17.4 Simulation and Control of Behaviors

This section addresses methods for validating hypotheses via modeling future scenarios or unobservable situations. It includes conventional rule-based, pattern-based, and planning-based techniques. Rule-based methods involve predefined rules for agent actions, offering clear interpretability but requiring extensive effort for adaptation to new situations. Pattern-based techniques identify behavioral patterns from data without explicit rules. Planning-based methods, often linked to reinforcement learning, focus on achieving long-term objectives through computed policies or paths.

17.4.1 Pattern-Based Approaches

Pattern-based approaches learn dynamics directly from data, requiring minimal human input. These methods, mainly applied in team sports, focus on simulating short-term multi-agent trajectories and long-term team outcomes. They often employ supervised learning, not only for feature and rule extraction but also for simulating and evaluating player behaviors.

Predicting short-term multi-agent trajectories in sports such as basketball and soccer is a significant challenge in machine learning. Techniques that have been used include recurrent neural networks (RNN) (Hauri et al., 2021; Zheng et al., 2016), variational RNNs (Fujii et al., 2024; Yeh et al., 2019; Zhan et al., 2019), and generative adversarial networks (Chen et al., 2018). Recent developments in graph neural networks (Yeh et al., 2019) have addressed the permutation problem that was described in the unsupervised learning subsection.

Additionally, tactical evaluation of predicted trajectories is crucial. For instance, optimizing defensive strategies in basketball (Sha, 2018), and using counterfactual predictions to validate models (Chen et al., 2023; Fujii et al., 2024; Yeh et al., 2019) have been considered, despite the lack of ground truth in such hypothetical scenarios.

17.4.2 Planning-Based Approaches

Planning-based methods in machine learning focus on the long-term movement goals of agents, and developing policies or paths to achieve these objectives. These methods can be broadly divided into inverse and forward planning approaches.

Inverse planning uses statistical learning to estimate action models or reward functions from observed data. It is akin to applying reinforcement learning in real-world scenarios, evaluating player or team actions and states for goal achievement, distinct from feature extraction or trajectory prediction. Examples include evaluating shooting efficiency (Cervone et al., 2014, 2016) and defensive action quality in basketball (Wang et al., 2018), as well as estimating state-action values (Chen et al., 2022; Yanai et al., 2022).

Forward planning assumes optimal criteria for agent movements, using predefined reward functions such as scoring in sports. It has been used in virtual competitions, for example, in a 3-vs-3 basketball simulator (Tang et al., 2018), which is currently available on a different platform at https://github.com/FuxiRL/DunkCityDynasty. Such reinforcement learning aims to outperform human players. While similar behaviors can be learned from sports data using imitation learning, combining inverse and forward planning remains a relatively unexplored area, and bridging this gap should be a key focus of future research.

17.5 Practical Applications and Future Directions

This section explores practical applications and future directions of sports analytics, focusing on the use of rules, features, and behaviors derived from machine learning. A direct application of play classification and score prediction is reducing the workload for manual play analysis. While defining specific plays for classification can be challenging, methods including unsupervised learning can facilitate play recommendation systems (Sha et al., 2018), akin to web search algorithms. Short-term future predictions, as discussed in Sect. 17.4, can visually demonstrate potential player movements and team responses in upcoming games. For long-term predictions, forecasting opponent team strategies and outcomes can aid in tactical planning.

Future research directions include addressing the high cost of location information, used primarily in professional sports. Advances in software (and hardware) technologies, for example, computer vision techniques (Yao, 2021), are expected to make data acquisition more accessible and accurate. Open-source video datasets (Cui et al., 2023) are contributing to this development. Additionally, while high performance is often demanded for practical applications, it is worth assessing whether these analytical methods offer more cost-effective and accurate results than manual human analysis.

References

Alcorn, M. A., & Nguyen, A. (2021). baller2vec: A multi-entity transformer for multi-agent spatiotemporal modeling. *arXiv preprint arXiv:2102.03291*.

Bourbousson, J., Sève, C., & McGarry, T. (2010). Space–time coordination dynamics in basketball: Part 2. The interaction between the two teams. *Journal of Sports Sciences, 28*(3), 349–358.

Bunker, R. P., Duy, V. N. L., Tabei, Y., Takeuchi, I., & Fujii, K. (2024). Multi-agent statistically discriminative sub-trajectory mining and an application to NBA basketball. Journal of Quantitative Analysis in Sports, (0).

Cervone, D., D'Amour, A., Bornn, L., & Goldsberry, K. (2014). Pointwise: Predicting points and valuing decisions in real time with NBA optical tracking data. In *Proceedings of the MIT Sloan Sports Analytics Conference*.

Cervone, D., D'Amour, A., Bornn, L., & Goldsberry, K. (2016). A multiresolution stochastic process model for predicting basketball possession outcomes. *Journal of the American Statistical Association, 111*(514), 585–599.

Chen, C.-Y., Lai, W., Hsieh, H.-Y., Zheng, W.-H., Wang, Y.-S., & Chuang, J.-H. (2018). Generating defensive plays in basketball games. In *Proceedings of the 26th ACM International Conference on Multimedia* (pp. 1580–1588).

Chen, X., Jiang, J.-Y., Jin, K., Zhou, Y., Liu, M., Brantingham, P. J., & Wang, W. (2022). Reliable: Offline reinforcement learning for tactical strategies in professional basketball games. In *Proceedings of the 31st ACM International Conference on Information & Knowledge Management* (pp. 3023–3032).

Chen, X., Wang, W.-Y., Hu, Z., Chou, C., Hoang, L., Jin, K., et al. (2023). Professional basketball player behavior synthesis via planning with diffusion. *arXiv preprint arXiv:2306.04090.*

Cui, Y., Zeng, C., Zhao, X., Yang, Y., Wu, G., & Wang, L. (2023). SportsMOT: A large multi-object tracking dataset in multiple sports scenes. In *Proceedings of the IEEE/CVF International Conference on Computer Vision* (pp. 9921–9931).

Fassmeyer, D., Fassmeyer, P., & Brefeld, U. (2022). Semi-supervised generative models for multiagent trajectories. *Advances in Neural Information Processing Systems, 35*, 37267–37281.

Fujii, K. (2021). Data-driven analysis for understanding team sports behaviors. *Journal of Robotics and Mechatronics, 33*(3), 505–514.

Fujii, K., Kawasaki, T., Inaba, Y., & Kawahara, Y. (2018). Prediction and classification in equation-free collective motion dynamics. *PLoS Computational Biology, 14*(11), e1006545.

Fujii, K., Takeishi, N., Hojo, M., Inaba, Y., & Kawahara, Y. (2020). Physically-interpretable classification of network dynamics for complex collective motions. *Scientific Reports, 10*, 3005.

Fujii, K., Takeishi, N., Kawahara, Y., & Takeda, K. (2024). Decentralized policy learning with partial observation and mechanical constraints for multi-person modeling. *Neural Networks, 171*, 40–52.

Gudmundsson, J., & Horton, M. (2017). Spatio-temporal analysis of team sports. *ACM Computing Surveys (CSUR), 50*(2), 1–34.

Hauri, S., & Vucetic, S. (2022). Group activity recognition in basketball tracking data–neural embeddings in team sports (nets). *arXiv preprint arXiv:2209.00451.*

Hauri, S., Djuric, N., Radosavljevic, V., & Vucetic, S. (2021). Multi-modal trajectory prediction of NBA players. In *Proceedings of the IEEE/CVF Winter Conference on Applications of Computer Vision* (pp. 1640–1649).

Hojo, M., Fujii, K., Inaba, Y., Motoyasu, Y., & Kawahara, Y. (2018). Automatically recognizing strategic cooperative behaviors in various situations of a team sport. *PLoS One, 13*(12), e0209247.

Lamas, L., Santana, F., Heiner, M., Ugrinowitsch, C., & Fellingham, G. (2015). Modeling the offensive-defensive interaction and resulting outcomes in basketball. *PLoS One, 10*(12), 1–14. https://doi.org/10.1371/journal.pone.0144435

McQueen, A., Wiens, J., & Guttag, J. (2014). Automatically recognizing on-ball screens. In *Proceedings of the MIT Sloan Sports Analytics Conference.*

Miller, A. C., & Bornn, L. (2017). Possession sketches: Mapping NBA strategies [Conference Proceedings]. In *Proceedings of the MIT Sloan Sports Analytics Conference.*

Miller, A., Bornn, L., Adams, R., & Goldsberry, K. (2014). Factorized point process intensities: A spatial analysis of professional basketball. In *International Conference on Machine Learning* (pp. 235–243).

Mortensen, J., & Bornn, L. (2019). From Markov models to Poisson point processes: Modeling movement in the NBA. In *Proceedings of the 2019 MIT Sloan Sports Analytics Conference.*

Perše, M., Kristan, M., Kovačič, S., Vučkovič, G., & Perš, J. (2009). A trajectory-based analysis of coordinated team activity in a basketball game. *Computer Vision and Image Understanding, 113*(5), 612–621.

Sarlis, V., & Tjortjis, C. (2020). Sports analytics—Evaluation of basketball players and team performance. *Information Systems, 93*, 101562.

Sha, L. (2018). *Representing and predicting multi-agent data in adversarial team sports*. Unpublished doctoral dissertation. Queensland University of Technology.

Sha, L., Lucey, P., Yue, Y., Carr, P., Rohlf, C., & Matthews, I. (2016). Chalkboarding: A new spatiotemporal query paradigm for sports play retrieval. In *International Conference on Intelligent User Interfaces* (pp. 336–347).

Sha, L., Lucey, P., Zheng, S., Kim, T., Yue, Y., & Sridharan, S. (2017). Fine-grained retrieval of sports plays using tree-based alignment of trajectories. *arXiv preprint arXiv:1710.02255*.

Sha, L., Lucey, P., Yue, Y., Wei, X., Hobbs, J., Rohlf, C., & Sridharan, S. (2018). Interactive sports analytics: An intelligent interface for utilizing trajectories for interactive sports play retrieval and analytics. *ACM Transactions on Computer-Human Interaction (TOCHI), 25*(2), 1–32.

Sicilia, A., Pelechrinis, K., & Goldsberry, K. (2019). Deephoops: Evaluating micro-actions in basketball using deep feature representations of spatio-temporal data. In *Proceedings of the 25th ACM SIGKDD International Conference on Knowledge Discovery & Data Mining* (pp. 2096–2104).

Stavropoulos, N., Papadopoulou, A., & Kolias, P. (2021). Evaluating the efficiency of off-ball screens in elite basketball teams via second-order Markov modelling. *Mathematics, 9*(16), 1991.

Supola, B., Hoch, T., & Baca, A. (2022). Modeling the extra pass in basketball–an assessment of one of the most crucial skills for creating great ball movement. *International Journal of Computer Science in Sport, 22*(1), 13–29.

Supola, B., Hoch, T., & Baca, A. (2023). Modeling the offensive-defensive interaction and resulting outcomes in basketball. *PLoS One, 18*(2), e0281467.

Tang, H., Hao, J., Lv, T., Chen, Y., Zhang, Z., Jia, H., et al. (2018). Hierarchical deep multiagent reinforcement learning with temporal abstraction. *arXiv preprint arXiv:1809.09332*.

Terner, Z., & Franks, A. (2021). Modeling player and team performance in basketball. *Annual Review of Statistics and Its Application, 8*, 1–23.

Tian, C., De Silva, V., Caine, M., & Swanson, S. (2019). Use of machine learning to automate the identification of basketball strategies using whole team player tracking data. *Applied Sciences, 10*(1), 24.

van der Maaten, L., & Hinton, G. (2008). Visualizing data using t-SNE. *Journal of Machine Learning Research, 9*, 2579–2605.

Wang, K.-C., & Zemel, R. (2016). Classifying NBA offensive plays using neural networks. In *Proceedings of the MIT Sloan Sports Analytics Conference.*

Wang, J., Fox, I., Skaza, J., Linck, N., Singh, S., & Wiens, J. (2018). The advantage of doubling: A deep reinforcement learning approach to studying the double team in the NBA. *arXiv preprint arXiv:1803.02940*.

Wu, Y., Deng, D., Xie, X., He, M., Xu, J., Zhang, H., et al. (2022). Obtracker: Visual analytics of off-ball movements in basketball. *IEEE Transactions on Visualization and Computer Graphics, 29*(1), 929–939.

Yanai, C., Solomon, A., Katz, G., Shapira, B., & Rokach, L. (2022). Q-ball: Modeling basketball games using deep reinforcement learning. In *Proceedings of the AAAI Conference on Artificial Intelligence* (Vol. 36, pp. 8806–8813).

Yao, P. (2021). Real-time analysis of basketball sports data based on deep learning. *Complexity, 2021*, 1–11.

Yeh, R. A., Schwing, A. G., Huang, J., & Murphy, K. (2019). Diverse generation for multi-agent sports games. In *Proceedings of the IEEE Conference on Computer Vision and Pattern Recognition* (pp. 4610–4619).

Yue, Y., Lucey, P., Carr, P., Bialkowski, A., & Matthews, I. (2014). Learning fine-grained spatial models for dynamic sports play prediction. In *2014 IEEE International Conference on Data Mining* (pp. 670–679). https://doi.org/10.1109/ICDM.2014.106

Zhan, E., Zheng, S., Yue, Y., Sha, L., & Lucey, P. (2019). Generating multi-agent trajectories using programmatic weak supervision. In *International Conference on Learning Representations*.

Zhang, Z., Takeda, K., & Fujii, K. (2022). Cooperative play classification in team sports via semi-supervised learning. *International Journal of Computer Science in Sport, 21*(1), 111–121.

Zhang, Z., Bunker, R., Takeda, K., & Fujii, K. (2023). Multi-agent deep-learning based comparative analysis of team sport trajectories. *IEEE Access, 11*, 43305–43315.

Zheng, S., Yue, Y., & Hobbs, J. (2016). Generating long-term trajectories using deep hierarchical networks. *Advances in Neural Information Processing Systems, 29*, 1543–1551.

Zheng, S., Yu, R., & Yue, Y. (2018). Multi-resolution tensor learning for large-scale spatial data. *arXiv preprint arXiv:1802.06825*.

Artificial Intelligence to Promote Physical Activity

18

Vahid Farrahi, Horacio Sánchez-Trigo, Marcus Schmidt, and Thomas Jaitner

Abstract

Despite the widely recognised benefits of physical activity (PA), the world has witnessed a pandemic of physical inactivity. A significant number of people spend most of their time being physically inactive, and there are no significant signs that this trend is reversing. In recent years, the rise of artificial intelligence (AI) and machine learning (ML) has opened up new ways to promote PA and active living in innovative ways. This chapter explores three key emerging areas relevant to the promotion of PA, namely (1) fitness tracking with wearable devices, (2) chatbots, and (3) gamification of PA, and shows how the integration of AI and ML in these areas can make PA interventions more effective and sustainable.

18.1 Background

The health benefits of PA are currently well documented. Recent studies provide clear evidence that any amount of PA, even in small doses, is associated with a lower risk of all-cause mortality (Chastin et al., 2021; Niemelä et al., 2023) and cardiovascular mortality (Walmsley et al., 2022) and cancer-related mortality (Garcia et al., 2023). PA has also been shown to be associated with healthy growth

V. Farrahi (✉) · M. Schmidt · T. Jaitner
Institute for Sport and Sport Science, TU Dortmund University, Dortmund, Germany
e-mail: vahid.farrahi@tu-dortmund.de; marcus2.schmidt@tu-dortmund.de; thomas.jaitner@tu-dortmund.de

H. Sánchez-Trigo
Department of Physical Education and Sport, University of Seville, Seville, Spain
e-mail: fstrigo@us.es

D. Memmert (ed.), *Artificial Intelligence and Machine Learning in Sports Science*, https://doi.org/10.1007/978-3-662-70155-3_18

and ageing (Lipnowski et al., 2012; Moreno-Agostino et al., 2020) and better cardiometabolic health (Farrahi et al., 2021, 2023), reduced symptoms of depression (Noetel et al., 2024) and the prevention of various other chronic diseases (Anderson & Durstine, 2019).

Despite the widely recognised benefits of PA, the world has been experiencing a pandemic of physical inactivity for years (Kohl et al., 2012; Pratt et al., 2020). Many people worldwide do not reach the minimum recommendations for PA and spend most of their time sitting down. The pandemic of physical inactivity is a global crisis that crosses borders and affects people of all ages and socioeconomic backgrounds (Guthold et al., 2018). The consequences include rising rates of obesity and major non-communicable diseases such as diabetes, cardiovascular disease and cancer (Lee et al., 2012) at an estimated cost of billions of dollars to healthcare systems worldwide (Ding et al., 2016). This results in an urgent need to increase and promote PA (Pratt et al., 2020).

Approaches to increasing PA can be divided into informational, behavioural, social, environmental and policy approaches (Kahn et al., 2002). PA interventions that rely on such approaches range from measures in schools, the workplace and healthcare to the creation of accessible and inclusive recreational spaces (Cooper et al., 2021; Jirathananuwat & Pongpirul, 2017; Moon et al., 2024; Woodhead et al., 2023). These efforts and measures also include policy adjustments, infrastructure improvements and the promotion of active transport (Chillón et al., 2011; Durão et al., 2023). Public awareness campaigns and initiatives aimed at changing social norms and policies to prioritise active transport have been implemented (den Braver et al., 2022). However, despite several reportedly effective programmes to promote active lifestyles and PA (Pratt et al., 2020), there is increasing recognition of the importance of tailoring PA interventions to the needs and circumstances of each individual to promote sustained positive behaviour change (Ma et al., 2021, 2022).

However, even the current guidelines for PA for different age groups and populations are developed and disseminated on a 'one-size-fits-most' basis (DiPietro et al., 2020), overlooking the diverse individual needs and circumstances that influence an individual's ability and motivation to be active (Tremblay et al., 2024). Personalised approaches to PA promotion can encompass a range of considerations, such as the individual's characteristics, preferences, needs, inclination and ability to engage in PA, health status and other factors (An et al., 2024; Ma et al., 2021, 2022; Tremblay et al., 2024). Personalised approaches can potentially influence PA behaviour and outcomes more effectively and promote long-term behaviour change.

► Personalised and sustainable approaches to promoting PA include tailored interventions that are tailored to the individual's needs and preferences, while ensuring that these strategies are effective and sustained over time.

18.2 The Rise of AI and ML

In recent years, the development and application of advanced AI and ML tools and algorithms in healthcare and medicine has increased significantly (see Chap. 1 for details). The trend towards the use of AI methods is particularly evident in applications where the focus is on personalisation and personalised data is available. One such example is the use of AI tools for precision and personalised medicine (König et al., 2017). These tools harness the power of AI to customise medical treatments and interventions based on the individual characteristics of each patient. The aim is to deliver the right treatment to the right patient, at the right dose and at the right time. Precision medicine and personalised approaches have already led to revolutionary changes in many research areas and fields and continue to evolve in PA research (Farrahi & Clare, 2024; Farrahi & Rostami, 2024; Tremblay et al., 2024).

With the advent of AI, the promotion of PA can be approached in a personalised and innovative way. Existing studies have shown that AI-driven interventions can lead to significant improvements in PA levels and outcomes (An et al., 2024; Hassoon et al., 2021), indicating that these technologies have the potential to promote health behaviour change and have a significant impact on public health. The potential of AI to promote and increase PA is evident in several ways. For example, the integration of AI in PA promotion and improvement can facilitate personalised and adaptive PA interventions, promote positive PA behaviours through real-time feedback systems, and provide data-driven insights that promote PA. Three key emerging areas relevant to the promotion of PA are (1) wearable fitness trackers, (2) chatbots, and (3) gamification of PA. This chapter shows how AI and ML applications have been used to increase the effectiveness and sustainability of PA interventions.

▶ The most important additional benefits of AI and ML for personalised and sustainable PA support are:

- Individualisation of the intervention, taking into account personal preferences, lifestyle and existing health conditions.
- Adaptive goal setting that evolves with the participant's progress and changing circumstances.
- Regular and timely follow-up to ensure that the measures remain effective.
- Progressive and motivational adjustments to the intervention based on ongoing assessment and feedback.

18.3 Fitness Tracking on the Wrist

Wearable fitness tracking is currently experiencing a huge upsurge in popularity among many people and offers a remarkable and unique opportunity to promote positive health behaviours. Recent technological advances have enabled the development of many types of wearable, low-cost and unobtrusive sensors that can be worn by the wearer to monitor a wide range of physiological parameters and track PA behaviour (Kaewkannate & Kim, 2016). Such wearable devices have already found their way into the lives of many people.

Utilising the features and data provided by fitness trackers offers various opportunities to promote and improve PA. Wearable fitness trackers serve as a platform for individuals to self-monitor their behaviour and initiate positive lifestyle changes (Ferguson et al., 2022; Kaewkannate & Kim, 2016). Studies have shown that fitness trackers can positively influence people's lifestyle and PA behaviours in several ways—by prompting users to set goals, by allowing them to monitor things that are important to them, by reminding them when they are not on track to achieve their PA goals, or by providing them with personalised recommendations and feedback (Ferguson et al., 2022; Lyons et al., 2014). In addition to these benefits, fitness trackers also foster a sense of community among users. Many fitness tracker apps include social features that allow users to socialise with friends, join groups of like-minded people and participate in challenges (Stragier et al., 2016). This social aspect can provide additional motivation and make the process of maintaining an active lifestyle more enjoyable. There is substantial evidence that fitness monitoring offer at least some benefits to the wearer. A recent review of the existing literature shows that in over 160,000 participants across all included studies, individuals wearing an activity tracker took approximately 1800 more steps each day, resulting in a weight loss of approximately 1 kg (Ferguson et al., 2022). Several other studies report that interventions to promote active lifestyles and PA using wearable activity trackers can lead to increased PA and reduced sedentary behaviour (Li et al., 2021; Szeto et al., 2023).

AI and ML play an important role in augmenting and enhancing PA through fitness trackers. At the core of fitness trackers, AI and ML provide these devices with the ability to learn from the collected data, recognise the unique patterns in the wearer's PA behaviour, and make predictions or classifications based on patterns and relationships in the data (Nahavandi et al., 2022). Unsupervised ML algorithms, for example, have demonstrated their effectiveness when applied to fitness tracker data. They are able to identify unique patterns specific to the wearer's PA behaviour (Li et al., 2018). Furthermore, these algorithms have shown that they are able to detect and identify changes in PA behaviour over time (Díaz et al., 2020; Sprint et al., 2016). This ability to track and analyse changes in behaviour is critical to providing timely and adaptive PA recommendations, thereby increasing the likelihood that PA goals will be achieved. Providing these trends and patterns to the users themselves can also be informative and helpful for users to understand and analyse their own PA behaviour.

Predictive modelling and algorithms have also been applied to fitness tracking data to predict future behaviours based on historical data (Dijkhuis et al., 2018; Zhou et al., 2019). Predictive algorithms can automatically analyse data to identify patterns in PA behaviour, making PA interventions both personalised and dynamic. These models are able to learn the unique characteristics of different individuals' PA behaviour, enabling tailored recommendations that adapt to the user's specific circumstances and needs (An et al., 2024). This individualised approach can lead to more effective interventions as recommendations are based on a comprehensive and continuous analysis of the user's PA behaviour rather than generic advice (Dijkhuis et al., 2018). In addition, the ability to dynamically update recommendations based on real-time data can help users stay motivated and engaged, improving adherence to PA routines (Dijkhuis et al., 2018). For example, if a user is typically less active on weekends, the model could predict this pattern and provide early recommendations to increase activity during this time. There is initial evidence that supervised ML algorithms, when applied to historical step and goal attainment data, are able to predict exercise relapse with high accuracy (85%) (Zhou et al., 2019). Such predictive models could be used to suggest specific workouts or activities that the user enjoys or to encourage them to stay active with motivational messages.

It is estimated that 40% of households in the United States now own a wearable device and the prevalence is similar in many other European and non-European countries (Wearable devices ownership in the United States in 2021, by age, Statista (2022). https://www.statista.com/statistics/1310364/wearable-devices-ownership-age-us/). Preliminary data suggests that wearable device owners are generally willing to share their data and use digital health technologies aimed at changing lifestyle and health-related behaviours (Shandhi et al., 2024). This willingness is promising as it emphasises the potential of integrating data-intensive AI and ML techniques into fitness trackers to enhance their capabilities and make them more effective tools for promoting and improving PA. Despite the potential limitations in the accuracy of fitness trackers for measuring PA (Kaewkannate & Kim, 2016), the personalized data they collect can be leveraged with AI and ML to enhance fitness outcomes. This combination can deliver timely nudges to promote PA, provide personalized exercise recommendations, adjust the frequency and timing of these recommendations based on individual needs in real time, monitor trends and progress, and boost motivation for maintaining PA.

18.4 AI Chatbots

A chatbot is essentially a computer programme or software application that simulates a conversation with human users, usually through text or voice-based interactions. Chatbots can be rule-based or limited to a certain number of responses. An AI chatbot is a type of chatbot that contains AI technologies at its core (Aggarwal et al., 2023). This integration enables the AI chatbot to understand context, learn from interactions and formulate responses that are not pre-programmed. These conversational agents play an important role in improving user experience, providing personalised assistance and automating interactions in various contexts, including health advice and education (Aggarwal et al., 2023; Zhang et al., 2020).

The application of AI chatbots to lifestyle change programmes and interventions is one of the promising areas for the development of cost-effective and feasible behavioural interventions to promote PA (Zhang et al., 2020). These tools engage users in natural, conversational interactions and provide tailored exercise programmes, motivation and guidance. AI chatbots have remarkable features that can overcome the limitations of promoting PA with traditional intervention strategies (Oh et al., 2021; Zhang et al., 2020). A key advantage is the elimination of traditional face-to-face visits. Traditional strategies often involve scheduled, face-to-face sessions with a fitness trainer or health coach. These can be both time-consuming and costly. However, with an AI chatbot accessible via a web or mobile application, these in-person visits become redundant (Oh et al., 2021; Zhang et al., 2020). Users can interact with the chatbot at any time, regardless of the time of day, whether early in the morning or late at night. Essentially, AI chatbots are able to provide personalised recommendations based on preferences and goals and offer advice and support in real time, helping to reduce both time and financial expenditure for participants. Another notable feature of AI chatbots is that they can introduce new communication modalities that allow individuals to receive, understand and utilise information, suggestions and support on a personalised level (Zhang et al., 2020). Personalised interaction can help overcome barriers such as lack of self-efficacy or social support (Maher et al., 2020).

Recent developments in the field of AI chatbots are promising; research suggests that these technologies can positively impact PA behaviour and outcomes (Oh et al., 2021; Singh et al., 2023). The use of AI-driven chatbots has demonstrated feasibility and efficacy in improving PA and body composition over a 12-week period (Maher et al., 2020). Studies examining the impact of chatbots on lifestyle behaviours also suggest that these technologies can effectively promote health-related behaviours and improve PA outcomes. For example, a recent systematic review and meta-analysis of chatbot interventions targeting PA (as well as other lifestyles such as sleep and diet) found that chatbots can positively influence PA behaviours and increase total PA, moderate to vigorous PA, and daily steps (Singh et al., 2023). The results were consistent and showed that chatbot interventions were effective across a range of populations and age groups, with benefits observed in both short- and long-term interventions.

A notable advance in recent years has been the emergence of large language models (LLMs)—a form of generative AI—which are characterised by their extensive knowledge and ability to participate in human-like dialogues (Thirunavukarasu et al., 2023). These models are able to understand and generate human language in a way that is remarkably similar to human communication. In particular, LLMs have been shown to have a comprehensive understanding of medicine and healthcare (Thirunavukarasu et al., 2023), although more research is needed to fully establish their capabilities and validity for these domains. Preliminary data suggest that existing LMM models have potential for use in personalised health advice (Birkun & Gautam, 2024), including PA prescription and exercise (Dergaa et al., 2023). A recent study examining OpenAI's Generative Pre-Trained Transformer 4 (GPT-4)—a widely used LLM model—suggests that it can create safe exercise prescriptions for patient profiles with different health conditions and fitness goals (Dergaa et al., 2023). Based on expert assessments, the exercise programmes developed by GPT-4 allow individuals with average fitness levels to maintain their current health status, although they do not appear to provide sufficient incentives for substantial health or fitness improvement.

Another important advance is the development of vision-language models in the field of chatbots (Christensen et al., 2024). These models combine the ability to understand both visual and textual content, improving the interaction experience. They can process and interpret images in conjunction with text to provide a more comprehensive and richer response (Christensen et al., 2024). Although still in their infancy, several potential applications of these techniques in terms of promoting PA are foreseeable. For example, these models are likely to be able to assess the mood of users based on their facial expressions and adapt motivational messages accordingly. For users with hearing loss or problems, models could be used to recognise sign language and provide personalised recommendations.

18.5 Gamification of the PA

Gamification is an effective strategy for promoting and improving PA by utilising elements of game design in non-game contexts (Patel et al., 2019). This approach taps into people's natural need for competition, achievement and social interaction and makes PA more interesting and enjoyable. AI-powered gamification can significantly increase motivation and engagement in PA by providing personalised challenges and rewards (Modi & Gochhait, 2023). These systems use AI and ML algorithms to analyse user data and tailor the gaming experience to the individual's fitness level, goals and preferences. For example, AI can adjust the difficulty level of challenges based on the user's performance and ensure that tasks are neither too easy nor too difficult to maintain an optimal level of challenge and motivation. Personalised rewards such as virtual badges, points or even tangible incentives can be designed to match the user's interests and progress to encourage sustained engagement (Tolks et al., 2024). Studies have shown that gamified interventions can significantly increase PA levels and improve overall health, highlighting the practical benefits of AI-driven gamification (Tolks et al., 2024).

Social interaction and competition are critical components of gamification that can be enhanced by AI. AI algorithms can create virtual communities, leaderboards and social challenges that foster a sense of camaraderie and healthy competition among users. By participating in group challenges or competing with friends, users are more likely to stay motivated and stick to their fitness goals (Sañudo et al., 2024). AI can also analyse social dynamics and make recommendations to increase user engagement in gamified PA (Hamari & Koivisto, 2015).

AI-powered gamification also includes behaviour change techniques to encourage long-term PA. By applying theories from behavioural science, AI can develop gamified experiences that encourage users to adopt and maintain healthy habits. Interventions to increase PA are likely to fail in achieving long-term adherence. However, AI methods can be adaptive, a capacity that can be utilized in gamification to make positive changes in PA behaviours sustainable. Techniques such as goal setting, positive reinforcement and habit formation can be integrated into the game experience, making it easier for users to achieve their fitness goals. For example, AI can set incremental goals that gradually increase in difficulty, helping users to improve their confidence and skills over time (Pérez et al., 2023).

Popular fitness games and apps such as Pokémon GO, Zombies, Run! and Ring Fit Adventure are an example of the potential of gamified PA. These applications utilise AI to create an engaging experience that combines PA with gaming elements. Pokémon GO, for example, encourages users to run and explore their surroundings to catch virtual creatures, while Zombies, Run! turns running into an adventure game where users have to complete missions and avoid virtual zombies. Ring Fit Adventure combines movement with a role-playing game where users perform physical activities to progress through the game. These examples show how AI can transform PA into an interactive and enjoyable experience (Pérez et al., 2023).

18.6 Challenges in the Use of AI in PA Promotion

Although AI and ML have considerable potential to advance PA, there are also challenges and limitations. The implementation of AI and ML techniques is usually complex and time-consuming and requires high computing power. The high costs associated with the development and use of AI tools to promote PA can limit their accessibility. Another major limitation lies in the inherent complexity of AI and ML, which makes it difficult to explain these techniques in terms that are easily understood by humans. The lack of transparency, often referred to as the "black box" of AI models, seems to hinder the full adaptation of AI applications in the medical and health-related fields (Martinho et al., 2021). For example, when making a recommendation for PA, one hopes to know the reasons for prescribing a particular treatment. Without clear explanations, it becomes difficult to justify and adhere to the prescribed plan, which could undermine the effectiveness of AI-guided interventions. Explainable AI (XAI) is an active area dedicated to improving the interpretability of AI and ML approaches (Confalonieri et al., 2021). These efforts should mitigate some of the limitations associated with the lack of transparency of AI and ML and favour their wider use in the promotion of PA.

The existing literature has also highlighted ethical concerns regarding the use of AI algorithms in medical and health research (Anom, 2020; Celi et al., 2022). When it comes to PA, the safety and appropriateness of AI-driven interventions must be critically evaluated and considered. The integration of AI in the promotion of PA raises the question of accountability and responsibility for the decisions made by these AI systems. When an AI-driven intervention leads to a negative outcome, it can be difficult to determine who is responsible—the developers, the healthcare providers or the AI system itself. This ambiguity can lead to legal and ethical dilemmas that need to be resolved to ensure responsible use of AI technologies. Other concerns, such as the potential for algorithmic errors and the impact on privacy and security, are also important and need to be considered.

AI and ML models can be biased because they are trained on large amounts of historical data that are likely to contain biases and inequalities in healthcare (Celi et al., 2022). The training data may contain an under- or over-representation of certain ethnic, gender or socioeconomic groups. When AI and ML models are trained on such biased data, they may unintentionally reflect these biases and even exacerbate existing inequalities (Celi et al., 2022). For example, if a dataset consists primarily of data from one demographic group, the resulting model may be less accurate or effective for people outside that group. This can potentially lead to unfair and inaccurate recommendations and exacerbate existing health inequalities.

18.7 Conclusion

In recent years, the rapid emergence of state-of-the-art AI and ML has opened up innovative and new ways to promote PA. Existing research data suggests that PA interventions driven by AI can positively and significantly influence PA behaviour and outcomes. AI have proven successful in fitness tracking, chatbots and gamified PA systems, enabling individualisation of interventions, dynamic goal setting and easier accessibility and adaptability of PA promotion. Although AI is promising for the promotion and improvement of PA, challenges and ethical considerations must also be taken into account. To realise the full potential of AI and ML in PA promotion, ethical considerations must be taken into account and the transparency and interpretability of AI and ML must be increased.

References

Aggarwal, A., Tam, C. C., Wu, D., Li, X., & Qiao, S. (2023). Artificial intelligence-based chatbots for promoting health behavioural changes: Systematic review. *Journal of Medical Internet Research, 25*, e40789.

An, R., Shen, J., Wang, J., & Yang, Y. (2024). A scoping review of methodologies for applying artificial intelligence to physical activity interventions. *Journal of Sport and Health Science, 13*(3), 428–441. https://doi.org/10.1016/j.jshs.2023.09.010

Anderson, E., & Durstine, J. L. (2019). Physical activity, exercise, and chronic diseases: A brief review. *Sports Medicine and Health Science, 1*(1), 3–10.

Anom, B. Y. (2020). Ethics of big data and artificial intelligence in medicine. *Ethics, Medicine and Public Health, 15*, 100568.

Birkun, A. A., & Gautam, A. (2024). Large language model-based chatbot as a source of advice on first aid in heart attack. *Current Problems in Cardiology, 49*(1 Pt A), 102048.

Celi, L. A., Cellini, J., Charpignon, M. L., Dee, E. C., Dernoncourt, F., Eber, R., et al. (2022). Sources of bias in artificial intelligence that perpetuate healthcare disparities—A global review. *PLOS Digital Health, 1*(3), e0000022.

Chastin, S., et al. (2021). Joint association between accelerometry-measured daily combination of time spent in physical activity, sedentary behaviour and sleep and all-cause mortality: A pooled analysis of six prospective cohorts using compositional analysis. *British Journal of Sports Medicine, 55*(22), 1277–1285.

Chillón, P., Evenson, K. R., Vaughn, A., & Ward, D. S. (2011). A systematic review of interventions for promoting active transportation to school. *International Journal of Behavioural Nutrition and Physical Activity, 8*, 1–17.

Christensen, M., Vukadinovic, M., Yuan, N., & Ouyang, D. (2024). Vision-language foundation model for echocardiogram interpretation. *Nature Medicine, 30*, 1481–1488.

Confalonieri, R., Coba, L., Wagner, B., & Besold, T. R. (2021). A historical perspective of explainable artificial intelligence. *Wiley Interdisciplinary Reviews: Data Mining and Knowledge Discovery, 11*(1), e1391.

Cooper, J., Murphy, J., Woods, C., Van Nassau, F., McGrath, A., Callaghan, D., et al. (2021). Barriers and facilitators to implementing community-based physical activity interventions: A qualitative systematic review. *International Journal of Behavioural Nutrition and Physical Activity, 18*, 1–13.

den Braver, N. R., Garcia Bengoechea, E., Messing, S., Kelly, L., Schoonmade, L. J., Volf, K., et al. (2022). The impact of mass-media campaigns on physical activity: A review of reviews through a policy lens. *European Journal of Public Health, 32*(Supplement_4), iv71–iv83.

Dergaa, I., Saad, H. B., El Omri, A., Glenn, J., Clark, C., Washif, J., et al. (2023). Using artificial intelligence for exercise prescription in personalised health promotion: A critical evaluation of OpenAI's GPT-4 model. *Biology of Sport, 41*(2), 221–241.

Díaz, C., Galy, O., Caillaud, C., & Yacef, K. (2020). A clustering approach for modeling and analyzing changes in physical activity behaviours from accelerometers. *IEEE Access, 8*, 224123–224134.

Dijkhuis, T. B., Blaauw, F. J., Van Ittersum, M. W., Velthuijsen, H., & Aiello, M. (2018). Personalised physical activity coaching: A machine learning approach. *Sensors, 18*(2), 623.

Ding, D., Lawson, K. D., Kolbe-Alexander, T. L., Finkelstein, E. A., Katzmarzyk, P. T., Van Mechelen, W., & Pratt, M. (2016). The economic burden of physical inactivity: A global analysis of major non-communicable diseases. *The Lancet, 388*(10051), 1311–1324.

DiPietro, L., Al-Ansari, S. S., Biddle, S. J., Borodulin, K., Bull, F. C., Buman, M. P., et al. (2020). Advancing the global physical activity agenda: Recommendations for future research by the 2020 WHO physical activity and sedentary behaviour guidelines development group. *International Journal of Behavioural Nutrition and Physical Activity, 17*, 1–11.

Durão, S., Burns, J., Schmidt, B. M., Tumusiime, D., Hohlfeld, A., Pfadenhauer, L., et al. (2023). Infrastructure, policy and regulatory interventions to increase physical activity to prevent cardiovascular diseases and diabetes: A systematic review. *BMC Public Health, 23*(1), 112.

Farrahi, V., & Clare, P. (2024). Artificial intelligence and machine learning-powerful yet underutilised tools and algorithms in physical activity and sedentary behavior research. *Journal of Physical Activity and Health, 21*(4), 320–322.

Farrahi, V., & Rostami, M. (2024). Machine learning in physical activity, sedentary, and sleep behaviour research. *Journal of Activity, Sedentary and Sleep Behaviors, 3*(1), 5.

Farrahi, V., Kangas, M., Walmsley, R., Niemelä, M., Kiviniemi, A., Puukka, K., et al. (2021). Compositional associations of sleep and activities within the 24-h cycle with cardiometabolic health markers in adults. *Medicine and Science in Sports and Exercise, 53*(2), 324.

Farrahi, V., Rostami, M., Nauha, L., Korpisaari, M., Niemelä, M., Jämsä, T., et al. (2023). Replacing sedentary time with physical activity and sleep: Associations with cardiometabolic health markers in adults. *Scandinavian Journal of Medicine & Science in Sports, 33*(6), 907–920.

Ferguson, T., Olds, T., Curtis, R., Blake, H., Crozier, A. J., Dankiw, K., et al. (2022). Effectiveness of wearable activity trackers to increase physical activity and improve health: A systematic review of systematic reviews and meta-analyses. *The Lancet Digital Health, 4*(8), e615–e626.

Garcia, L., Pearce, M., Abbas, A., Mok, A., Strain, T., Ali, S., et al. (2023). Non-occupational physical activity and risk of cardiovascular disease, cancer and mortality outcomes: A dose-response meta-analysis of large prospective studies. *British Journal of Sports Medicine, 57*(15), 979–989.

Guthold, R., Stevens, G. A., Riley, L. M., & Bull, F. C. (2018). Worldwide trends in insufficient physical activity from 2001 to 2016: A pooled analysis of 358 population-based surveys with 1·9 million participants. *The Lancet Global Health, 6*(10), e1077–e1086.

Hamari, J., & Koivisto, J. (2015). "Working out for likes": An empirical study on social influence in exercise gamification. *Computers in Human Behaviour, 50*, 333–347.

Hassoon, A., Baig, Y., Naiman, D. Q., Celentano, D. D., Lansey, D., Stearns, V., et al. (2021). Randomised trial of two artificial intelligence coaching interventions to increase physical activity in cancer survivors. *npj Digital Medicine, 4*(1), 168.

Jirathananuwat, A., & Pongpirul, K. (2017). Promoting physical activity in the workplace: A systematic meta-review. *Journal of Occupational Health, 59*(5), 385–393.

Kaewkannate, K., & Kim, S. (2016). A comparison of wearable fitness devices. *BMC Public Health, 16*, 1–16.

Kahn, E. B., Ramsey, L. T., Brownson, R. C., Heath, G. W., Howze, E. H., Powell, K. E., et al. (2002). The effectiveness of interventions to increase physical activity: A systematic review. *American Journal of Preventive Medicine, 22*(4), 73–107.

Kohl, H. W., Craig, C. L., Lambert, E. V., Inoue, S., Alkandari, J. R., Leetongin, G., & Kahlmeier, S. (2012). The pandemic of physical inactivity: Global action for public health. *The Lancet, 380*(9838), 294–305.

König, I. R., Fuchs, O., Hansen, G., von Mutius, E., & Kopp, M. V. (2017). What is precision medicine. *European Respiratory Journal, 50*(4), 1700391. https://doi.org/10.1183/13993003.00391-2017

Lee, I. M., Shiroma, E. J., Lobelo, F., Puska, P., Blair, S. N., & Katzmarzyk, P. T. (2012). Effect of physical inactivity on major non-communicable diseases worldwide: An analysis of burden of disease and life expectancy. *The Lancet, 380*(9838), 219–229.

Li, Z., Das, S., Codella, J., Hao, T., Lin, K., Maduri, C., & Chen, C. H. (2018). An adaptive, data-driven personalised advisor for increasing physical activity. *IEEE Journal of Biomedical and Health Informatics, 23*(3), 999–1010.

Li, C., Chen, X., & Bi, X. (2021). Wearable activity trackers for promoting physical activity: A systematic meta-analytic review. *International Journal of Medical Informatics, 152*, 104487.

Lipnowski, S., LeBlanc, C. M., & Canadian Paediatric Society, & Healthy Active Living and Sports Medicine Committee. (2012). Healthy active living: Physical activity guidelines for children and adolescents. *Paediatrics & Child Health, 17*(4), 209–210.

Lyons, E. J., Lewis, Z. H., Mayrsohn, B. G., & Rowland, J. L. (2014). Behaviour change techniques implemented in electronic lifestyle activity monitors: A systematic content analysis. *Journal of Medical Internet Research, 16*(8), e192.

Ma, J. K., Floegel, T. A., Li, L. C., Leese, J., De Vera, M. A., Beauchamp, M. R., et al. (2021). Tailored physical activity behaviour change interventions: Challenges and opportunities. *Translational Behavioural Medicine, 11*(12), 2174–2181.

Ma, J. K., Ramachandran, S., Sandhu, A., Tsui, K., Hoens, A. M., Hu, D., & Li, L. C. (2022). Tailored interventions for supporting physical activity participation in people with arthritis and related conditions: A systematic review. *Current Treatment Options in Rheumatology, 8*(4), 117–142.

Maher, C. A., Davis, C. R., Curtis, R. G., Short, C. E., & Murphy, K. J. (2020). A physical activity and diet programme delivered by artificially intelligent virtual health coach: Proof-of-concept study. *JMIR mHealth and uHealth, 8*(7), e17558.

Martinho, A., Kroesen, M., & Chorus, C. (2021). A healthy debate: Exploring the views of medical doctors on the ethics of artificial intelligence. *Artificial Intelligence in Medicine, 121*, 102190.

Modi, T., & Gochhait, S. (2023, March). Impact of artificial intelligence on gamification: Current applications. In *2023 International Conference on Innovative Data Communication Technologies and Application (ICIDCA)* (pp. 287–290). IEEE.

Moon, J., Webster, C. A., Stodden, D. F., Brian, A., Mulvey, K. L., Beets, M., et al. (2024). Systematic review and meta-analysis of physical activity interventions to increase elementary children's motor competence: A comprehensive school physical activity programme perspective. *BMC Public Health, 24*(1), 826.

Moreno-Agostino, D., Daskalopoulou, C., Wu, Y. T., Koukounari, A., Haro, J. M., Tyrovolas, S., et al. (2020). The impact of physical activity on healthy ageing trajectories: Evidence from eight cohort studies. *International Journal of Behavioural Nutrition and Physical Activity, 17*, 1–12.

Nahavandi, D., Alizadehsani, R., Khosravi, A., & Acharya, U. R. (2022). Application of artificial intelligence in wearable devices: Opportunities and challenges. *Computer Methods and Programmes in Biomedicine, 213*, 106541.

Niemelä, M., Kiviniemi, A., Ikäheimo, T. M., Tulppo, M., Korpelainen, R., Jämsä, T., & Farrahi, V. (2023). Compositional association of 24-h movement behaviour with incident major adverse cardiac events and all-cause mortality. *Scandinavian Journal of Medicine & Science in Sports, 33*(5), 641–650.

Noetel, M., Sanders, T., Gallardo-Gómez, D., Taylor, P., del Pozo Cruz, B., Van Den Hoek, D., et al. (2024). Effect of exercise for depression: Systematic review and network meta-analysis of randomised controlled trials. *BMJ, 384*, e075847. https://doi.org/10.1136/bmj-2023-075847

Oh, Y. J., Zhang, J., Fang, M. L., & Fukuoka, Y. (2021). A systematic review of artificial intelligence chatbots for promoting physical activity, healthy diet, and weight loss. *International Journal of Behavioural Nutrition and Physical Activity, 18*, 1–25.

Patel, M. S., Small, D. S., Harrison, J. D., Fortunato, M. P., Oon, A. L., Rareshide, C. A., et al. (2019). Effectiveness of behaviorally designed gamification interventions with social incentives for increasing physical activity among overweight and obese adults across the United States: The STEP UP randomized clinical trial. *JAMA Internal Medicine, 179*(12), 1624–1632.

Pérez, J., Castro, M., & López, G. (2023). *Serious games and ai: Challenges and opportunities for computational social science*. IEEE Access.

Pratt, M., Varela, A. R., Salvo, D., Kohl, H. W., III, & Ding, D. (2020). Attacking the pandemic of physical inactivity: What is holding us back? *British Journal of Sports Medicine, 54*(13), 760–762.

Sañudo, B., Sanchez-Trigo, H., Domínguez, R., Flores-Aguilar, G., Sánchez-Oliver, A., Moral, J. E., & Oviedo-Caro, M. Á. (2024). A randomised controlled mHealth trial that evaluates social comparison-oriented gamification to improve physical activity, sleep quantity, and quality of life in young adults. *Psychology of Sport and Exercise, 72*, 102590.

Shandhi, M. M. H., Singh, K., Janson, N., Ashar, P., Singh, G., Lu, B., et al. (2024). Assessment of ownership of smart devices and the acceptability of digital health data sharing. *npj Digital Medicine, 7*(1), 44.

Singh, B., Olds, T., Brinsley, J., Dumuid, D., Virgara, R., Matricciani, L., et al. (2023). Systematic review and meta-analysis of the effectiveness of chatbots on lifestyle behaviours. *npj Digital Medicine, 6*(1), 118.

Sprint, G., Cook, D. J., & Schmitter-Edgecombe, M. (2016). Unsupervised detection and analysis of changes in everyday physical activity data. *Journal of Biomedical Informatics, 63*, 54–65.

Stragier, J., Abeele, M. V., Mechant, P., & De Marez, L. (2016). Understanding persistence in the use of online fitness communities: Comparing novice and experienced users. *Computers in Human Behaviour, 64*, 34–42.

Szeto, K., Arnold, J., Singh, B., Gower, B., Simpson, C. E., & Maher, C. (2023). Interventions using wearable activity trackers to improve patient physical activity and other outcomes in adults who are hospitalised: A systematic review and meta-analysis. *JAMA Network Open, 6*(6), e2318478–e2318478.

Thirunavukarasu, A. J., Ting, D. S. J., Elangovan, K., Gutierrez, L., Tan, T. F., & Ting, D. S. W. (2023). Large language models in medicine. *Nature Medicine, 29*(8), 1930–1940.

Tolks, D., Schmidt, J. J., & Kuhn, S. (2024). The role of AI in serious games and gamification for health: Scoping review. *JMIR Serious Games, 12*(1), e48258.

Tremblay, M. S., Duncan, M. J., Kuzik, N., Silva, D. A. S., & Carson, V. (2024). Towards precision 24-hour movement behaviour recommendations—The next new paradigm? *Journal of Sport and Health Science, 13*(6), 743–748. https://doi.org/10.1016/j.jshs.2024.05.003

Walmsley, R., Chan, S., Smith-Byrne, K., Ramakrishnan, R., Woodward, M., Rahimi, K., et al. (2022). Reallocation of time between device-measured movement behaviours and risk of incident cardiovascular disease. *British Journal of Sports Medicine, 56*(18), 1008–1017.

Woodhead, G., Sivaramakrishnan, D., & Baker, G. (2023). Promoting physical activity to patients: A scoping review of the perceptions of doctors in the United Kingdom. *Systematic Reviews, 12*(1), 104.

Zhang, J., Oh, Y. J., Lange, P., Yu, Z., & Fukuoka, Y. (2020). Artificial intelligence chatbot behaviour change model for designing artificial intelligence chatbots to promote physical activity and a healthy diet. *Journal of Medical Internet Research, 22*(9), e22845.

Zhou, M., Fukuoka, Y., Goldberg, K., Vittinghoff, E., & Aswani, A. (2019). Applying machine learning to predict future adherence to physical activity programmes. *BMC Medical Informatics and Decision Making, 19*, 1–11.

Machine Learning for Injury Prediction 19

Matthias Kempe

Abstract

The sports industry emphasizes injury prevention, but current methods struggle due to data complexity. Machine learning (ML) offers promise by synthesizing athlete data for informed decisions. Research proposes ML-based decision support systems, simplifying variables for better insights. Classification algorithms dominate injury prediction, with ML showing promise but lacking consistency in predictors. Short-term models aid coaching staff, while long-term models benefit performance and operational departments. Challenges include addressing temporal dependencies and data leakage during preprocessing. Thorough data inspection and domain knowledge are crucial to prevent overconfidence in model predictions. ML presents an opportunity to advance injury prediction and athlete management in sports, but it's important to address key challenges and ensure robustness before using them in practice.

19.1 Introduction

A common saying in sport is "the best ability is availability", highlighting the importance of injuries for teams and players. Sports teams are losing millions of dollars in actual and potential revenue each year due to athlete injuries, despite significant investment in sports medicine, strength and conditioning, and analytics. Still athlete monitoring and management presently fails to meaningfully forecast injury risk (Buchheit, 2017) due to a lack in data analytics tools that better represent the complexity and variety of interactions across factors.

M. Kempe (✉)
Faculty of Medical Sciences, Universität Groningen, Groningen, Netherlands
e-mail: m.kempe@umcg.nl

D. Memmert (ed.), *Artificial Intelligence and Machine Learning in Sports Science*,
https://doi.org/10.1007/978-3-662-70155-3_19

Using Machine Learning (ML) to synthesize the vast amount of data enabling coaches, trainers, and support staff to make educated decisions might be the most influential use in sports practice. By now, a professional sports team monitors their athletes extensively, recording variables of training prescription and performance (external load), as well as the physiological and mental strains of these training sessions (internal load). Next to that, they also perform regular performance testing and record individual variables like age, injury history, growth spurt, etc. The complexity and possible interactions of these variables are hard to grasp which is why Robertson made the argument to in-stall decision support systems to create recommendations for practitioners based on "historical" data using ML (Robertson, 2020). Such a system creates simplified variables and/or visualization by synthesizing various information to make better-informed decisions. Especially in terms of athlete monitoring and injury prevention, different approaches were introduced in the last decade to serve this purpose. While some of them proved to be too simplistic others, while being accurate aren't endorsed by practitioners.

External (training) **load** can be simply defined as 'what the athlete does' and can be observed, i.e. the physical work actually performed by the athlete during the training. External training load can be measured by quantifying what the athlete does for example using GPS derived units (e.g. speed, accelerations, distance), force, level of resistance, work, etc. **Internal** (training) **load** typically refers to the internal (to the body) responses experienced by an athlete during the exercise or the training session. The internal responses can be measured using psychological (e.g. rating of perceived exertion), physiological (e.g. heart rate and EMG) or other kinds of measures representing internal responses during exercise (Jeffries et al., 2022).

19.2 "Simplistic" Models for Athlete Monitoring and Injury Prediction

As mentioned earlier the goal of a decision support system for athlete monitoring is to maximize possible performance gains, while at the same time minimizing the probability of an injury. It needs to be easy to implement for daily use and easy to interpret for the people working within. Gabbett introduced the idea of the acute to chronic workload ratio (ACWR) which could, given his argumentation, solve this problem (Gabbett, 2016). As can be seen in Eq. 19.1, ACWR standardizes the most recent training stress (often last 7 days) by a midterm training load (e.g. 28 days). Using this approach, several studies showed that if the ACWR stays between 0.8–1.3 ("sweet spot") the risk of injury is low while exceeding workload ratios of ≥1.5 ("danger zone"), dramatically increasing the risk of injury.

$$\mathrm{ACWR} = \frac{\mathrm{Fatige}\left(\mathrm{Acute\ Workload}\right)}{\mathrm{Fitness}\left(\mathrm{Chronic\ Workload}\right)} \tag{19.1}$$

Given the positive evidence and the simplicity of the approach, ACWR was quickly accepted as the "gold standard" for injury prevention and athlete monitoring and got endorsed by various sports federations. The flaws of this approach, a consequence of its simplicity, were pointed out in recent years by Impellizeri and colleagues (2021). In their paper, they use various regression analyses to show that ACWR is not a normalization of acute load but rather a rescaling this way magnifying the effect of acute load and adding additional noise and artifacts. By doing so, they prove the lack of predictive value of ACWR as it is as accurate as a simple intercept model based on chronic load. They conclude that the idea of ACWR should be dismissed and other models are needed. In a follow-up, Impellizeri and colleagues (2020) suggested that models for injury prediction should expand from load variables, expand sample sizes, use more complex modeling, and improve generalizability with e.g. out of sample testing. Based on these recommendations, ML are a logical step to try to create improved decision-making systems, as they enable to incorporate all the aforementioned shortcomings.

19.3 Machine Learning for Injury Prediction: A Success Story?

Supervised machine learning, particularly classification algorithms, are predominantly utilized in this field, as the variables to be predicted are typically binary (injury vs. no injury) or categorical (severity of an injury on a specific scale). In its simplest form, the task of any classification model is to accurately predict injuries (a positive class) and non-injuries (a negative class). Commonly used classification algorithms include logistic regression, decision trees, random forests, k-nearest neighbours (KNN), support vector machines (SVM), artificial neural networks (ANN), and "ensemble methods" (e.g., bagging and boosting). Due to the setup of these algorithms, some are easier to interpret (white-box) than others (black-box). White-box algorithms directly link input to outputs, thereby providing easy-to-use formulas (e.g., regressions) or cutoff values (e.g., decision trees). This information can then be directly used to inform practitioners and adapt training regimes. Black-box algorithms do not offer this direct interpretability. However, for most of them, post-hoc analyses are available (e.g., SHAP values) to provide information on the relative importance of specific input features.

Two recent review articles highlighted that ML is a very promising approach for injury prediction with some studies achieve strong prediction performance (Majumdar et al., 2022; Van Eetvelde et al., 2021). They also showed that the field focusses on white-box methods, especially tree-based algorithms (e.g. Decision Tress, Random Forrest or XGBoost) to allow for easier interpretation. Included studies used a wide range of predictors, from training-related (e.g. training load, psychological and neuromuscular assessment, stress level) to rather stable (e.g. demographics, genetic markers, anthropometric measurements, previous injury) factors. Because of the wide variety of features used over the different papers, not much consistency was found in the reported important predictors. The features that were reported as important were previous injuries, higher training load, and higher body size (in youth players only). This variety shows that there is no general solution for injury prediction by now. In addition, one can also see that there are two different general ideas that are linked to the timeline of the prediction. The first approach is providing a general idea of injury prevalence of an athlete based on general characteristics and periodic testing (long term models). These models answer the question:

1. "Will my athlete get injury this season/in next 2 months?".

The second approach is more time sensitive (short term models) and often related to training load variables answering the question:

2. "Will my athlete get injury in the next training session/completion?"

19.4 Short-Term Vs Long-Term Models

As indicated by the two types of questions short and long-term models are trying to answer, there results might be also used by different personal. Short-term models are more useful for the coaching staff and can be used for load management, line-up configurations or game strategy. On the other hand, long-term models can provide insights for the performance/sport science staff via suggestions for training adjustments, injury prevention strategies and further assessments as well as for the analytics and operational department for performance projections and contract negotiations.

19.4.1 Features

Depending on the model's use case, different features are required as input. Long-term models utilize more stable variables such as demographics (e.g., age, maturation, height, weight), testing data (e.g., yo-yo test, counter movement jumps, fat percentage), and aggregated load measures (e.g., distance covered in the last month). Typically, short-term models rely on aggregated load measures over a shorter time frame (last 3 days or last 7 days) and incorporate self-reported measures (e.g., readiness to perform or sleep quality). Particularly for short-term models, it's crucial to include features that are sensitive to day-to-day variations or capture potential overuse that could lead to an injury. During an initial exploratory analysis of the data, this can be checked using, for example, coefficients of variation.

19.4.2 Model Validation

To evaluate the performance of an ML model, it is common to split the data randomly into a training and test set or to perform this split repeatedly using a cross-validation approach. However, these typical train-test-split schemes are often not applicable for injury prediction. Data points within a dataset may not be independent from each other, making it challenging to randomly assign them to the training or test set. One prerequisite for evaluating ML models is that the test data needs to be "unseen" by the training data or future data. For instance, internal load data of players exhibits temporal dependency, as fatigue from yesterday's training might influence reported Rate of Perceived Exertion or heart rate. If these observations are then split into the training and test sets, the test set may still hold information from the training set. Similarly, if two performance measures (e.g., counter movement jump) of one athlete are spread across the training and test sets, the model might optimize for specific individuals rather than an underlying factor. This can lead to overconfidence (overfitting) of the model. This phenomenon, where data splits are not independent, is also known as data leakage.

Data leakage is when information about the test data is introduced into the model during training, nullifying the benefits of separating training and test data (Kaufman et al., 2012).

Figure 19.1 provides an overview on methods to overcome the aforementioned issues of temporal dependency and subject consistency.

Differentiating between the two types of injury prediction approaches, long-term models have the advantage of observing more athletes as they often do not rely on daily measurements, which are harder to conduct with a larger group. Due to this sample size advantage, they offer the potential for more generalizable results, such as application in different contexts. However, achieving this requires careful consideration of subject consistency to prevent the model from overfitting to specific individuals. This becomes especially critical if some athletes have more observations in the dataset due to longer team tenure compared to others who may have joined more recently. Similarly, including multiple teams in the model can introduce bias if one

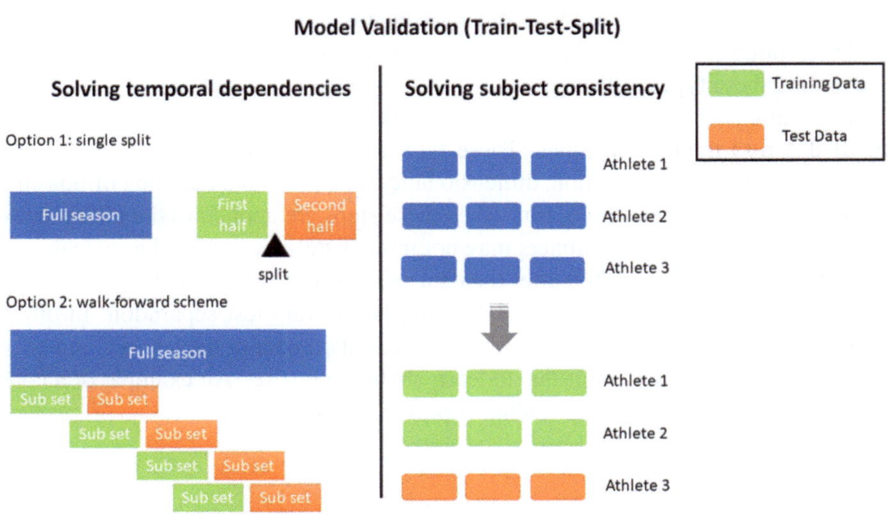

Fig. 19.1 Schemes split the data sample into a train and test set to improve the performance and generalization of machine learning models

team has a higher number of injuries than another. In such cases, it's essential to ensure that one team is either entirely in the test or train set. For injury prediction models, this necessitates creating train-test splits rather than relying solely on off-the-shelf functions to ensure the correct distribution of athletes.

Regarding temporal dependencies, long-term models commonly employ a single split (see Fig. 19.1) due to their larger sample size and focus on prediction. On the other hand, short-term models are often developed for smaller groups of athletes, such as individual teams. They are frequently adapted and retrained to accommodate new training stimuli. Therefore, a walk-forward scheme is recommended for such models. More advanced approaches include prequential evaluation or distributed k-fold evaluation, which are used to assess classifiers of data streams. For an overview and open-source implementation, please refer to Haug et al. (2022).

19.4.3 Data Leakage During Pre-processing

One of the main issues that was pointed out by Van Eetvelde and colleagues (2021) in their review were data leakage issues in the data preprocessing. As mentioned earlier data leakage creates an overconfidence in the model and leads to worse results when the model gets implemented in practice.

Sasse and colleagues (2023) provide an in-depth overview of common examples of data leakage when setting up an ML pipeline. A typical mistake, often made during feature selection, is called *target leakage*. This occurs when information about the prediction target is inadvertently included in the input data. For instance, the number of visits to a physiotherapist might be included as a feature, which could indicate that an athlete is recovering from an injury or experiencing symptoms. However, the goal is typically to make predictions before symptoms occur, so including such a feature means it already contains information indicating that the athlete is injured. Thorough data inspection and domain knowledge are essential for preventing these types of leakage.

In addition to feature selection, several preprocessing steps often need to be applied across multiple samples. These typically include feature scaling, feature normalization, feature selection, dimensionality reduction, and handling of missing values. Test-to-train leakage occurs when these steps are applied to the entire dataset before splitting it. Practitioners may not immediately recognize this as leakage, since the ML model is trained after the data split. However, estimating preprocessing parameters on the entire dataset invalidates the train-test separation. In other words, data in the training set is transformed based on data in the test set, and crucially, ML models can exploit this to learn about the test set. An example of a correct implementation of feature scaling is provided in Fig. 19.2.

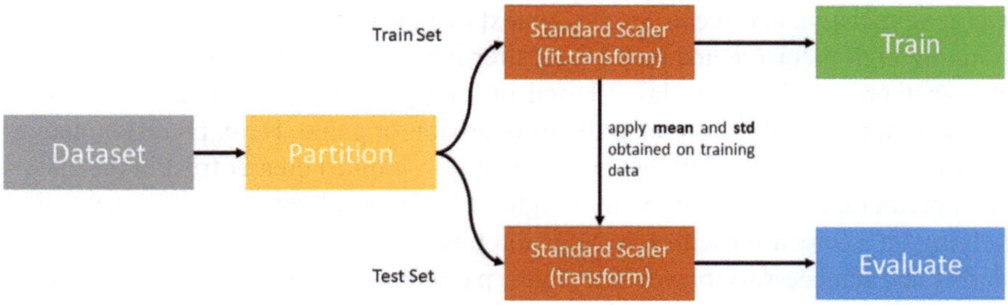

Fig. 19.2 Example on how to avoid data leakage during data pre-processing with the example of data scaling using StandardScaler from scikit-learn library

19.5 Working with Imbalanced Data Sets

Fortunately, athletes are more often healthy than injured. This normally leads to an imbalanced dataset, meaning we have a large number of one observation (majority class), in our case, not injured, and a small number of the other observation (minority class), injured. If we randomly allocate data to either the train or test set, it could happen that none of the cases of the minority class are in the training set. We would, therefore, just train our model to predict being not injured, which is not our goal. Even in less extreme examples, a skewed distribution of both classes can lead to insufficient prediction. Therefore, the majority and minority class should be evenly distributed in the training and test set; this is achieved by stratifying the data.

One further issue of class imbalance is that it might diminish the prediction performance of the model as it will tent to overemphasize the majority class. This means the model would be good at predicting not getting injured and poorly at getting injured. A common approach to deal with this is implementing under and oversampling strategies. The open-source project Imbalanced-learn (Lemaître et al., 2017) provides an overview on update methods for both options. It needs to mentioned, that under and oversampling should just be used in the training set and never in the test set.

▶ Inspecting, visualizing, and comparing your data before and after splitting it into a train and test set is essential to minimize mistakes in data (pre-) processing.

19.6 Measures to Validate Injury Prediction Models

Evaluating the effectiveness of a classifier commonly relies on accuracy, yet this measure falls short when confronted with imbalanced data, where one class outweighs the other. Achieving a high accuracy score may be as simple as guessing the majority class. For assessing injury prediction models, alternative performance metrics prove more apt, such as the area under the receiver operating curve (ROC AUC), average precision, and recall. In the context of injury forecasting, the focus lies on attaining high precision and recall values for injury class 1. Imagine a coach deciding whether to sideline a player based on injury forecasts—where a forecast of injury prompts the player to skip the next session or game. Here, precision gauges the reliability of the predictions: higher precision implies greater trust in the classifier's predictions, correlating with a higher likelihood of actual injury occurrence. Relying on a forecaster with low precision poses risks, resulting in numerous false alarms and unnecessary player stoppages, particularly detrimental for key players. Conversely, recall measures the forecaster's ability to identify injuries, with higher recall indicating a greater proportion of detected injuries. A forecaster with low recall misses many injuries, leading to players participating in sessions or games and potentially sustaining injuries. Trusting a forecaster with low recall increases the likelihood of misclassifying actual injuries as non-injuries. Given these examples it is important to optimize and evaluate an injury prediction model on precision and recall on the injury class.

One still needs to note that all the aforementioned evaluation methods are binary, therefore labeling injury or non-injury. However, classification model does actually provide a class probability and then provides a class label based on a cut-off value (normally 50%). Measures like AUC Scores, precision, and recall don't account for a decision being made by probability of 51% being injured or 95% being injured. In a recent study Carey and colleagues (2018) showed that this could lead to higher false discovery rate, especially when using training load data. They therefore advise to use evaluation methods take class probabilities into account. Commonly used examples for such evaluation methods are Brier and logarithmic scoring (e.g. Log loss).

19.7 Conclusion

To sum up, sports medicine faces significant challenges in accurately forecasting athlete injuries despite substantial investments in technology and analytics. The complexity of athlete monitoring and the multifaceted nature of injury risk necessitate sophisticated data analytics tools, particularly those leveraging Machine Learning (ML). Notably, ML algorithms offer the potential to discern patterns and interactions across various factors influencing injury risk, ranging from external training load to physiological and psychological responses. However, the efficacy of ML models relies heavily on proper validation and handling of data issues such as imbalanced datasets and data leakage during preprocessing. Additionally, it's important to consider the field of application of the ML model, as both short-term and long-term models for injury prediction serve different purposes within sports organizations. Short-term models cater to immediate coaching decisions, while long-term models inform broader injury prevention strategies and performance projections.

Despite advancements in ML-based injury prediction, several challenges persist, including model validation and the interpretation of evaluation metrics. Notably, the focus on precision and recall underscores the importance of accurately identifying both injured and non-injured athletes, thereby minimizing false alarms and missed injuries. Moving forward, efforts to refine ML models for injury prediction should prioritize robust validation methods, account for the nuances of athlete monitoring, and consider the practical implications for sports organizations. By leveraging ML effectively, sports teams can enhance injury prevention strategies and optimize athlete performance, ultimately realizing the adage that "the best ability is availability" in the competitive arena of sports.

References

Buchheit, M. (2017). Houston, we still have a problem. *International Journal of Sports Physiology and Performance, 12*(8), 1111–1114. https://doi.org/10.1123/ijspp.2017-0422

Carey, D. L., Crossley, K. M., Whiteley, R., Mosler, A., Ong, K.-L., Crow, J., & Morris, M. E. (2018). Modeling training loads and injuries: The dangers of discretization. *Medicine & Science in Sports & Exercise, 50*(11), 2267–2276. https://doi.org/10.1249/MSS.0000000000001685

Gabbett, T. J. (2016). The training-injury prevention paradox: Should athletes be training smarter and harder? *British Journal of Sports Medicine, 50*(5), 273–280. https://doi.org/10.1136/bjsports-2015-095788

Haug, J., Tramountani, E., & Kasneci, G. (2022). *Standardized evaluation of machine learning methods for evolving data streams*. http://arxiv.org/abs/2204.13625

Impellizzeri, F. M., McCall, A., Ward, P., Bornn, L., & Coutts, A. J. (2020). Training load and its role in injury prevention, part 2: Conceptual and methodologic pitfalls. *Journal of Athletic Training, 55*(9), 893–901. https://doi.org/10.4085/1062-6050-501-19

Impellizzeri, F. M., Woodcock, S., Coutts, A. J., Fanchini, M., McCall, A., & Vigotsky, A. D. (2021). What role do chronic workloads play in the acute to chronic workload ratio? Time to dismiss ACWR and its underlying theory. *Sports Medicine, 51*(3), 581–592. https://doi.org/10.1007/s40279-020-01378-6

Jeffries, A. C., Marcora, S. M., Coutts, A. J., Wallace, L., McCall, A., & Impellizzeri, F. M. (2022). Development of a revised conceptual framework of physical training for use in research and practice. *Sports Medicine, 52*(4), 709–724. https://doi.org/10.1007/s40279-021-01551-5

Kaufman, S., Rosset, S., Perlich, C., & Stitelman, O. (2012). Leakage in data mining. *ACM Transactions on Knowledge Discovery from Data, 6*(4), 1–21. https://doi.org/10.1145/2382577.2382579

Lemaître, G., Nogueira, F., & Aridas, C. K. (2017). Imbalanced-learn: A Python toolbox to tackle the curse of imbalanced datasets in machine learning. *Journal of Machine Learning Research, 18*, 1–5. http://jmlr.org/papers/v18/16-365.html

Majumdar, A., Bakirov, R., Hodges, D., Scott, S., & Rees, T. (2022). Machine learning for understanding and predicting injuries in football. *Sports Medicine—Open, 8*(1), 73. https://doi.org/10.1186/s40798-022-00465-4

Robertson, P. S. (2020). Man & machine: Adaptive tools for the contemporary performance analyst. *Journal of Sports Sciences, 38*(18), 2118–2126. https://doi.org/10.1080/02640414.2020.1774143

Sasse, L., Nicolaisen-Sobesky, E., Dukart, J., Eickhoff, S. B., Götz, M., Hamdan, S., Komeyer, V., Kulkarni, A., Lahnakoski, J., Love, B. C., Raimondo, F., & Patil, K. R. (2023). *On leakage in machine learning pipelines*.

Van Eetvelde, H., Mendonça, L. D., Ley, C., Seil, R., & Tischer, T. (2021). Machine learning methods in sport injury prediction and prevention: A systematic review. *Journal of Experimental Orthopaedics, 8*(1), 27. https://doi.org/10.1186/s40634-021-00346-x

Event Data Based Action Valuation: Possibilities and Limitations

20

Leo Martins de Sá-Freire

Abstract

This chapter explores the evolving field of action valuation metrics in soccer analytics, which significantly enhance the depth of quantitative analysis beyond traditional aggregate statistics. By providing a detailed assessment of each action's value, these metrics address the specificity and granularity often missing in traditional statistical approaches. We will examine the VAEP metric as a case study, highlighting its utility and the insights it can offer into player and team performance over a season. We also discuss the limitations of these metrics, particularly the lack of spatial context in event data. Despite these challenges, action valuation metrics remain a potent tool, with ongoing developments aimed at refining their accuracy and applicability in professional sports. This chapter underscores the importance of using these metrics judiciously and explores potential enhancements to overcome their inherent limitations.

20.1 Introduction

Data Science in team sports encompasses, among various tasks, the challenge of measuring the impact of individual actions on the final outcome of a game. This is crucial for enabling different analyses, from searching for efficient strategies to identifying high-performing athletes.

L. M. de Sá-Freire (✉)
Departamento de Ciencia da Computação, Universidade Federal de Minas Gerais, Belo Horizonte MG, Brazil

Clube Atletico Mineiro, Belo Horizonte MG, Brazil

D. Memmert (ed.), *Artificial Intelligence and Machine Learning in Sports Science*, https://doi.org/10.1007/978-3-662-70155-3_20

Clearly, developing strategies to perform this task in a useful and applicable manner is intrinsically related to understanding the nature of the sport being analyzed. In the unique realm of soccer, characterized by its infrequent scoring, assessing individual actions becomes both a challenge and a necessity. If most of what happens in a match is not directly related to scoring a goal, how can we measure the impact and the importance that each action has on a match?

Over the past few decades, the task of considering what athletes and teams do during matches essentially amounted to an exercise in counting (Mackenzie & Cushion, 2013; Zhou et al., 2021). Statistics such as the number of passes, shots, tackles, aerial duels, dribbles, and interceptions were the usual way of characterizing the performance of the players involved in a game. Over time, more specific statistics emerged. For example, the simple count of passes evolved into more granular metrics such as passes to the final third and key passes.

This approach falls under what is now termed as aggregate statistics: delineating which metrics are relevant and tabulating the frequency of each occurrence for every player throughout a match. Naturally, this has expanded the range of data collected on individual performance in soccer, opening up possibilities for more intricate analyses. An example is the radar plot, which remains widely used in the

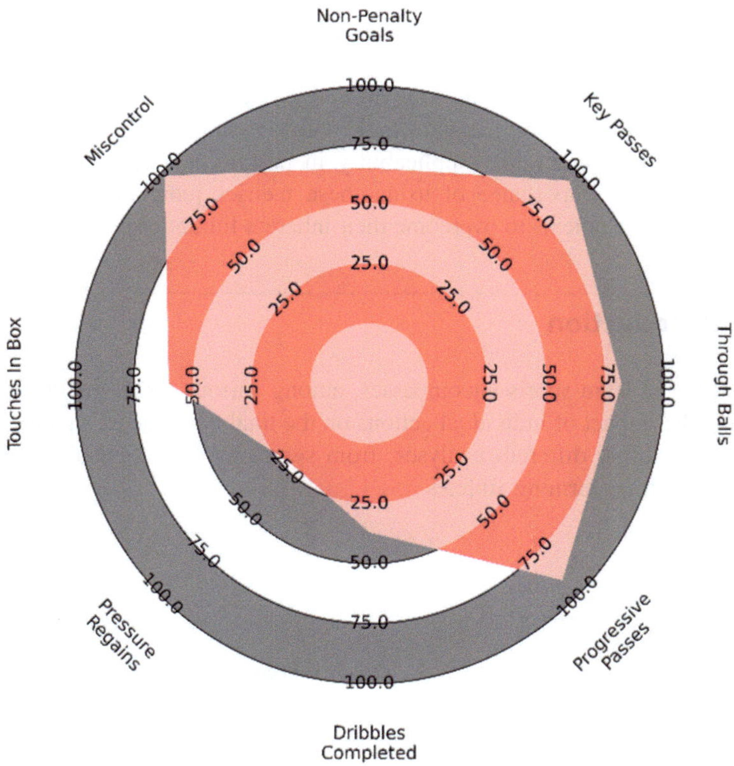

Fig. 20.1 Radar plot example

sports industry for characterizing and comparing players. The radar plot presented in Fig. 20.1 clearly illustrates that this specific player demonstrates excellence in passing the ball but shows limitations in dribbling and pressuring the opponent.

Although with aggregate statistics we can group and count player actions according to the statistics of interest, there is a substantial loss in this process: two actions that fit within the same statistic can be very different from each other. Take, for example, a pass to the final third. A short pass from midfield to the right wing would be counted the same as a long pass from the defensive field to the opponent's area that puts the receiver in a position to shoot, even though the latter seems to have a much higher value for the team that executed it.

This limitation can lead to distortions. A merely frequency-based approach may favor a midfielder who made five passes of the first type during a game over one who made four passes of the second type. This justifies the need for an approach that can consider the peculiarities of each player's actions on the ball, beyond the macro categories defined by the previously established statistics.

The combination of more refined data collection with the use of computational techniques and advanced statistics has enabled the development of such an approach, called action valuation. In the next section, we will briefly present the data formats currently collected in soccer matches: event and tracking data. Additionally, we will introduce VAEP, one of the key metrics of action valuation that has been widely used.

20.2 Granular Data Types

Apart from aggregate statistics, most of the quantitative research on soccer relies on two different data formats: event data and tracking data. Event data represent each match as a sequence of on-ball actions, where each action (pass, dribble, tackle, etc.) is registered as a row in a table, with attributes stored in different columns (type of action, player executing it, initial and final coordinates, part of the body, outcome, etc.). This offers a tabular representation of soccer matches that encompasses what happens with the ball throughout the game, yet it overlooks aspects related to off-ball players.

Tracking data, on the other hand, consists of capturing in constant frames the location of all properly identified players on the field as well as the ball. This data format provides the spatial context of the game in its entirety. However, it involves a more expensive and complex collection process, is heavier, and has reduced availability.

The existence of refined quantitative representations for soccer matches as described above has enabled the development of techniques to qualify actions considering their peculiarities. The task of action valuation consists of assigning a value to each action during a match, representing its contribution to the team's success (or failure). In this context, success can be assessed as the occurrence of a shot or, more commonly, a goal. Next, we will present one of the main approaches to valuing actions based solely on event data. Finally, we will present results from their implementations for a complete season and discuss limitations related to the lack of spatial context.

20.3 Valuing Actions by Estimating Probabilities (VAEP)

VAEP is a metric introduced by Decroos et al. (2019). Its framework is based on estimating the probabilities of the team in possession of the ball scoring or conceding a goal within the next 10 actions relative to the current game state. From these estimated probabilities, a numerical value is assigned to each action based on how much that action moved their team closer to scoring and further from conceding.

To estimate these probabilities, two models are trained in an identical structure. These models are constructed through a supervised learning process, where features extracted from sequences of three consecutive events (game states) are introduced, and binary labels indicating the occurrence of a goal (for or against the team in possession) within the next 10 actions are associated. This generates two probability estimators (*pscores* and *pconcedes*) for the occurrence of goals in new unlabeled data. Thus, it becomes possible to evaluate each action a_i according to the equation below:

$$v(a_i) = \Delta\, pscores(a_i) - \Delta\, pconcedes(a_i)$$

In order to present a concrete example of the metric's application, we used a free sample of event data provided by Wyscout (Pappalardo et al., 2019), which includes the top five European leagues for the 2017/18 season. We used matches from La Liga, Ligue 1, Serie A, and the Premier League to train the models and set the Bundesliga matches as the test dataset. In the tables below, it is possible to observe the performance of the two trained models evaluated through the Normalized Brier Score (NBS) (Decroos & Davis, 2019) and the Expected Calibration Error (ECE).

The Normalized Brier Score measures the accuracy of probabilistic predictions, with lower values indicating better performance, and it is normalized to account for the baseline (average of true labels) prediction accuracy. A lower NBS indicates that the model's predicted probabilities are closer to the actual outcomes. The Expected Calibration Error assesses how well the predicted probabilities match the actual outcomes, with lower values indicating better calibration. A lower ECE means that the predicted probabilities are more reliable and accurately reflect the true likelihood of events (Table 20.1).

Additionally, the following table highlights the top 5 players who generated the most value in the mentioned Bundesliga season (Table 20.2).

Table 20.1 Performance of trained VAEP models

Model	Normalized brier score	Expected calibration error
pscores	0.81	0.0010
pconcedes	0.88	0.0007

Table 20.2 Top 5 players in VAEP value (Bundesliga 17/18)

Position	Player	Position	VAEP value	Goals + Assists
1	Andrej Kramaric	Striker	12.71	21
2	Robert Lewandowski	Striker	12.64	31
3	Julian Brandt	Midfielder	12.48	14
4	Leon Bailey	Winger	12.38	15
5	Kevin Volland	Striker	11.82	16

20.4 Discussion

The estimation of calibrated probabilities and an NBS indicating a reduction in the expected error of the predictions compared to a naive baseline suggests the practical utility of the VAEP framework. Moreover, by observing the players with the best evaluations in the aggregate of all on ball actions throughout the 17/18 season, we see that there is a certain compatibility between the model's results and the intuition of those who follow the sport in identifying the top players of the championship.

Although somewhat aligned with expectations, the results provide notable additions. First, the assurance that the evaluation considered all on-ball actions of each player throughout the championship. This is somewhat different from almost all subjective evaluations, which likely ignore part of what the player did that went unnoticed by aggregate statistics. Second, the comparative gain that quantifying the values of actions allows. A numeric performance indicator not only lets us rank the performance of different athletes but also observe the distances between various players.

Furthermore, the framework offers a multitude of analysis possibilities. For instance, it is possible to analyze the value generated by each player in a specific type of action (such as passes). Another feasible scenario would be to examine the values generated in specific parts of the field. VAEP also allows us to analyze which 2-to-2 interactions (passer and receiver) generate the most value for a team. Also, more collective analyses can be developed from questions like: which side of the attack is more dangerous for a team? In which zones of the field does the opponent concede more value to their opponents? The possibilities are extensive, and, with real questions and sufficient creativity, the metric becomes a very powerful tool.

However, the VAEP framework also comes with significant limitations. Foremost among these limitations are those stemming from the data format it utilizes. As previously mentioned, event data represent matches through the sequence of on-ball actions that occurred without recording the positions of off-ball players in each instance. When watching a soccer game, we intuitively look for free players, large spaces without defenders, and we form expectations when attackers have few or no opponents between them and the goal. It is clear that, at least from our perception, spatial occupation and the relationships between the positions of attackers and defenders are fundamental aspects for estimating the probabilities of goal occurrences. Since the VAEP features are built from data that does not contain information about the positions of off-ball players, they are not taken into account in the probability estimation and, therefore, in the value attributed to actions.

To illustrate this limitation, imagine a diagonal pass from the midfield to the opponent's final third. Such a pass could occur in different game contexts. A possibility is in a structured build-up, as seen in the left diagram of Fig. 20.2, where the left defender finds the right winger with some space to control the ball, but with the opponent's defense still in position. Another possible context is during a counter-attack, seen in the right diagram of Fig. 20.2, where such a pass effectively removes almost all opponents from the play and puts the receiver in a position to progress

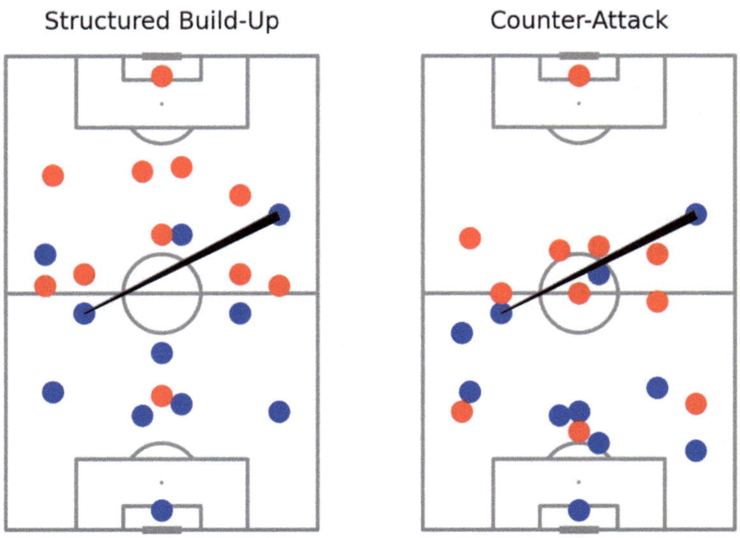

Fig. 20.2 Two different scenarios for a diagonal pass

freely towards the goal. Naturally, these two situations would have distinct probabilities of resulting in a goal.

However, the events recorded in the dataset would be exactly the same (Fig. 20.3). Thus, the features constructed by the VAEP framework would differ only in the attributes related to the two actions prior to the pass that make up the gamestate. This would lead the metric to assign a similar value to both passes, even though in reality they contribute to a different increase in the actual chance of scoring a goal.

Aware of this limitation, previous research have shown that, for example, the absence of spatial context hinders the VAEP framework and event data-based action valuation metrics in distinguishing passes that resulted in goals in the initial third of the field from those that did not (de Sá-Freire & Vaz-de-Melo, 2023). Moreover, there are examples of studies that have sought to add spatial context to the task of action valuation, either from metrics originally based on event data (Everett et al., 2022; Robberechts et al., 2023) or from metrics already structured with tracking data (Fernández et al., 2019, 2021).

While the issue is apparent, its significance varies depending on how the framework is intended to be utilized. One notable instance is the utilization of model-estimated probabilities to forecast the likelihood of subsequent goals after particular actions, wherein the effect of under-representation is notably profound. However, this is not the general use of VAEP. Most of the time, the values assigned by the metric are aggregated over considerable intervals to evaluate athletes and teams across multiple matches. The primary focus of the analysis is to identify players/teams that consistently increase the chances of scoring goals and decrease the chances of conceding. This process tends to minimize the impacts that some underestimations or overestimations might have.

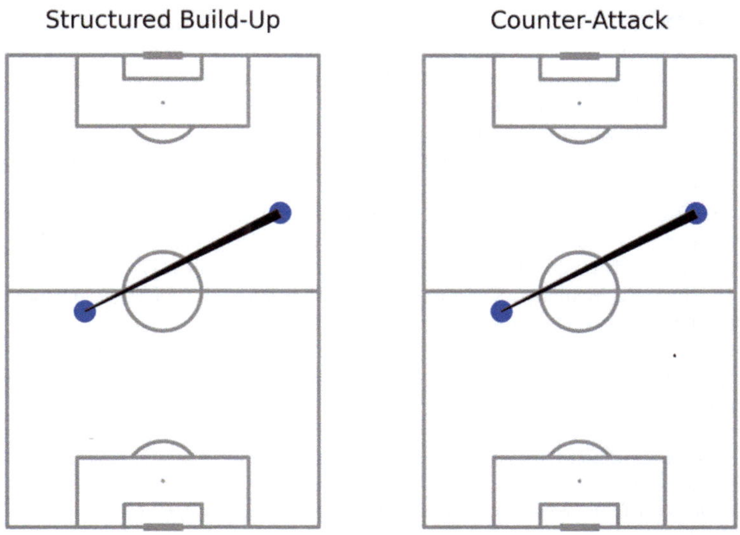

Fig. 20.3 Two different scenarios for a diagonal pass represented only with event data

In summary, this VAEP model implementation exemplifies the practical benefits and inherent limitations of current event-based action valuation metrics in soccer analytics. While these metrics are able to quantify action values, providing insights into game dynamics and player contributions, it is crucial to use them with an understanding of their limitations, particularly the absence of spatial context. Nevertheless, by aggregating these values over multiple games, the potential for nuanced insights into team strategies and player effectiveness remains, albeit with the understanding that some inaccuracies may persist. As the field progresses, exploring more comprehensive data and refining analytical techniques could provide a pathway to enhance the precision and relevance of these metrics.

20.5 Conclusion

Action valuation metrics significantly advance quantitative soccer analysis by overcoming the primary limitation of aggregate statistics—lack of specificity. These metrics, exemplified by VAEP, unlock valuable individual and collective insights useful in both competitive data analysis and academic research.

However, these metrics are not free of limitations. For instance, VAEP and similar event data-based metrics lack spatial context because they do not account for the positions of off-ball players. This can cause different game situations to be represented identically in the framework, leading to averaged values.

Despite these shortcomings, the metrics remain valuable when used appropriately. Awareness of their limitations is crucial for responsible usage and for justifying further development to mitigate these downsides. Practitioners continuously refine these methods, enhancing the impact of sports analytics on professional sports decision-making.

References

de Sá-Freire, L. M., & Vaz-de-Melo, P. O. (2023, September). Field depth matters: Comparing the valuation of passes in football. In *International Workshop on Machine Learning and Data Mining for Sports Analytics* (pp. 64–73). Springer.

Decroos, T., & Davis, J. (2019, December). Interpretable prediction of goals in soccer. In *Proceedings of the AAAI-20 Workshop on Artificial Intelligence in Team Sports.*

Decroos, T., Bransen, L., Van Haaren, J., & Davis, J. (2019, July). Actions speak louder than goals: Valuing player actions in soccer. In *Proceedings of the 25th ACM SIGKDD International Conference on Knowledge Discovery & Data Mining* (pp. 1851–1861).

Everett, G., Beal, R., Matthews, T., Norman, T., & Ramchurn, G. (2022). Contextual expected threat using spatial event data. In *Statsbomb Conference 2022.*

Fernández, J., Bornn, L., & Cervone, D. (2019, March). Decomposing the immeasurable sport: A deep learning expected possession value framework for soccer. In *13th MIT Sloan Sports Analytics Conference.*

Fernández, J., Bornn, L., & Cervone, D. (2021). A framework for the fine-grained evaluation of the instantaneous expected value of soccer possessions. *Machine Learning, 110*(6), 1389–1427.

Mackenzie, R., & Cushion, C. (2013). Performance analysis in football: A critical review and implications for future research. *Journal of Sports Sciences, 31*(6), 639–676.

Pappalardo, L., Cintia, P., Rossi, A., Massucco, E., Ferragina, P., Pedreschi, D., & Giannotti, F. (2019). A public data set of spatio-temporal match events in soccer competitions. *Scientific Data, 6*(1), 236.

Robberechts, P., Van Roy, M., & Davis, J. (2023, August). un-xPass: Measuring soccer player's creativity. In *Proceedings of the 29th ACM SIGKDD Conference on Knowledge Discovery and Data Mining* (pp. 4768–4777).

Zhou, C., Calvo, A. L., Robertson, S., & Gómez, M. Á. (2021). Long-term influence of technical, physical performance indicators and situational variables on match outcome in male professional Chinese soccer. *Journal of Sports Sciences, 39*(6), 598–608.

Part V

Outlook

21 Responsible Use of Machine Learning in Sports Science

Fabian Wunderlich

Abstract

Artificial Intelligence (AI) in general and Machine Learning (ML) in particular have brought substantial change to society, business and science, which of course also applies to the field of sports. While research has initially concentrated on methodological questions and opportunities for application, the rise of ML also creates new problems, yielding increased discussion on the responsible use of ML. The present chapter discusses the development of ML from fundamental research to a broad societal usage and recent political efforts to create guidelines as well as a legal framework for the use of AI. Several issues connected with ML-based solutions, such as discrimination, missing explainability, methodologically incorrect use of models and application for questionable purposes are discussed domain-independently. Moreover, the chapter aims to promote awareness of ML responsibility in the domain of sports by addressing specific sports-related ethical, economic and legal issues, but also issues of applicability and questions related to the way we want our society to be in the future.

21.1 Development of Societal and Political Discussion on ML

The rise of Machine Learning (ML) is a success story that has been made possible through scientific advances as well as societal changes. The foundation for the methods in various ML approaches have been laid decades ago in fundamental research from mathematics and computer science (Ho, 1995; Hochreiter & Schmidhuber, 1997; Rumelhart et al., 1986). The full potential of ML methods,

F. Wunderlich (✉)
Institute of Exercise Training and Sport Informatics, German Sport University Cologne, Köln, Germany
e-mail: f.wunderlich@dshs-koeln.de

D. Memmert (ed.), *Artificial Intelligence and Machine Learning in Sports Science*, https://doi.org/10.1007/978-3-662-70155-3_21

however, only became clear some time later, partially because aspects like rising computational power and data availability naturally favored the possibilities of ML. For quite a long time ML has been developed and discussed by experts in the field, while the impact on everyday life and societal discussion was rather limited. Important steps towards an increased social awareness and broader discussion of ML have been successful applications in domains like image recognition (Redmon et al., 2016), speech recognition (Sak et al., 2014) and recommendation systems (Zhang et al., 2019). The recent rise of large language models like ChatGPT can be considered another massive step in the social awareness of the power of ML-based solutions and their capability to solve everyday problems and interact with their users. ML-based tools that accelerate our ability to access knowledge, save us time and effort through image or speech recognition, or enrich our lives with intelligent recommendations have become fully integrated into our everyday life. At the same time societal and political discussion starts to increasingly focus on the disadvantages resulting from this development, including ethical, societal or legal issues of ML.

21.1.1 Ethics Guidelines of the European Union

In 2019, the European Union published guidelines for Trustworthy AI (European Union, 2019), being further evidence that discussion on ethical aspects of ML included in the even broader concept of Artificial Intelligence (AI) have reached the highest political levels. Moreover, it shows that the scientific focus on methodological aspects in the initial phase of AI has changed to a more holistic assessment of how humans develop, use and interact with AI-based systems in a broader social context. The guidelines state three prerequisites for AI to be trustworthy, namely that it should be "lawful, complying with all applicable laws and regulations", "ethical, ensuring adherence to ethical principles and values" as well as "robust, both from a technical and social perspective" (European Union, 2019, p. 5). Robustness specifically refers to the avoidance of unintentional damage caused by AI-systems. The guidelines further specify seven requirements that should be ensured by all stakeholders involved during the life cycle of AI systems from development to usage. These requirements are listed below (see European Union, 2019, p. 14).

- Human agency and oversight
- Technical robustness and safety
- Privacy and data governance
- Transparency
- Diversity, non-discrimination and fairness
- Societal and environmental wellbeing
- Accountability

The need to formulate such guidelines emphasizes the existence of numerous potential issues that arise or could arise from the use of AI in society.

21.1.2 Legal Framework of the European Union

In 2021 the European Commission proposed the *Artificial Intelligence Act* (AI Act) as a first regulatory approach to establish a legal basis for dealing with risks caused by AI-based systems. Its purpose is to ensure that AI-based systems do not violate fundamental rights and are consistent with the values of the European Union, with regard to aspects of development, marketing and use (European Union, 2021). The difficult balancing act between the advantages and disadvantages of AI is underlined by mentioning "the twin objective of promoting the uptake of AI and of addressing the risks associated with certain uses of such technology." (European Union, 2021, p. 1). Prohibited uses of AI are defined and distinguish between "unacceptable risk", "high risk" and "low or minimal risk" (European Union, 2021, p. 12), which may include "practices that have a significant potential to manipulate persons through subliminal techniques beyond their consciousness or exploit vulnerabilities of specific vulnerable groups such as children or persons with disabilities in order to materially distort their behavior in a manner that is likely to cause them or another person psychological or physical harm." (European Union, 2021, p. 13). This initiative of the European Union emphasizes that downsides of AI are not only an ethical problem to be discussed in society, but actually are regarded as serious enough to require legal regulations. The initiative of the European Union for legal regulation can certainly be regarded as unchartered territory. In this respect, it must be noted that in this very dynamic environment, the above information on regulation may no longer be completely up to date at the time of reading this book and at the same time the legal regulations may already be lagging behind the latest AI-related developments as soon as they are enacted.

21.2 Potential Disadvantages of Machine Learning

This section provides examples on potential downsides of ML that have been discussed in the scientific literature independent of any specific domain of application. Obviously, only a very incomplete selection of current and future downsides of the use of ML-based solutions can be covered.

21.2.1 Discrimination

When talking about discrimination, we refer to an unfair act of treating different groups unequally dependent on characteristics such as gender, race, sexual orientation, religion, physical or mental illnesses, social status, etc. Interestingly, it could be argued that discrimination is a human trait based on the very subjective character of human decisions, while nothing could be less subjective than data-based decisions. As such, algorithmic and data-based models could even be helpful to objectivize decisions and prevent discrimination. This line of reasoning, however, becomes flawed in the era of AI, where algorithms are deliberately developed to imitate human behavior including human decision making, which consequently might rebuild human weaknesses or flaws.

Researchers have started to identify several aspects of discrimination in AI-based applications, explain the reasons behind discrimination and identify strategies to avoid these pitfalls in future work (Zou & Schiebinger, 2018). Application cases that fall under the risk of discrimination or discrimination-based misjudgement are diverse. For example, ML methods for facial analysis used to identify the gender of a person have been shown to be highly accurate for lighter-skinned males, while failing to accurately recognise darker-skinned females (Buolamwini & Gebru, 2018). Using the concept of word embeddings, researchers have investigated the closeness of words and found that human race and gender biases evidenced in psychological experiments such as the association of certain skills or professions with a certain gender can also be found in purely ML-based algorithmic methods trained on textual data (Caliskan et al., 2017). Further example domains that have been argued to be susceptible to discrimination are healthcare applications such as the prediction of pediatric sepsis, where existing racistic tendencies are at the risk of being transferred to ML solutions (Sveen et al., 2022) or interviews to evaluate the hireability of individuals, where it has been shown that ML-based methods cause bias if no appropriate mitigation strategies are used (Booth et al., 2021).

When discussing the reasons behind algorithmic discrimination, careful evaluation and precise explanation is needed. Stating that ML in general or a certain ML method is racistic or designed to be racistic, for example, is a highly simplified view. AI as a broader concept including rule-based algorithms may indeed be designed in a racistic way if the rules inherently reflect racistic biases. ML methods as a part of AI, however, are inherently designed to learn from data, which means that ML methods confronted with human-generated data are at the risk of rebuilding human subjective decisions with all their potential flaws. As such, ML methods themselves would not be considered racistic, but will inevitably be prone to learn to be racistic if the data and labels used for training reflects racistic biases. As such ML methods can often retain racistic tendencies and the same is true for other issues like sexism or discrimination of any other sort. For this reason, avoiding imbalances and biases in the training data are one of the most straightforward approaches to avoid discrimination in ML (see Zou & Schiebinger, 2018). Another approach is to better understand existing discrimination in trained ML models by visually illustrating and analysing discrimination in the specific method (Wang et al., 2020).

21.2.2 Black Box Character

With regard to disadvantages of ML methods, one of the most frequently raised points is their black box character. On the input side of an ML model, we have a clearly defined feature space and on the output side, we have a clearly defined output in relation to the task the model is expected to fulfill, i.e. some classification or regression task. However, the exact processes of how inputs are transferred into outputs can be widely unknown and difficult to unravel from a human perspective. This is the reason for using the black box analogy, particularly if contrasting ML methods to more classical statistical models like regression models, which are designed to very clearly inform about how outputs are derived from inputs by using a mathematical model with limited complexity and providing its user with interpretable coefficients related to each specific input.

The limited transparency about how a ML model comes up with its decision becomes an ethical issue, once potentially far-reaching decisions are based on the assessment of the model. As such black box machine learning models are discussed very controversially, for example, in the domain of sports medicine (Bullock et al., 2022). Another relevant issue raised by the black box character concerns the use of ML methods in research. Applied science is interested in the capabilities to fulfill a task, which also involves the invention of ML methods. However, theory-based research is aiming to understand processes and validate theories by data, which can become highly difficult if not having a deep understanding of the processes the ML model uses.

Another possible negative side-effect of the increased capabilities, success and usage in critical decision-making domains might be the potential loss of human knowledge. If, in future, ML is capable of outperforming doctors in crucial tasks like detecting skin cancer (Haenssle et al., 2018), it can be assumed that doctors and patients will more and more rely on algorithmic evaluation than on human evaluation. This, however, will be in line with a decreased need for doctors to make own judgements and potentially even a discontinuation of training doctors in this task. The consequence would be a dependency on algorithmic evaluation and a loss of human capability to perform such tasks, which is even aggravated if the ML method itself is a black box and as such the reasons for its decisions are hardly traceable for humans.

In computer science, serious effort is recently made to unravel the black box character and better understand why and how ML models obtain decisions. Such approaches are summarized under the terms of explainability and interpretability, which, however, are not always separated clearly in the literature (Gunning et al., 2019; Preece, 2018). Usually, explainability refers more to the question of why a result was obtained (i.e. which inputs were responsible for this output) while interpretability refers more to the question of how a method obtained a specific result (i.e. how inputs were transferred to outputs in this method). Example approaches are the usage of feature importance in Random Forest or SHAP values (Lundberg & Lee, 2017).

21.2.3 Misuse of ML Methods

One of the rationales of the AI Act discussed in the first section is to avoid harm to persons through the use (or rather the misuse) of ML. In that sense, another downside of ML is that it can be misused either deliberately or through a lack of sufficient knowledge (see Volovici et al., 2022 for the domain of clinical research). The aforementioned authors argue that complex models should be tested carefully enough in order to avoid the promotion of models that owe their success to overfitting of data. Moreover, simpler statistical models should be used routinely as benchmarks in order to have a realistic estimation of the advantages of more complex models, if existent at all. This helps to avoid an overuse, unconsciously overconfident view on the abilities of ML models compared to simpler models or even a deliberate overselling of such models. If not having a significant benefit over simple models in terms of accuracy, the use of ML constitutes a bad trade-off given the advantages of simple models in terms of explainability (see also Sect. 21.2.2) and replicability of results. Another misuse of ML may stem from the usage of biased data (see also Sect. 21.2.1) or insufficiently large datasets inherently inappropriate for the data-intensive training of ML models.

In addition to methodological unsound use (or misuse), methodologically correct models can also be misused if they are used to pursue questionable objectives or even pose a danger to social cohesion and democracy. The extent of the influence that algorithmic solutions can have on human perception and opinion can be demonstrated empirically. Kramer et al. (2014) report that it was possible to change the emotional reactions Facebook users expressed in own posts by deliberate manipulation of the emotional content in the feed of these users. Epstein and Robertson (2015) found that deliberate biases introduced to online search rankings are sufficient to change the preferences of undecided voters without them being aware of any manipulation.

These examples illustrate that the influence of algorithmic solutions including ML models on our behavior needs to be observed critically and the objectives pursued when applying ML need to be questioned carefully. Recommendation systems used in social media might amplify political opinions of users, even without deliberate manipulation. If content matching the existing political opinion of a user is increasingly proposed, users are at the risk of no longer perceiving contrary opinions. As a consequence, existing opinions can be assumed to be reinforced, leading to a polarization of opinion in society in the long term. Recommendation systems used in online shops and streaming services can be seen as smart advertising and assistance for the customer, but at the same time might make the persons exposed to them spend more money or time than would be considered positive for them. Further alarming examples are related to the generation of fake news including videos and spoken statements hardly distinguishable from truth.

21.3 Responsibility Issues in Sports

So far, this chapter has discussed the background of responsibility issues related to ML and has given several examples. However, these were widely independent of the domain of application. In line with the topic of this book, it is worth discussing specific challenges for applications of ML in sports science and sports practice. As in virtually every other domain, ML has opened up new opportunities and research domains in sports science. Examples are performance analysis including analysis of tactical behaviour (Dick & Brefeld, 2019; Memmert et al., 2017), quantification of the risk for injuries (Rossi et al., 2018), forecasting of match outcomes (Horvat & Job, 2020; Wunderlich et al., 2025), analysis of sports-related social media data (Kolbinger & Knopp, 2020), applications of computer vision in sports (Theiner et al., 2022) or sports medicine (Ramkumar et al., 2022).

The current literature on ML in sports seems to be focused on opportunities rather than risks. Despite rare examples (Bullock et al., 2022), critical discussion on ethical or legal aspects of ML is highly limited, if existent. This might be partly driven by the fact that, with the exception of sports medicine, there is a rather limited impact of the use of ML-based solutions in sports to the physical or psychological health of humans, if at all. For experimental research in sports science, it is well-established practice to discuss ethical question and how to follow ethical principles before even carrying out a research project. Applications of ML, however, have been highly data-driven so far and hardly involve humans in a direct way in the sense of human-ML interaction. This might change as soon as direct interaction between humans and ML-based solutions become more common, which is likely to be the case in the near future, e.g. through the use of large language models.

Given the relative novelty of this topic and the missing standards of how to responsibly use ML in sports, this section is intended to raise questions rather than answer them. Certainly, as a first step towards responsibility, increased awareness of this important aspect will be needed. In that sense, before applying ML techniques to sports-related questions, researchers should spend some thoughts on possible downsides of doing so. Discussion on such attempts should not only consider results and future opportunities, but also existing and possible future issues arising from such research. Table 21.1 summarises several examples, which includes ethical, economic and legal aspects, but also aspects of applicability or questions related to the way we want our society to be in the future. While it is focused on the use of ML in sports, it also relates to broader issues of using data and technology. Table 21.1 is neither sorted by importance nor aims for completeness.

It will be interesting to see how these questions or similar questions are answered in future research and practice and how the rapidly advancing methods in ML and social familiarisation with it will change the way we answer those questions in the future. As the importance of responsibility in ML in sports will increase, scientists could work towards guidelines in the near future to avoid negative consequences of ML in sports, as has been already done in other domains (Wiens et al., 2019).

Table 21.1 Example questions that may arise from using data, technology and ML methods in the domain of sports or that might arise in the near future

Topic	Question	Case example
Ethical usage of results	How do we deal with situations where the opportunities of ML models are inconsistent with ethical concepts of our society?	Do we want to tell a child that it will not be successful in sports based on a decision of a ML model on talent identification?
Participation in sports	What if technical and/or methodological progress hinders the participation of humans in sports events?	Do we want to deprive line judges or umpires of the opportunity to participate in tennis competitions as their decisions can be replaced by the hawk-eye system?
Data protection	To what extent do we want to protect personal data of athletes?	If data with personal or medical background may increasingly get available through media coverage of sports events, do we need to protect athletes from analysis of such data?
Black box models	How do we deal with model decisions, that we are unable to explain and understand?	How will a coach give feedback on a decision to an athlete, if the decision was based on a ML model and the coach does not even understand the reasons for this decision?
Trust	How do we deal with possible mistrust towards ML-based decision support?	Will club officials keep trusting on human intuition despite objective evidence in favour of ML-based solutions?
Security	How can we ensure model security?	Do we want to trust a ML model for far-reaching decisions if we have to be afraid of it being subject to errors either unintentionally or through malicious influence by third parties
Discrimination	How can we avoid that ML models reinforce unequal treatment of groups due to personal characteristics?	If talents chosen for a youth team have been selected in a biased or discriminatory way in the past, how can we avoid this to be reinforced in algorithmic solutions in the future?
Data fairness	How do we deal with inequality in data sets?	If athletes with a certain gender, race or social status are over-represented in a training set for injury prediction, will we obtain generalisable results?
Legal issues	How can we avoid legal violations caused by the use of ML models?	Are ML-based models in relation to nutrition, planning of training programmes or sports medicine decisions at the risk of being prohibited by law due to potential psychological or physical harm caused?
Undesirable consequences	What if data-driven solutions have negative side effects on sports?	Will people still be interested in and entertained by watching soccer if referee decisions, player transfers and decisions by coaches are largely automated model-based processes?

(continued)

Table 21.1 (continued)

Topic	Question	Case example
Teaching methods	To what extent do we want to include digital tools in teaching?	Do we want to include digital tools in sports lessons at school are do we want sports to be a counterweight to digitisation?
Employment	To what extent do we want human work to be replaced by ML-based solutions?	If data-driven player scouting is good enough to replace humans, do we want to deprive persons from the possibility to work in sports?
Addiction	How do we deal with the risk of ML provoking addictive behaviour?	Are smart recommendation systems desirable, that might entice persons to consume sports events, comment sports on social media, spend money on sports or engage in sports betting in an unhealthy manner?
Social interaction	How do we deal with the fact that the use of ML-based solutions might replace social interaction?	If ML-based tools are able to give feedback on how to do exercises correctly based on video data, might we still want a coach to do it in order to maintain social interaction?

21.4 Conclusion

As with probably all technological advances in the past, AI- and ML-based solutions involve downsides that require responsible use of these techniques. ML models may involve the risk of incorrect use, undesirable consequences, ethical issues or even manipulative misuse in practically every domain of application. Sport is certainly not one of the domains in which responsibility issues have been discussed most intensively to date, but nevertheless numerous issues in sports-related usage of ML exist. Being to early for providing guidelines on how to deal with these issues, the chapter summarises diverse questions related to the use of ML in sports. The topic of ML responsibility creates the need for societal and political discussion, scientific guidelines and a legal framework to avoid or minimise downsides of ML (in sports) in the near future.

References

Booth, B. M., Hickman, L., Subburaj, S. K., Tay, L., Woo, S. E., & D'Mello, S. K. (2021). Bias and fairness in multimodal machine learning: A case study of automated video interviews. In *Proceedings of the 2021 International Conference on Multimodal Interaction.*

Bullock, G. S., Hughes, T., Arundale, A. H., Ward, P., Collins, G. S., & Kluzek, S. (2022). Black box prediction methods in sports medicine deserve a red card for reckless practice: A change of tactics is needed to advance athlete care. *Sports Medicine (Auckland, N.Z.), 52*(8), 1729–1735. https://doi.org/10.1007/s40279-022-01655-6

Buolamwini, J., & Gebru, T. (2018). Gender shades: Intersectional accuracy disparities in commercial gender classification. In *Conference on Fairness, Accountability and Transparency. Symposium Conducted at the Meeting of PMLR.*

Caliskan, A., Bryson, J. J., & Narayanan, A. (2017). Semantics derived automatically from language corpora contain human-like biases. *Science, 356*(6334), 183–186.

Dick, U., & Brefeld, U. (2019). Learning to rate player positioning in soccer. *Big Data, 7*(1), 71–82. https://doi.org/10.1089/big.2018.0054

Epstein, R., & Robertson, R. E. (2015). The search engine manipulation effect (SEME) and its possible impact on the outcomes of elections. *Proceedings of the National Academy of Sciences of the United States of America, 112*(33), E4512–E4521. https://doi.org/10.1073/pnas.1419828112

European Union. (2019). *Ethics guidelines for trustworthy AI*. https://digital-strategy.ec.europa.eu/en/library/ethics-guidelines-trustworthy-ai

European Union. (2021). *Proposal for a regulation of the European Parliament and of the Council laying down harmonised rules on artificial intelligence (Artificial Intelligence Act) and amending certain union legislative acts*. https://eur-lex.europa.eu/legal-content/EN/TXT/?uri=CELEX:52021PC0206

Gunning, D., Stefik, M., Choi, J., Miller, T., Stumpf, S., & Yang, G.-Z. (2019). Xai-Explainable artificial intelligence. *Science Robotics, 4*(37), eaay7120. https://doi.org/10.1126/scirobotics.aay7120

Haenssle, H. A., Fink, C., Schneiderbauer, R., Toberer, F., Buhl, T., Blum, A., et al. (2018). Man against machine: Diagnostic performance of a deep learning convolutional neural network for dermoscopic melanoma recognition in comparison to 58 dermatologists. *Annals of Oncology: Official Journal of the European Society for Medical Oncology, 29*(8), 1836–1842. https://doi.org/10.1093/annonc/mdy166

Ho, T. K. (1995). Random decision forests. In *Proceedings of the third International Conference on Document Analysis and Recognition: August 14-16, 1995, Montréal, Canada* (pp. 278–282). IEEE Computer Society Press. https://doi.org/10.1109/ICDAR.1995.598994

Hochreiter, S., & Schmidhuber, J. (1997). Long short-term memory. *Neural Computation, 9*(8), 1735–1780. https://doi.org/10.1162/neco.1997.9.8.1735

Horvat, T., & Job, J. (2020). The use of machine learning in sport outcome prediction: A review. *WIREs Data Mining and Knowledge Discovery, 10*(5), e1380. https://doi.org/10.1002/widm.1380

Kolbinger, O., & Knopp, M. (2020). Video kills the sentiment—Exploring fans' reception of the video assistant referee in the English Premier League using Twitter data. *PLoS One, 15*(12), e0242728. https://doi.org/10.1371/journal.pone.0242728

Kramer, A. D. I., Guillory, J. E., & Hancock, J. T. (2014). Experimental evidence of massive-scale emotional contagion through social networks. *Proceedings of the National Academy of Sciences of the United States of America, 111*(24), 8788–8790. https://doi.org/10.1073/pnas.1320040111

Lundberg, S. M., & Lee, S.-I. (2017). A unified approach to interpreting model predictions. In I. Guyon, U. V. Luxburg, S. Bengio, H. Wallach, R. Fergus, S. Vishwanathan, & R. Garnett (Eds.), *Advances in neural information processing systems 30* (pp. 4765–4774). Curran Associates, Inc. http://papers.nips.cc/paper/7062-a-unified-approach-to-interpreting-model-predictions.pdf

Memmert, D., Lemmink, K. A. P. M., & Sampaio, J. (2017). Current approaches to tactical performance analyses in soccer using position data. *Sports Medicine (Auckland, N.Z.), 47*(1), 1–10. https://doi.org/10.1007/s40279-016-0562-5

Preece, A. (2018). Asking 'Why' in AI: Explainability of intelligent systems—Perspectives and challenges. *Intelligent Systems in Accounting, Finance and Management, 25*(2), 63–72. https://doi.org/10.1002/isaf.1422

Ramkumar, P. N., Luu, B. C., Haeberle, H. S., Karnuta, J. M., Nwachukwu, B. U., & Williams, R. J. (2022). Sports medicine and artificial intelligence: A primer. *The American Journal of Sports Medicine, 50*(4), 1166–1174. https://doi.org/10.1177/03635465211008648

Redmon, J., Divvala, S., Girshick, R., & Farhadi, A. (2016). You only look once: Unified, real-time object detection. In *2016 IEEE Conference on Computer Vision and Pattern Recognition (CVPR)* (pp. 779–788). IEEE. https://doi.org/10.1109/CVPR.2016.91

Rossi, A., Pappalardo, L., Cintia, P., Iaia, F. M., Fernàndez, J., & Medina, D. (2018). Effective injury forecasting in soccer with GPS training data and machine learning. *PLoS One, 13*(7), e0201264. https://doi.org/10.1371/journal.pone.0201264

Rumelhart, D. E., Hinton, G. E., & Williams, R. J. (1986). Learning representations by back-propagating errors. *Nature, 323*(6088), 533–536.

Sak, H., Senior, A. W., & Beaufays, F. (2014). Long short-term memory recurrent neural network architectures for large scale acoustic modeling. In *INTERSPEECH: 15th Annual Conference on the International Speech Communication Association.*

Sveen, W., Dewan, M., & Dexheimer, J. W. (2022). The risk of coding racism into pediatric sepsis care: The necessity of antiracism in machine learning. *The Journal of Pediatrics, 247*, 129–132.

Theiner, J., Gritz, W., Müller-Budack, E., Rein, R., Memmert, D., & Ewerth, R. (2022). Extraction of positional player data from broadcast soccer videos. In *Proceedings of the IEEE/CVF Winter Conference on Applications of Computer Vision.*

Volovici, V., Syn, N. L., Ercole, A., Zhao, J. J., & Liu, N. (2022). Steps to avoid overuse and misuse of machine learning in clinical research. *Nature Medicine, 28*(10), 1996–1999. https://doi.org/10.1038/s41591-022-01961-6

Wang, Q., Xu, Z., Chen, Z., Wang, Y., Liu, S., & Qu, H. (2020). Visual analysis of discrimination in machine learning. *IEEE Transactions on Visualization and Computer Graphics, 27*(2), 1470–1480.

Wiens, J., Saria, S., Sendak, M., Ghassemi, M., Liu, V. X., Doshi-Velez, F., et al. (2019). Do no harm: A roadmap for responsible machine learning for health care. *Nature Medicine, 25*(9), 1337–1340. https://doi.org/10.1038/s41591-019-0548-6

Wunderlich, F., Biermann, H., Yang, W., Bassek, M., Raabe, D., Elbert, N., Memmert, D. & Garnica Caparrós, M. (2025). Assessing machine learning and data imputation approaches to handle the issue of data sparsity in sports forecasting. *Machine Learning, 114*(2), 1–28.

Zhang, S., Yao, L., Sun, A., & Tay, Y. (2019). Deep learning based recommender system. *ACM Computing Surveys, 52*(1), 1–38. https://doi.org/10.1145/3285029

Zou, J., & Schiebinger, L. (2018). *AI can be sexist and racist—It's time to make it fair.* Nature Publishing Group UK.

Opportunities and Challenges of Artificial Intelligence in Sports Science

22

Chris Richter and Enda King

Abstract

Imagine a world in which the full potential of AI is realised. In this world, we will be able to predict critical events such as injuries in sports—without interfering with athletes' daily routines. AI will play a key role in making data more accessible and this in turn will improve our ability to predict, prevent and improve injuries. By analysing vast amounts of data, AI can uncover patterns and correlations that may go unnoticed using traditional methods, leading to better prediction of injuries and more tailored prevention and rehabilitation plans, as well as improved athletic performance. However, the path to integrating AI into sport is fraught with challenges. We have to deal with issues such as ensuring AI accuracy, compliance with regulations, understanding data changes and the responsible use of AI. Despite these hurdles, the gradual introduction of AI into sport is paving the way for an era of more informed decision-making. It promises a future where decisions are based on solid data, leading to training and coaching that is both safer and more effective for athletes.

22.1 Introduction

Sport is on the cusp of a revolutionary era that is being driven by breakthroughs in artificial intelligence (AI), which is developing at an unprecedented rate. AI offers many opportunities for the development of systems to diagnose, treat and

C. Richter (✉)
Richter Analytics UG, Grafing, Germany
e-mail: chris@richtics.com

E. King
Head of Elite Performance and Development Aspetar, Doha, Qatar

D. Memmert (ed.), *Artificial Intelligence and Machine Learning in Sports Science*,
https://doi.org/10.1007/978-3-662-70155-3_22

rehabilitate sports injuries. Although AI also offers opportunities to improve athletic performance through customised training and optimisation strategies, the focus of this chapter is on injury prevention and rehabilitation. We have focussed on injuries as they have a profound impact on athletes' careers and well-being.

AI models have the potential to revolutionise injury prevention by accurately predicting the likelihood of both acute and chronic injuries before they occur—enabling early intervention strategies tailored to the individual athlete's risk factors and physical condition. Although much effort has been invested in injury prediction, the general consensus is that "there is no screening test for the prediction of sports injuries with adequate test characteristics and no intervention study that supports screening for injury risk" (Bahr, 2016). We find ourselves in this situation because testing protocols fail to assess the full range of properties required for effective injury prediction, focusing mainly on one dimension (e.g. strength or symmetry) only. This is also the case in rehabilitation, where traditional approaches and tests have typically followed a standardised, time-centered protocol. Here, there are often fixed timelines for milestones such as returning to running or starting sport-specific training. These conventional methods often overlook athletes' individual recovery pathways and do not take into account the complex nature of their physical readiness. For example, assessing readiness for high-impact activities based solely on time since surgery, strength or symmetry ignores critical aspects such as range of motion and functional ability. AI technologies can address these shortcomings by supporting a more nuanced, comprehensive approach to injury prediction, rehabilitation and testing. By facilitating regular, detailed assessments, AI enables a broader, multi-faceted analysis of an athlete's current health status that is not limited to a single test or metric.

AI has historically been associated with advanced scientific processes and science fiction films that can only be used by the most skilled and specialised scientists and programmers. However, due to the benefits of AI and democratisation (especially Large Language Models), the use of AI now permeates almost every aspect of our lives. Sports science is no different, and those working in sports need to educate themselves on the ever-advancing applicability of AI in their profession. Sports scientists who understand the direction in which AI is heading will not only ensure that they are providing the best possible support to their athletes to enhance their performance as efficiently as possible, but they will also improve the way in which they record and analyse information (data). Unfortunately, there is no universal/clear definition of AI, and there are many buzzwords around AI (machine learning, deep learning, natural language processing or big data, to name a few). These terms and AI are also often used interchangeably as they are closely related. This leads to confusion in terms of meaning and applicability. The term AI encompasses any predictive algorithm (model) that enables a machine to perform a task that we would call "smart" or intelligent. AI can be a simple rule-based model (defined by a human) or a very complex and hard-to-understand deep learning model (that can learn to estimate or predict a dependent variable from a database of many independent variables). Between a simple rule-based model and a deep learning model, there is a wide range of statistical, machine learning and data science models. In general, most AI models are designed to solve a single, well-defined problem and predict or generate an outcome based on an input (text, numbers, images, etc.). The more complex a task becomes, the more important the data set used to train the model (training data) becomes. A prediction / answer from an AI should be considered as the most probable answer to a given input based on the training data. Very complex tasks are often not handled by a single AI model, but by an ensemble of models that work together (AI framework). These models break down a complex task into smaller, more manageable components that can be effectively solved by smaller specialised models.

Now that we have developed a common understanding of AI, let's look at how AI could change sports science. We believe that AI will help us to bring technologies to the playing field or gym that can only be used in laboratories today. With this development, we will develop better and more personalised training interventions that will lead to better performance in elite athletes and better prediction of injuries.

AI is often marketed as magical, but it is not magic. It's maths, and it's important to understand that developing an effective AI application requires skills and expertise. Before we can realise the full potential of AI, we need to overcome three major hurdles:

Hurdle 1—Sample size: A key challenge in sports science research, particularly in studies aimed at understanding sports injuries, is the limited number of samples collected. This limitation is often due to the high costs associated with data collection, cleaning and processing, which require a lot of expertise, technology and time. However, AI has the potential to drastically reduce these costs and enable the collection of larger samples. Thus, AI could bring big data to sports science and significantly improve data availability and the depth of our insights.

Hurdle 2—Frequency of recording: In sports science research focussed on understanding sports injuries, the frequency of data collection is a major hurdle. Many longitudinal studies only collect data at the beginning of the season but track injuries throughout the season, resulting in a blind spot between initial data collection and injury occurrence. During this blind spot, numerous variables can change. This methodology leads to ambiguity in our data sets and poses a significant problem. AI has the potential to simplify data collection and allow researchers to not only collect a greater number of samples, but also increase the frequency of data collection, improving the accuracy and relevance of our datasets.

Hurdle 3—Dimensionality is another challenge in sports science research, especially in studies that aim to understand sports injuries, as the data sets usually have low dimensionality. Many studies focus narrowly on individual aspects, such as movement patterns during a jump. This ignores the multifaceted nature of sports injuries (risk factors can include muscle imbalances, previous injuries, fatigue levels and anatomical features that are critical to understanding injury mechanisms). AI can improve our research by integrating different areas through the use of specialised AI models. These models will incorporate the knowledge and experience of experts from different fields into our studies and enrich the analysis with diverse, interdisciplinary insights.

Dimensionality refers to the number of areas recorded in the data set. Each dimension contains the knowledge for a specific area (e.g. exercise, nutrition or sleep) related to an athlete. High dimensionality provides a more comprehensive overview of an athlete, but is also a challenge to analyse due to the complexity and computational effort involved in processing large sets of variables.

Big data refers to data that is generated from numerous sources with high velocity, variety and veracity. In the context of sports science, this can include everything from athlete performance data, video analyses of movements, physiological data, social media interactions and more. These data sets are characterised by their volume, which exceeds the capabilities of traditional data processing tools to collect, manage and process within a reasonable timeframe. Big data in sports science offers unparalleled opportunities for insights into performance improvement, injury prevention and personalised training plans.

Having overcome these three major hurdles, AI has the potential to improve our understanding of sports injuries to the point where we can question Bahr's (2016) statement that no screening test can predict injury with sufficient accuracy. Although AI has immense potential to revolutionise sports science, it cannot drive change alone. We must actively contribute to pushing the boundaries of the field. While not every sports scientist needs to have an in-depth knowledge of how AI works, a basic awareness of the key challenges in implementing AI frameworks is crucial. In the following two sections, we will take a closer look at the opportunities that AI offers for sport and discuss the key challenges associated with each opportunity.

22.2 Possibilities

In this section, we focus on discussing in more detail the opportunities that AI offers in sport. It is an exercise in "sky-blue thinking", where we identify and discuss the potential and promising aspects of AI without thinking about the constraints of current technological requirements or conceptual limitations. The challenges to be overcome for each opportunity are explored in a later section. Progress within one opportunity can significantly impact progress in another opportunity. This interdependence reflects the dependency between the different opportunities and the cumulative impact of AI on different aspects of sports science.

It is important to recognise that the possibilities of AI are enormous and constantly evolving. Despite the extensive review and projection in this chapter, it is likely that the rapidly advancing technologies in the field of AI and its possibilities are beyond our current horizon. Although this section aims to be exhaustive, the list of possibilities mentioned here is likely to be incomplete.

To achieve the ultimate goal of predicting sports injuries, we need to recognise that this daunting task needs to be broken down into more manageable segments to be tackled by specific AI models.

22.2.1 Opportunity: Availability of Data

One segment is ensuring good data availability. Critical factors are sufficient sample size to capture different injury mechanisms, adequate frequency of data collection to track incremental changes, and capturing dimensions that contribute to injury (mental health, nutrition or sleep). By implementing a comprehensive data strategy, we not only create the basis for injury prediction, but are also able to conduct continuous monitoring of athletes, which also allows for immediate feedback.

22.2.1.1 Opportunity: Reducing Costs, Time and the Need for Expertise

Data collection, even if limited to one dimension, is a time-consuming and costly process that requires the expertise of experts. Although an institution is able to hire a variety of experts and allocate sufficient resources for advanced technology to collect data across multiple dimensions, collecting a multidimensional dataset of an athlete is currently an overly time-consuming process. Furthermore, creating a comprehensive data set is also not very feasible as it could significantly disrupt an athlete's training and competition schedule. Every scientist's dream is to have data collected at a high frequency (e.g. daily or weekly). This would leave athletes with less time for training or competition, as they would have to spend hours in a lab during the day. This logistical challenge explains the lack of multidimensional data sets in sports science. However, AI models provide a solution as they offer the possibility to create such comprehensive data sets without compromising the athletes' time.

Example of the recording of biomechanical data: To record human movement, it is currently common practice to take an athlete to a biomechanics lab. A biomechanics lab is usually equipped with a series of infrared cameras and a force plate. The researcher must apply reflective markers to the athlete (approximately 30–45 min) before the athlete can perform the movement recording (approximately 15 min). It currently takes about 60 min to perform a movement recording with six exercises, with a large part of the time being used for attaching the markers to the athlete. The size of the laboratory limits the movements that can be recorded (recording a 100 metre run would require considerable effort). In addition, the recorded data must also be processed after the motion capture, which takes between 15 min and 2 h (depending on the lab setup and expertise in the lab). AI is already challenging the state of the art and will change the way we capture motion. Already today, companies are offering three-dimensional motion capture via a video recorded with a mobile phone. In other words, a biomechanics lab is already in your pocket. This is made possible by an AI method called "pose estimation" that identifies key points on the human body, which is basically the same as tracking markers in a lab facility, except no markers need to be attached and no post-processing is required.

Pose estimation is a task in the field of computer vision and artificial intelligence that involves recognising and tracking the position and orientation of human body parts in images or videos.

For the interested reader, there is a good open source example from the Max Planck Institute for Intelligent Systems called: SMPL. SMPL stands for: A Skinned Multi-Person Linear Model (SMPL) and is a realistic 3D model of the human body. SMPL is based on data from thousands of 3D body scans (https://smpl.is.tue.mpg.de/).

The force plates integrated into the floor of a motion capture laboratory are used to record ground reaction forces, which enable the calculation of joint torques and forces in subsequent analysis steps. AI models that estimate ground reaction forces (GRF) from video data have also already been presented by researchers. In a systematic review, Ancillao et al. (2018) reported that vertical GRFs can be accurately predicted in single-leg interactions with the ground, while estimating forces in two-legged tasks is still a challenge. The combination of pose estimation and GRF-AI modelling could move movement assessment from the laboratory to the training and competition venues without the need for athlete preparation. Furthermore, the appropriate infrastructure (video cameras) is already in place and would enable a high data collection frequency without any interaction with the athletes. This future may not be so far away, as there are already practical examples of such a future, at least in tennis (https://www.wingfield.io/).

Example of nutritional data collection: One dimension that is of great interest to a sports scientist is the dimension of nutrition. To record an athlete's diet, researchers are currently limited to questionnaires or protocols. This type of data collection is prone to errors and memory lapses. An athlete could forget to enter data and simply fill something in, manipulate the data to cover "cheat days", or simply not enter any data. It is therefore not surprising that the dimension of nutrition is not a regular feature in sports science studies. Again, AI could help us to capture data accurately and facilitate the collection of such data (in terms of time and cost). Recent advances in computer vision have enabled the development of tools that are able to estimate the calorie content of meals based on an image alone.

Computer vision is an area of artificial intelligence that enables computers to derive meaningful information from images. It involves automatically extracting, analysing and understanding useful information from a single image or a sequence of images.

SnapCalorie, for example, is an AI-powered app that estimates a calorie count and provides a breakdown of a meal's macronutrients based on a single photo taken with a smartphone (https://www.snapcalorie.com/). It is conceivable that in the future, AI will eliminate the need to manually record calorie intake and macronutrients by simply wearing a camera that can automatically record calorie and macronutrient content. Currently, mobile applications for food recognition and quantity estimation are still under development (Amugongo et al., 2023).

As can be seen, AI will enable researchers to collect multi-dimensional data sets in a more cost-effective and time-efficient manner, without the need for a large number of experts and with little or no intervention in the athlete. This will lead to the collection of more diverse and larger data sets, which in turn will enable advances in injury prediction. In addition, the reduction in cost could also increase the availability of open/shared datasets, which will further drive progress and enable testing of models with other data sources.

22.2.1.2 Opportunity: Enrichment of Data

Another opportunity offered by AI models is the ability to enrich data (add new features) that would previously have required a lot of computing time. In biomechanical analyses, for example, muscle activation and muscle forces can be calculated from movement recordings. This is difficult to do in real time, as many mathematical optimisation steps are required. With AI models, however, it would be possible to determine these variables much more quickly. An AI model can be trained to estimate muscle activation and muscle forces from joint angle and moment data. Insights into muscle activation and muscle forces could be crucial for understanding injuries or identifying movement deficits (e.g. movement patterns that place more stress on tissues than others). In addition, such enriching AI models could also be found in other areas. Imagine your athlete is a marathon runner. Here, you won't be able to record video throughout the competition or training. However, you could use sensors that only measure the kinematics of the tibia. Even though this doesn't look like meaningful data, AI models have already been used to translate such sensor readings into GRF signals (Ancillao et al., 2018), which would be very meaningful for them.

22.2.1.3 Opportunity: Synthetic Data

The final possibility discussed in relation to data is the ability of AI to generate synthetic data. Synthetic data is data created by AI that mimics real-world data sets. The generation of synthetic data can be very useful to us as it can help us improve data privacy, combat data scarcity and train and test models. In terms of improving data privacy, by generating data that mirrors real athletes without containing actual personal data, researchers can analyse sensitive topics while fully protecting athletes' privacy. In terms of data scarcity, in some cases it is difficult to collect a large enough sample, even if AI will bring improvements. For rare diseases, limited access to athletes or certain events that are difficult to observe, synthetic data could fill the gap and provide the data needed to improve current knowledge. Synthetic data will also help to improve the training of AI models as it can be used to balance unbalanced data sets. An unbalanced data set is when certain events (injuries) occur far less frequently in the data than other events (no injuries), which can distort the AI model and affect the accuracy of predictions.

22.2.2 Data Analysis

As AI will help sports scientists overcome the three biggest hurdles in terms of data availability, many new opportunities are opening up, paving the way for transformative advances in the field. In this section, we look at how AI's ability to process complex multidimensional datasets will not only improve our understanding of injury prediction, but also revolutionise the methods we use to gain meaningful insights by enabling deeper, more nuanced analysis that was previously unattainable. This will set a new standard for precision and efficiency in sports research.

22.2.2.1 Opportunity: Data Processing

As data collection in sport has improved, we now have huge amounts of complex multidimensional information at our disposal. This may sound like a treasure, but it also presents us with a a new challenge: how can we extract the information (characteristics) from this mountain of data that is important for predicting injuries? In order to extract meaningful features, we need to convert the raw data into a structured database. A structured database can be thought of as an organised cabinet containing a large amount of information, but divided into different categories or sections, similar to dividing a large amount of sports statistics into separate, easily accessible folders or lists. It is similar to organising a huge, messy room filled with many items into neatly labelled boxes. This step is crucial and often tedious, error-prone and requires a deep understanding of the subject areas in the database. A well-structured database allows both humans and AI models to efficiently perform analyses that are essential for predicting and understanding injuries.

A characteristic is defined as a specific, measurable attribute or property of an athlete that can come from different dimensions. For example, "height" and "weight" are characteristics from the dimension of physical measurement, while "speed", "endurance" and "power" are characteristics from the dimension of performance measurement. Similarly, "heart rate" and "VO_2 max" are features of physiological data, and "reaction time" and "decision speed" are cognitive indicators. These features are important inputs for AI models that enable the analysis, prediction and understanding of athletes' performance and health status.

If we intelligently combine AI, especially machine learning, with human expertise, we can almost completely automate this task. To tackle the complexity of organising and understanding our data, we should use 'domain-specific expert models' (AI assistants), which are specialised tools for specific tasks. In the example of motion capture, an AI assistant to organise a long motion record could organise the motion data by identifying and separating each jump or sprint step within the data. Another assistant could recognise errors in the execution of the movement, such as a "bad" landing after the jump. Another assistant could search through this organised data to find important features (e.g. maximum knee flexion). This can be done in an expert-driven way, where the AI assistant learns from previous research which features we should look out for. Another AI assistant could extract features in a data-driven way, where the AI uses dimensionality reduction techniques to identify important features based on the data features. The combination of an expert and a data-driven approach will help us to utilise knowledge from previous research and not miss important information.

> Dimension reduction is a data analysis technique in which the number of variables is reduced while retaining only the most important information. This process simplifies a data set with many dimensions (characteristics) to a manageable size while retaining the essence of the original data. By focusing on the most important features, dimensionality reduction helps visualise data, speeds up calculations and improves the performance of machine learning models by eliminating redundant or irrelevant information.

By transforming raw data into a clear and structured format that is ready for analysis, AI assistants will prove invaluable as they drastically reduce the time, effort and expertise normally required. This paves the way for a breakthrough in injury prediction.

22.2.2.2 Opportunity: Data Insights

Having transformed raw data into a structured and manageable database, we are now at a point where we can report on opportunities that AI brings to the data analysis phase. AI techniques can discover complex patterns and provide insights that go beyond the capabilities of traditional analysis.

Clustering models can be used to identify groupings of athletes based on various characteristics that can be used to tailor training and rehabilitation programmes more efficiently. This in turn allows us to separate or enrich the data set with subgroups, enabling the identification of specific athlete needs. For example, these AI models could differentiate between “short distance sprinters” and “long distance marathon runners” based on their performance and physiological characteristics. Further segmentation within these groups could be based on performance categories (e.g. juniors, adults or masters), leading to tailored training programmes that take into account the unique requirements and characteristics of each subgroup. This approach allows for personalised training that is optimised for the specific needs of each athlete, while taking into account the particular challenges and benefits that the subgroup’s profile brings. While AI will help identify subgroups, AI will not be able to do all the work. When using clustering models, the experience of the sports scientist is required to decide how many groups are present in the data and how they can be translated for the athletes and coaches.

Clustering models are machine learning algorithms that are designed to categorise data into meaningful groups (clusters) without prior knowledge of the group assignment. These models recognise patterns and similarities in the data and allow athletes with common characteristics to be grouped together in a cluster. The main goal is to ensure that the data points within a cluster are more similar to each other than the points in other clusters.

Natural Language Processing (NLP) models *will change the* way we interact with the data we have about an athlete. By leveraging the ability to analyse and interpret both text and other types of data, NLP opens the door to a variety of ground-breaking applications, such as automated written feedback, personalised training programme design and AI-powered interactive data conversations. In automated written feedback, NLP models examine a database to generate detailed written feedback. Based on the instructions given to the computer in natural language, the analysis can highlight strengths, identify areas for improvement or recommend adjustments to training approaches—all formulated in natural language. This means that even less experienced sports scientists without analysis experience can produce expert-level analyses, ensuring that athletes have access to world-class guidance at all times. The individual training programme design section describes how NLP training plans can be tailored to the specific goals and limitations of individual athletes. By analysing an athlete’s data against proven best practice principles, NLP can formulate plans that maximise progress while taking into account the guidance provided by the user. Finally, the points of an AI-powered interactive data conversation provide for the ability to upload an athlete's data and have a dialogue with the computer about their performance, potential injury risks and training directions. NLP facilitates this by enabling queries and advice in everyday language, with the system responding like a colleague or external counsellor. This ability allows for a deep dive into various concerns, such as assessing injury risks based on actual data or adapting training strategies.

Another use of AI is the detection of anomalies, to provide information about-Vortrag risk of injury. Anomaly detection can be understood as the identification of rare behaviours that deviate from the majority of data and do not conform to a clearly defined notion of normal behaviour. For all time series data in the field of sports science, there should be no such thing as an anomaly—especially in exercise science. Any abnormal pattern in an athlete's data over time or in the team as a whole should indicate a factor that is disrupting the current state. If the pattern is related to abnormal strength measurements, for example at the beginning of the season, it could be discarded as a change due to the athlete's strength abilities. However, if there is no such event, an abnormal pattern should be considered a red flag, as it could indicate that the athlete is fatigued, is at the beginning of developing a chronic condition or is ingesting unauthorised substances (doping). In general, AI models are excellent at recognising patterns and dependencies over time that could be overlooked by conventional statistical methods. This is because AI models can manage large amounts of complex data and understand "normal fluctuations" (by learning the underlying patterns that might be missed by traditional statistical methods). The ability of AI models to continuously learn and adapt to the data is particularly beneficial in sports.

Time series analysis involves analysing data points collected or recorded at successive points in time to identify trends, seasonal effects and long-term patterns. Time series analysis is particularly important in sport for monitoring injury risk and physiological metrics over time. It helps to understand how variables such as training load, recovery times and performance metrics evolve and provides insights that can lead to training adaptations and periodisation strategies to optimise performance and reduce the risk of injury.

22.2.2.3 The Opportunity: Predicting Future Events

Predictive modelling in AI will be a transformative force in sports science as it can bridge the gap between data analysis and future injury/injury prevention. AI models, especially deep learning models, are very good at understanding the complexity required to predict future injuries. This can be particularly important considering that important factors in human movement, such as the interplay between joints (coordination), movement variability and tissue stress, are currently not well understood.

Deep learning refers to AI models that learn from data in a way that mimics the way the human brain solves problems. It uses algorithms known as neural networks to process large data sets, recognise patterns and make predictions. This allows deep learning models to perform complex tasks such as image recognition, natural language processing and predictive analytics with remarkable accuracy and minimal human guidance.

AI models are therefore much better at predicting future injuries. However, if we want to predict future injuries, we need to distinguish between chronic and acute injuries. Acute injuries happen suddenly (e.g. torn ligaments or fractures), while chronic injuries are likely to result from overuse of a muscle or joint over time. Predicting chronic injuries is therefore easier than predicting acute injuries, as they are related to tissue overload and changes in movement (abnormalities) can give a good indication of this. Acute injuries, on the other hand, are much more complicated as an athlete with poor movement patterns but good anatomical features (e.g. thick ACL) may not sustain an injury, whereas an athlete with good movement patterns but poor anatomical features (e.g. thin ACL) may sustain an injury. In such cases, many more data samples and greater dimensionality are needed. However, by utilising data from athletes, AI models and their unique ability to process sequential time series data, we will be much better able to predict an athlete's current injury risk. These predictions will then impact the training programme we develop for our athletes to not only proactively minimise injury potential but also optimise performance. In addition, AI models can also point us to previous 'similar' cases in our data sets, which will help us in the decisions we need to make. By analysing patterns and results of previous similar cases, we can adjust the features, feed them back to the AI and evaluate whether the prediction is in our favour. There should be a human element to the decision-making process to ensure that the data manipulation and resulting decisions are well controlled.

22.3 Challenges

The challenges associated with the use of AI in daily practice are complex and range from the need for good and robust performance, regulatory compliance, managing data discrepancy, understanding AI decisions and eliminating bias in AI models. These challenges are not only related to AI, but are also linked to the fields of statistics and data science. This link is based on the dependence of AI on mathematical concepts and the crucial role that data quality plays in determining model efficiency (for example, poor data quality inevitably leads to poor models).

This section examines the challenges associated with AI in detail. You will gain the insights you need to identify and implement effective strategies to overcome these challenges before implementing solutions. As a result, you will be able to utilise AI models more effectively and reliably in the future.

22.3.1 Challenge: What Is Good Enough?

It is crucial to ensure the accuracy and reliability of AI models, especially in high-stakes scenarios such as injury prevention and restart decisions. Therefore, it is important to define how good "good enough" is. When we talk about the performance of an AI model, it's easy to say "the model is right 95% of the time". When it comes to injury prediction, stating the percentage of accuracy can be misleading. For example, King et al. presented a model that predicted ipsilateral ACL reinjury with an accuracy of 76%, which may seem impressive at first (King et al., 2020). However, they also described that a baseline model that consistently predicted the most common outcome (no injury) achieved 96% accuracy. This startling contrast shows that a model predicting 'no injury' by default would achieve 96% accuracy because only 4% of athletes in the study cohort experienced re-injury, highlighting the importance of considering baseline data when assessing model performance. This example emphasises the need to consider baseline metrics when evaluating AI model performance and highlights two key challenges: understanding the baseline metrics against which a model is measured and determining the most appropriate performance metric. Overcoming the first challenge involves understanding the baseline. In order to determine the required performance of these models, you need to establish base or reference values (a baseline). The second challenge is choosing a suitable performance metric. Percentage accuracy may seem straightforward, but it is often not the most appropriate metric. Choosing the right performance metric depends on the specific requirements of the task to be solved. Each of the options described above requires a customised approach to define what is "good". This shows that the most appropriate performance metrics and acceptable error sizes vary widely across the different applications of AI in sports science.

The performance of an AI model can be measured in various ways (to name a few: Accuracy, precision, recall [sensitivity], f1 score, area under the ROC curve (AUC-ROC), mean absolute error, root mean square error, root mean square error, confusion matrix, R-squared, BLEU score, precision-recall curve). Each of these measures has its own use case, and the commonly used measure of precision is not ideal for every task.

Regarding the discussed opportunities in terms of data availability (e.g. image data on GRF signals), our goal is to reduce the error between the predictions of the AI model and a gold standard. Metrics such as: Mean Absolute Error, Standard Error of the Mean or Root Mean Squared Error would be best suited for such an evaluation. The accepted performance of the AI models should be closely aligned with the measurement error of the gold standard.

The measurement error of a device is the discrepancy between the measured value of the device and the actual true value of the measured quantity. This error includes both systematic errors, e.g. constant and predictable deviations from the true value, and random errors, which are due to unpredictable variations in the measurement process. Factors such as the precision of the device, its calibration, the way the user operates it and the environmental conditions can all affect the measurement error and ultimately influence the accuracy and reliability of the device's measurements.

In terms of the opportunities discussed in relation to data insights and predictions, the performance metrics of precision-recall or AUC (Area Under the Curve) should be utilised—with a higher value indicating a better model. However, it is important to realise that due to the complexity of factors influencing injury predictions, it is unlikely to achieve a perfect result. For example, certain movement patterns may increase the risk of injury, while other variables such as anatomical characteristics (e.g. cruciate ligament thickness) may decrease this risk, resulting in individuals with high movement pattern risk remaining uninjured. When evaluating prediction models, it is critical to compare them to an established gold standard, which may initially be human performance, to facilitate comparison and iterative improvement.

22.3.2 Challenge: Invisible Distortions (Bias)

The term bias refers to systematic errors within the AI model that lead to unfair or distorted predictions. Bias can arise from assumptions, choices or simplifications made during the development of an AI model and influence the results of the model in a distorted way. Bias can enter an AI model in two ways: through the process of data collection or during the creation of the AI model itself. Bias occurs when the data we use to train the AI does not accurately reflect the world or the target audience. Bias can result from the way we collect, select or handle the data, so that the AI performs well on data from athletes similar to the data from other athletes it was trained on, but not as well on other groups of athletes. For example, if an AI model for injury prediction was trained on data from young male football players, it is likely to perform poorly on older men, women or non-football players. Bias can also refer to problems in prediction due to assumptions made when developing the AI model. This may be due to the choice of model and features selected for training, which may unintentionally favour certain patterns or outcomes over others (e.g. maximizing the AI models performance without considering group membership). For example, during the training of the AI model, some features of the model could be optimised so that the model specialised towards a certain sport without being noticed by the developers. As a result, the model would provide less accurate results for athletes from other sports. In this scenario, the bias is not due to the data itself, but to how the model to optimized during the training.

To find out whether a model is biased in a way that is helpful for our use case or not, you can perform a validation study. For some opportunities, performing a validation study does not require much time or resources and quickly tells us whether the AI model is suitable for our needs (e.g. motion detection). However, for other tasks, checking for bias through a validation study is not as straightforward (e.g. injury prediction). We often don't know exactly what data the AI model has been trained on, and if we wanted to make sure that it knows all the edge cases that are important to us, we should create a database ourselves or dive deep into the trainings data, which means a lot of work. When it comes to understanding and predicting injuries using AI, determining whether the model is biased toward a case / athlete group requires a lot of effort. We would need to keep track of our athletes' injuries and check if the AI has missed some injuries or made wrong predictions. Analysing this information to find biased patterns in the predictions is difficult.

22.3.3 Challenge: Transparency, Explainability and Interpretability

Understanding the biases within an AI model is crucial, and this understanding is largely facilitated by understanding the basis on which the model makes its decisions. This brings into focus the importance of transparency, explainability and interpretability in the context of AI. In general terms, transparency sheds light on the entire development process of an AI model, explainability provides a clear understanding of the model's results, and interpretability provides direct insight into the model's decision-making process.

Transparency refers to the openness and clarity about the functioning and development process of an AI model. It encompasses the entire AI model development workflow, including data collection, model selection, training processes and deployment strategies. Transparency means that an outsider can understand the workflow of an AI model and the decisions made at each step, from conception to output.

Explainability refers to the degree to which a human can understand the result of an AI model. It is about providing humanly comprehensible reasons for how and why a model has arrived at its decision. Explainability is crucial for trust and accountability.

Interpretability is the extent to which a cause and effect can be observed within an AI model. It refers to the ability of the model to represent its processes in a way that is understandable to a human. An interpretable model allows users to understand why certain decisions or predictions are made based on the input data being processed. While all interpretable models are inherently explainable, not all explainable models are also interpretable, especially when it comes to complex models such as deep neural networks, which may require additional methods to achieve explainability.

Ideally, any AI model we use should be transparent, e.g. we know where the data comes from, how the model was created, how it was trained and what it will be used for. Transparency is crucial to determine whether the model is suitable for a particular task and to recognise any inherent biases. Depending on the intended use of an AI model, the need for explainability and interpretability will vary. In some scenarios, it is essential to fully understand how a model has made a decision, while in other cases a detailed understanding of the model is not necessarily required. For models that convert a signal into another measurement (e.g. predicting GRF from images), understanding the inner workings of the model may not have a large impact on its usefulness, as long as the accuracy of the data meets our requirements. The main focus in these scenarios is on the accuracy of the model. The situation is different when using AI for injury prediction. Here, understanding the decision-making process of a model is important for its usability. For injury prediction models in particular, it is not enough to know that an athlete has a high risk of injury without understanding the underlying reasons or knowing what steps can be taken to mitigate that risk and avoid injury. A model that predicts an injury without providing insight into how to prevent it is not really helpful.

Ultimately, the value of an AI model lies in its ability to effectively solve the tasks at hand. In the context of injury prediction, a model that only accurately predicts injuries but does not provide clear guidance on how to prevent an injury is invaluable.

22.3.4 Challenge: Changing Worlds (Data Drift)

Ensuring that AI models is performing, unbiased and understandable is critical, but it is equally important to understand and address data drift. The term data drift refers to changes in the underlying patterns that an AI model has learnt over time, resulting in a decrease in the accuracy and relevance of the model. Recognising, adapting and preparing for data drift are important skills to ensure that a model remains effective and reliable. This section looks at the impact of data drift on AI models and strategies to overcome this inevitable challenge with evolving data sets in sports science.

A good example of data drift in sports science is the evolution of high jump techniques—from the scissor jump to the Fosbury flop (Fig. 22.1). Initially, athletes used the scissor technique, but over time they switched to the Fosbury flop, a method that allows a better jump height with the same amount of force due to its more efficient technique. If an AI model trained on historical data is not updated to reflect such developments, it would continue to optimise for the outdated scissor technique and become obsolete as coaches and athletes adopt more efficient methods. This scenario emphasises how data drift can cause the predictions of a previously accurate AI model to become inadequate and highlights the need to continually update the AI model with new data to prevent it from becoming outdated and irrelevant.

Data drifts pose a particular challenge for AI injury prediction models, as numerous external factors can influence an athlete's underlying injury mechanism.

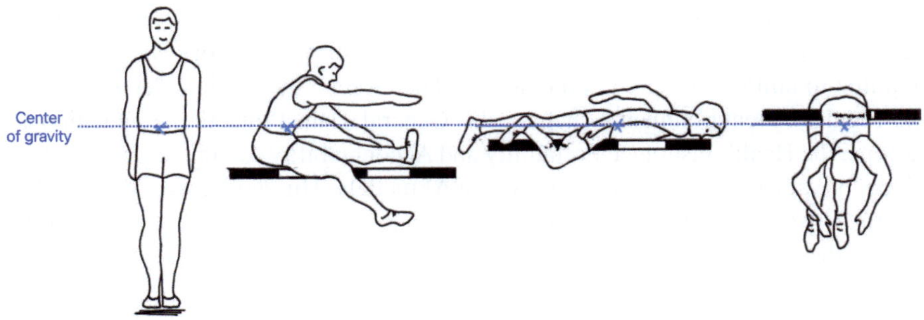

Fig. 22.1 Illustration of body centre of mass and jump height for the scissor, western roll and Fosbury flop technique

Changes in training history, sports practice, diet or fatigue can influence the factors that favour an injury over time. Therefore, injury prediction models need to be regularly re-evaluated and updated.

The impact of data drift is not the same for all AI applications in sports. While data acquisition models rely on consistent physical principles and are therefore less susceptible to data drift, data processing models can be moderately affected as incoming data changes and motion pattern adjustments slightly alter the signals they analyse. Nevertheless, data quality frameworks can recognise unnoticed data patterns and alert users to changes, making data drift in data processing models manageable and less consequential.

In summary, while data drift presents a nuanced challenge for various AI applications in sports science, a deep understanding of its dynamics and the implementation of continuous model refinement strategies are critical to maintaining the accuracy and usefulness of AI models amidst change.

22.3.5 Challenge: What Am I Allowed to Do?

Another challenge we face when working with AI models is navigating regulations. This challenge highlights the complexity of dealing with sensitive athlete data, which comes with significant privacy and security concerns. Regulatory compliance is a significant hurdle for the use of AI models in both clinical and professional sports. While advanced data analytics tools undoubtedly enhance our ability to monitor and improve injury management, they also impose a critical responsibility on us to protect the confidentiality and security of such data.

An athlete's data can include health, medical and nutritional data as well as performance metrics. Not only is this information private by nature, but it is also highly coveted in the competitive sports industry. In clinical and professional sports, each data point represents an individual, so health data cannot simply be aggregated from public sources. Obtaining the data required for AI models involves considerable effort, which emphasises the importance of careful data management and an appropriate security framework. A breach or mishandling of this data could have serious consequences, including legal action, loss of trust and potential damage to an athlete's career. Most of the data contains sensitive personal identifiers, making it difficult for providers to collect comprehensive information from various fields required to build injury prediction models. Therefore, a thorough understanding of the regulatory environment—such as the General Data Protection Regulation in Europe, the Health Insurance Portability and Accountability Act in the US and other global privacy laws—is critical for users of AI models. This emphasises the ongoing efforts to use AI effectively and ensure compliance with legal and ethical standards, despite the inherent challenges that sports science data brings.

22.4 Conclusion

This chapter has explored the opportunities that AI presents for sport science, but also the corresponding challenges. In describing the opportunities, we have embarked on a 'sky blue thinking' journey, identifying and considering the transformative possibilities of AI without the limitations of technology or conceptual barriers. We realised that the impact of AI on sports science is profound.

In the opportunities section, we uncovered the need for robust data collection strategies for AI models to eventually predict injuries, emphasising the importance of sufficient sample size, appropriate frequency of data collection and the inclusion of multiple injury-relevant dimensions. This area offers a number of opportunities for AI as it promises to revolutionise data collection by efficiently generating comprehensive multidimensional datasets without interfering with athletes' schedules. For us humans, these comprehensive, multidimensional data sets are difficult to understand, and here lies another opportunity for AI, as it is able to process large, multidimensional data sets that allow us to gain new insights and improve our understanding of sports injuries. From improving data processing to enabling sophisticated data insights through cluster modelling and natural language processing, AI will enable deeper, more nuanced analysis than is possible with traditional tools.

In the challenges section, we discussed a complex web of problems that we need to overcome if we want to use AI. The pursuit of modelling accuracy, the fight against bias, the quest for transparency, explainability and interpretability, and the daunting task of managing data drift—they all require our attention and ingenuity. In addition, regulatory compliance and the ethical handling of personal data emphasise the need for a conscientious approach to the integration of AI in sports science.

The integration of AI into sports science will revolutionise the field over the next 10 years. By 2034, we can expect to see a landscape where AI not only improves our understanding of sports injuries, but also enhances athletic performance through personalised training plans. Looking to the future, it is crucial that the development and implementation of AI in sports science is guided by ethical standards and equal access. If these challenges are successfully overcome, AI has the potential to not only transform the way athletes train and recover, but also to democratise sports science at a high level and make advanced training and health methods accessible to a wider range of athletes. This is because AI models could be "seamlessly" integrated into practical strategies to bring the latest research into everyday practice.

References

Amugongo, L. M., Kriebitz, A., Boch, A., & Lütge, C. (2023). Mobile computer vision-based applications for food recognition and volume and calorific estimation: A systematic review. *Healthcare, 11*, 59. https://doi.org/10.3390/healthcare11010059

Ancillao, A., Tedesco, S., Barton, J., & O'Flynn, B. (2018). Indirect measurement of ground reaction forces and moments by means of wearable inertial sensors: A systematic review. *Sensors (Basel), 18*(8), 2564. https://doi.org/10.3390/s18082564

Bahr, R. (2016). Why screening tests to predict injury do not work-and probably never will...: A critical review. *British Journal of Sports Medicine, 50*(13), 776–780.

King, E., Richter, C., Jackson, M., Franklyn-Miller, A., Falvey, E., Myer, G. D., et al. (2020). Factors influencing return to play and second anterior cruciate ligament injury rates in level 1 athletes after primary anterior cruciate ligament reconstruction: 2-Year follow-up on 1432 reconstructions at a single centre. *The American Journal of Sports Medicine, 48*(4), 812–824.

Index

D. Memmert (ed.), *Artificial Intelligence and Machine Learning in Sports Science*,
https://doi.org/10.1007/978-3-662-70155-3

FSC
www.fsc.org
MIX
Papier aus verantwortungsvollen Quellen
Paper from responsible sources
FSC® C105338

If you have any concerns about our products,
you can contact us on
ProductSafety@springernature.com

In case Publisher is established outside the EU,
the EU authorized representative is:
Springer Nature Customer Service Center GmbH
Europaplatz 3, 69115 Heidelberg, Germany

Printed by Libri Plureos GmbH
in Hamburg, Germany